T0134954

Machine Learning in Complex Networks

Thiago Christiano Silva • Liang Zhao

Machine Learning in Complex Networks

 Springer

Thiago Christiano Silva
Departament of Research (Depep)
Central Bank of Brazil
Brasília, Distrito Federal, Brazil

Liang Zhao
Computer Science and Mathematics
University of São Paulo (USP)
Ribeirão Preto, São Paulo, Brazil

ISBN 978-3-319-79234-7   ISBN 978-3-319-17290-3   (eBook)
DOI 10.1007/978-3-319-17290-3

Springer Cham Heidelberg New York Dordrecht London
© Springer International Publishing Switzerland 2016
Softcover reprint of the hardcover 1st edition 2016

Printed on acid-free paper

Springer International Publishing AG Switzerland is part of Springer Science+Business Media (www.
springer.com)

Preface

Machine learning stands as an important research area that aims at developing computational methods capable of improving their performances with previously acquired experiences. Although a large amount of machine learning techniques has been proposed and successfully applied in real systems, there are still many challenging issues that need to be addressed. In the last years, an increasing interest in techniques based on complex networks (large-scale graphs with nontrivial connection patterns) has been verified. This emergence is explained by the inherent advantages that the data representation as networks provides. They allow for capturing spatial, topological, and functional relations of the data. This book presents the features and possible advantages offered by complex networks in the machine learning domain. In the first part, we give an introduction to the machine learning and complex networks areas, supplying necessary background materials. Then, we present a comprehensive description on network-based machine learning. In the second part, we describe some specific techniques based on complex networks for supervised, unsupervised, and semi-supervised learning as case studies with the purpose of showing detailed know-how on network-based machine learning. Particularly, we explore a particle competition technique for both unsupervised and semi-supervised learning using a stochastic nonlinear dynamical system. We also walk through analytical aspects of the competitive system, enabling us to predict the behavior of the technique. Additionally, we deal with the problem of imperfect learning, exploring data reliability issues in semi-supervised learning and adapting the competitive system to withstand flawed training sets. Identifying and preventing error propagation have practical importance and are found to be of little investigation in the literature. Still in the second part of this book, we present a hybrid supervised classification technique that combines both low and high orders of learning. The low-level term is implemented by traditional classification techniques, while the high-level term is realized by extracting features of the underlying network constructed from the input data. The general idea of the model is that the low-level term classifies test instances by their physical features, while the high-level term measures the compliance of test instances with the pattern formation of the data. We show that the high-level technique can realize classification according to the semantic meaning of the data.

This book intends to bridge two widely studied research areas: machine learning and complex networks. Therefore, we hope it will generate broad interests to the scientific community. This book is intended to be employed by researchers and students who are interested in machine learning and complex networks. To accomplish that, not only have we included classic knowledge but also recent research results. This book is aimed to be self-contained and to give interested readers insights on modeling, analysis, and applications of network-based machine learning techniques. We also provide pointers to the literature for further reading on each explored topic. Moreover, numerous illustrative figures and step-by-step examples help readers to understand the main ideas and implementation details.

Acknowledgments

We would like to thank Dr. Marcos Gonçalves Quiles, Dr. Fabricio Aparecido Breve, Dr. João Roberto Bertini Jr., Dr. Thiago Henrique Cupertino, Dr. Andrés Eduardo Coca Salazar, Dr. Bilzã Marques de Araújo, Dr. Thiago Ferreira Covões, Dr. Elbert Einstein Nehrer Macau, Dr. Alneu Andrade Lopes, Dr. Xiaoming Liang, Dr. Zonghua Liu, Mr. Antonio Paulo Galdeano Damiance Junior, Ms. Tatyana Bitencourt Soares de Oliveira, Ms. Lilian Berton, Mr. Jean Pierre Huertas Lopez, Mr. Murillo Guimarães Carneiro, Mr. Leonardo Nascimento Ferreira, Mr. Fabio Willian Zamoner, Mr. Roberto Alves Gueleri, Mr. Fabiano Berardo de Sousa, Mr. Filipe Alves Neto Verri, and Mr. Paulo Roberto Urio for their collaboration in this research area along the last years. We thank Dr. Jorge Nakahara Jr. for carefully reviewing the book and for the continuous support given to us in the entire publication process. We appreciate Dr. Ying-Cheng Lai for introducing us to the fascinating research area of complex networks. We thank Mr. Hamlet Pessoa Farias Junior and Mr. Victor Dolirio Ferreira Barbosa for the useful discussions. We also thank Mr. João Eliakin Mota de Oliveira for providing us two figures. Further, we wish to thank the Institute of Mathematics and Computer Science (ICMC) and the Faculty of Philosophy, Sciences, and Literatures at Ribeirão Preto (FFCLRP) of the University of São Paulo (USP), in Brazil, and the Central Bank of Brazil for their institutional support. Finally, we would like to thank the São Paulo Research Foundation (FAPESP), the National Council for Scientific and Technological Development (CNPq), and the Coordination for the Improvement of Higher Level Education (CAPES), all in Brazil, for their financial support on our research works.

Brasília (DF) and Ribeirão Preto (SP), Brazil   Thiago Christiano Silva
November 2015   Liang Zhao

Contents

Biography

Dr. Thiago Christiano Silva received his B.E. degree (with first place Honors) in Computer Engineering from the University of São Paulo in 2009. In 2012, he concluded his Ph.D. degree in Mathematics and Computer Sciences within the same university, where he was the recipient of three distinctive awards due to his thesis. He was the winner of the "CAPES Thesis Award Contest in the area of Computer Science", the winner of the "University of São Paulo Thesis Competition," and the winner of the "International BRICS-CCI Ph.D. Thesis Competition," all in 2013. In 2014, he completed a 1-year postdoctoral research in machine learning and complex networks in the University of São Paulo. Since 2011, he works as a researcher at the Central Bank of Brazil, Brasília. His research interests include machine learning, complex networks, financial stability, systemic risk, and banking.

Dr. Liang Zhao received his B.S. degree from Wuhan University in China, and M.Sc. and Ph.D. degrees from the Aeronautic Institute of Technology (ITA) in Brazil, in 1988, 1996, and 1998, respectively, all in computer science. He joined, as a faculty member, the Institute of Mathematics and Computer Science (ICMC) of University of São Paulo (USP) in 2000. Currently, he is a Full Professor and head of the Department of Computer Science and Mathematics (DCM), Faculty of Philosophy, Sciences, and Literatures in Ribeirão Preto (FFCLRP) of USP. From 2003 to 2004, he was a Visiting Researcher with the Department of Mathematics, Arizona State University in the USA. His current research interests include machine learning, complex networks, artificial neural networks, and pattern recognition. He has published more than 180 scientific articles in refereed international journals, books, and conferences. Dr. Zhao is a recipient of the Brazilian Research Productivity Fellowship. He is currently an Associate Editor of the *Neural Networks* and he was an Associate Editor of the *IEEE Transactions on Neural Networks and Learning Systems* from 2009 to 2012. Dr. Zhao is a Senior Member of IEEE and a Member of International Neural Network Society (INNS).

List of Symbols

$\mathcal{G}$	Graph or network
$\mathcal{V}$	Set of vertices
$\mathcal{E}$	Set of edges
$\mathcal{L}$	Set of labeled training data
$\mathcal{U}$	Set of unlabeled data
$\mathcal{Y}$	Set of labels, targets, or labels
$\mathcal{K}$	Set of particles
$\mathcal{N}(v)$	Set of neighbors of vertex v
$\mathcal{X}$	Set of vector-based data items
$\mathcal{X}_L$	Set of vector-based labeled data items
$\mathcal{X}_U$	Set of vector-based unlabeled data items
$\mathcal{X}_{\text{training}}$	Training set
$\mathcal{X}_{\text{test}}$	Test set
$\mathcal{O}(.)$	Big O notation: states the worst time complexity of an algorithm
V	Number of vertices
E	Number of edges
L	Number of labeled training data
U	Number of unlabeled data
Y	Number of labels, targets, or labels
K	Number of particles
P	Number of attributes (dimensionality) of each data item
M	Number of groups or communities in a data set
λ	Counterweighting factor between random and preferential rules
ρ	Compliance term
$\mathbf{A}$	Weighted or non-weighted adjacency matrix
$\mathbf{P}$	Transition matrix
$\mathbf{R}$	Potential or fundamental matrix
$\mathbf{S}$	Similarity matrix
$\mathbf{D}$	Dissimilarity matrix
k_v	Degree of vertex v

$k_v^{(\text{in})}$	In-degree of vertex v
$k_v^{(\text{out})}$	Out-degree of vertex v
s_v	Strength of vertex v
$s_v^{(\text{in})}$	In-strength of vertex v
$s_v^{(\text{out})}$	Out-strength of vertex v
CC_v	Clustering coefficient of vertex v
$\bar{k}$	Average network degree or network connectivity
r	Network assortativity
Q	Network modularity
CC	Network average clustering coefficient
d_{uv}	Shortest path length from vertex u to v

Chapter 1
Introduction

Abstract When we say "learning," one of the words that comes into our mind may be "mystery"; when we talk about "large scale networks," we may associate it to the word "complexity." What happens when we put together these two concepts? In this chapter, we present an overview on complex network-based machine learning. Throughout the entire book, we show the diversity of approaches for treating such a subject.

1.1 General Background

Human beings are born with the fascinating gift of learning. With the aid of such ability, they absorb and assimilate knowledge throughout their entire life. In an attempt to simulate such notorious characteristic in a computational environment, the research area entitled *machine learning* arose, which aims at developing computational methods that are capable of "learning" with past accumulated experiences [11, 19, 33].

Based on the computational data representation obtained from a wide range of domains, machine learning techniques can generate models apt to organize the existing knowledge or, yet, mimic the behavior of a human expert in an automatically manner. Generally speaking, these techniques are traditionally divided into two paradigms: the supervised and the unsupervised learning [11, 19]. In the *supervised learning* case, the goal is to infer concepts regarding the data using both the data distribution and external knowledge in the form of labels. In essence, the supervised learning process aims to construct a mapping function conditioned to the provided training data set. When these labels comprise discrete values, the inference problem is denominated *classification*, whereas when label values are continuous, *regression*. In contrast, in the *unsupervised learning* case, the main task consists in finding intrinsic data structures. The learning process, in this case, is solely guided by the data relationships as no external knowledge about the existing labels is required [33].

Supervised learning requires a labeled data set for training. However, the task of manual labeling is, in the majority of the cases, a cumbersome and expensive process, which usually involves the work of human experts. In order to soften this shortcoming, a third learning paradigm denominated *semi-supervised learning*

© Springer International Publishing Switzerland 2016

T.C. Silva, L. Zhao, *Machine Learning in Complex Networks*,

DOI 10.1007/978-3-319-17290-3_1

has emerged. The main distinct feature of semi-supervised algorithms is that they employ both labeled and unlabeled data in the prediction process by means of, for example, a diffusive labeling process. Frequently in real-world situations, data sets consist of a great amount of unlabeled data and a few labeled data. In this way, semi-supervised learning can considerably reduce human experts' efforts. Moreover, empirical results have shown that the usage of unlabeled data can improve the performance of fully supervised classification [15, 50].

Over the last decade, there has been an increasing interest in network research, with the focus drifting away from analyzing small graphs to considering large-scale graphs, called *complex networks*. Such networks have emerged as a unified representation of complex systems in various branches of science. In general, they are used to model systems that have nontrivial topologies and are composed of a large amount of vertices [1, 8, 36].

The study of networks began with the development of the graph theory, inaugurated by Leonhard Euler in 1736 with the solution of the seven bridges problem of Königsberg, today Kaliningrad, Russia. The problem, much discussed at the time, recorded that there were seven bridges crossing the river Pregel, with two intermediate islands. The residents desired to know whether it was possible to cross all of these seven bridges, without repetition, and then return to the starting point. Euler demonstrated, in an analytical manner, for the Russian Academy of Sciences in St. Petersburg, that it was not possible to complete such a walk. For this end, he made use of a graphical representation consisting of points and curves connecting these points. It was the beginning of the formal representation of a graph or network,[1] known until today, with vertices and edges. Thereafter, many researchers began studying this branch of research in search of new theories and applications [36].

In fact, the first major step in the study of complex networks was driven by Paul Erdös and Alfréd Réyni, who analyzed a certain type of network called *random networks* in their work published in 1959 [20]. That investigation opened doors to a novel area of study termed the theory of random networks, which represents a mixture of the graph and probabilistic theories to generate and analyze large-scale graphs.

Following the chronology, in 1967, Stanley Milgram decided to accept the challenge posed by Frigyes Karinthy, which, inspired by the conjectures of Guglielmo Marconi in 1909, dared one to find another person to whom he/she could not be transitively connected by using at most five intermediate people [32]. And it was exactly because of this problem that the concept denominated *separation in six degrees* was born, which was the first seed for the study of the *small-world networks*. To address this challenge, Milgram conducted experiments in order to try to discover the probability of any two arbitrary people to know each other. For this, letters were sent to random people living in predetermined regions of the United States, whose inner content dealt with information about any other arbitrary person. If the person

---

[1]Since graph and network share the same definition, these two terms are interchangeable in this book.

referred to in the letter was known by the reader, then he/she would mail the letter for the recipient. On the other hand, if he/she did not know, then the letter should be sent to someone else known. At the end of experiment, Milgram found that the average number of referrals of one person to another reached 5.5 persons. He was, therefore, empirically discovering the property of small world, which states that even though there are millions of persons interconnected in a social network, the average distance between them is only a small amount, in the example, 5.5 persons [32].

Despite of the findings of Milgram, it was only in the late 1990s that surveys on this field were retaken. In 1998, Watts and Strogatz found that the average shortest paths in a network can be drastically reduced by a random alteration of few links, starting from a regular network [56]. The resulting network is called *small-world network*, which, as we have seen, had already been empirically discovered by Milgram. In 1999, Barabási and Albert discovered that many real networks have a degree distribution of vertices that obeys a power law: $P(k) \sim k^{-\gamma}$, where k is the number of connections of a randomly chosen vertex and γ is a scaling exponent [7]. This heterogeneous distribution describes the existence of a small number of vertices that has a large number of connections, while the majority only shares few connections. Such networks are called *scale-free networks*.

Driven by the technological advances and also by the increasing number of data to be jointly analyzed, the complex networks area has emerged as a unifying topic in complex systems and is present in various branches of science [13]. Structurally, complex networks are represented by a graph $\mathcal{G} = \langle \mathcal{V}, \mathcal{E} \rangle$, where $\mathcal{V}$ represents the set of vertices and $\mathcal{E}$, the set of edges. Complex networks can be conceived as a general modeling scheme for heterogenous systems with arbitrary sizes [3], as they naturally incorporate the usual nontrivial aspects of the system agents.

Evidences of complex networks in real-world settings abound. Among some examples of real-world systems in which the network representation is perfectly plausible, we can highlight: the *Internet* [22], the *World Wide Web* [2], biological neural networks [52, 57], financial networks [14, 43, 51], information networks [59], social networks among individuals [27, 46] and between companies and organizations [34], food webs [35], metabolic networks [16, 26] and distribution as the bloodstream [58], protein-protein networks [55], postal delivery and electricity distribution networks [3] etc.

The data representation in complex networks inherently presents some positive characteristics of which we can cite [53]:

- *The structural complexity*—which translates into the heterogeneous and nontrivial data connections that are shared between vertices in the network. This feature can be easily understood by taking into account the difficulty in visualizing the network properties.
- *The evolution*—which marks the constant changes in the network structure due to the inclusion and removal of vertices and connections [18].
- *The diversity of connections*—because the connections between vertices can have various physical meanings, such as capacity, length, width and direction. These features are often operationalized via multilayer networks [12]. These networks

have several network layers, each of which representing different aspects of the possible connections.

- *The dynamical nature*—which affects, at a large scale, the states of a network, as can be construed as traffic information [63], occurrences of failures in communications [60, 61, 63], similarity relations between vertices, the distribution of functions [36], among others.

1.2   Focus of This Book

Machine learning and data mining techniques using complex networks have triggered increased attention. This is because networks are ubiquitous in nature and everyday life, as several real-world problems can be directly represented in terms of networks. Moreover, many other kinds of data sets can be transformed into network representations using a suitable network formation technique. For instance, a set of items described by feature (or attribute) vectors can be transformed into a network by simply connecting each sample to its k nearest neighbors. In addition, the complex network representation unifies the structure, dynamics, and functions of the system that it represents. It does not only describe the interaction among vertices (structure) and the evolution of such interactions (dynamics), but also reveals how the structure and dynamics affect the overall functions of the network [39]. For example, it is well known that there is a strong connection between the structure of protein-protein interaction networks and protein functions [42]. The main motivation of network research is the ability of describing topological structures of the original system. In the machine learning domain, it has been shown that the topological structure is quite useful to detect clusters of arbitrary forms in data clustering [23, 29].

This book brings as main goals the review and the development of machine learning techniques that are based on complex networks. Our intention is twofold:

1. To provide a thorough coverage on this topic. In this way, the book not only gives a general vision on this topic but also facilitates the effort of readers in finding out relevant materials for their own development. For this purpose, comprehensive reviews on complex networks and network-based machine learning are explored in the first part of this book.
2. To describe and bridge the connection between machine learning and complex networks. To this end, we focus on several up to date developments of these two topics in the second part of the book (the last three chapters). Here, the three branches of the machine learning area are explored, i.e., the unsupervised, the semi-supervised, and the supervised learning areas.

In the first part of the book, we give an overview of relevant concepts and some technical details on network-based machine learning. Since both complex networks and machine learning are well developed and multidisciplinary research areas, we would like to follow this feature and present diversity of ideas. However, due to the vast research results developed so far, we must make choices on the content of the book. For example, synchronization in complex networks has not been reviewed in this book. We have not gone to technical details on some topics, for examples, epidemic spreading and graph kernels. The interested readers may get further knowledge with specific materials cited in this book.

In the review of complex networks in Chap. 2, we first introduce basic concepts and notations on graphs, then we present the classic complex network models. The focus of this chapter is to provide a comprehensive review on network measures in a categorized way. Many such measures have been applied to develop machine learning techniques, while others may be used for future developments probably by the readers of this book and can be easily found in classical materials.

Then we go to the review of machine learning in Chap. 3, where we introduce the three traditional paradigms: supervised, unsupervised, and semi-supervised learning. When exploring them, we pay attention to the basic concepts and characterization of machine learning instead of technical details, which are out of the scope of this book.

Many data sets are not already in a network format. In order to apply network-based techniques for data analysis, it is necessary to transform the original data sets into networks. Intuitively, a single data set can be represented by different networks, which may lead to different qualities of final results. Therefore, network construction from original data sets is a crucial issue for data mining and machine learning. For this reason, Chap. 4 is dedicated to reviewing and compiling various methods that can be employed for network construction. It is worth mentioning that network construction is still at its infant stage and there is a large space for exploration. The readers are invited to contribute to this interesting topic.

Given the background knowledge of complex networks and machine learning, we start to present to the readers an overview of network-based machine learning. As we will see, the majority of the supervised and semi-supervised learning techniques is within-network methods, i.e., probabilistic or deterministic inference of class labels of some vertices takes place through finding out the "best" route from the labeled to unlabeled vertices within the network. (In the second part of this book, we will present an across-network supervised learning technique, in which global information of the underlying network is taken into account). We will also show that network-based unsupervised learning is really a community detection task. The presence of *communities* is a striking phenomena of complex networks. The notion of community in networks is straightforward: each community is defined as a subnetwork whose vertices are densely connected within itself, but sparsely connected with the rest of the network. Community detection in complex networks has turned out to be an important topic in graph mining and in data mining [17, 23, 30, 40]. In graph theory, community detection corresponds to graph partition, which is an NP-complete problem [23]. For this reason, a lot of efforts has been spent to developing efficient but suboptimal solutions, such as the spectral method [38], the technique based on the "betweenness" measure [40], modularity optimization [37], community detection based on the Potts model [45], synchronization [6], information theory [24], and random walks [64].

As community detection is directly related to unsupervised learning, a review of this topic will be presented in Chaps. 2 and 6.

The second part of this book is devoted to presenting concrete ideas and technical details on specific realizations of the three machine learning paradigms. When investigating these developments, we are concerned in addressing:

- The development of novel computational methods in machine learning are expected to be equipped, whenever possible, with an underlying mathematical framework. The construction of models that rely on mathematical grounds opens way to characterizing the short- and long-run behaviors of these techniques in a concise and formal manner. We believe that this is an important step to better understand the dynamics of the models and, as a consequence, to better enable one to perceive the potentialities and the shortcomings offered by them. Nonetheless, whenever it is possible, empirical studies are likewise conducted to consolidate and to confirm the validity of the analytical predictions.
- The design of these machine learning techniques, in contrast to traditional methods, is expected to provide alternative and novel ways to solve the challenging problems posed by the machine learning area. In this way, novel and efficient algorithms modeled for tasks of clustering and classification are explored. In other words, we intend to investigate new features and possible advantages offered by complex networks in the machine learning domain. In fact, we do show that the network-based approach really brings interesting features for machine learning.
- Having in mind the possibility of utilizing the methods described herein in real-world applications, we also take into account the design of techniques that are complementary in terms of performance and computational complexity. Consequently, we must find equilibrium between quality and efficiency when designing network-based machine learning techniques.

In order to give a concrete vision and present some know-hows on designing complex network-based machine learning techniques, this book details the following up to date developments:

- With regard to the unsupervised learning area:

  1. The fundamental mechanism and technical details of the particle competition model in complex networks are presented. The particle competition model was originally proposed and applied in community detection in [44]. Then, it was reformulated in a formal dynamical system and extended to solve data clustering problems by [49]. The model consists of several particles walking within the network and competing with each other to occupy as many vertices as possible, while attempting to reject intruder particles. The particle's walking rule is composed of a stochastic combination of random and preferential movements. As we will see in Chap. 9, computer simulations reveal that this model presents high community detection rates, as well as low computational complexity. One strong argument for delving into the process of particle competition is that of its similarity to many social and

natural processes, such as: competition among animals, territorial exploration by humans (animals), election campaigns, among others. Furthermore, the random-preferential movement incorporated into the particles' movement policy can substantially improve the model's performance, as we will see in Chap. 9 and 10. This model corroborates the importance of the randomness role in evolutionary systems whose primary function is to prevent particles from falling into local traps in an automatic manner. Besides that, it endows particles with the ability to explore unknown territories. Therefore, a certain amount of randomness is essential for the learning process. This randomness is charged with representing the state "I do not know" and lends itself as an effective "explorer of new features."

2. The particle competition model is constructed under a stochastic nonlinear dynamical system. In this regard, we provide an analytical analysis of the model, deriving probabilistic expressions that are able to predict the model's behavior as time progresses. A numerical validation confirms the theoretical predictions. In addition, we show that the model generalizes the process of single random walks to multiple interacting random-preferential walks in a competitive way. Such generalization is realized by calibrating the parameters of the model. A convergence analysis of the particle competition model is supplied. Therein, we show that the model does not converge to a fixed point, but instead it is confined within a certain region with a finite diameter. Furthermore, an upper bound of this region is estimated. Such a feature is similar to real-world systems due to the presence of noises and other uncontrolled variables.

3. A fuzzy index for detecting overlapping cluster or community structures in the network is explored. Most of the traditional community detection methods aim at assigning each vertex to a single community [23]. However, in real networks, vertices are often shared among different communities [23]. For example, in the language network composed of words as vertices, the word "bright" might be a member of several communities, such as those representing words related to the following subjects: "light," "astronomy," "color," "intelligence," and so on [42]. In a social network, each person naturally belongs to the communities of the company where he/she works and also to the community of his/her family at the same time. Therefore, uncovering overlapping community structure is important not only for network mining, but also for data analysis in general, once a data set is transformed into a network [21, 31, 41, 42, 47, 54, 62]. A drawback of traditional techniques that identify overlapping communities is that the overlap detection procedure is performed as a separated or dedicated process apart from the standard community detection technique. In this way, additional computational time is often required. As a result, the entire process may have high computational complexity. As we will see, the particle competition technique detects overlap community structures during the community detection procedure by using the dynamic variables generated by the particle competition process. As a result,

the particle competition method does not increase the overall model's time complexity order when employed to identify overlapping structures.

- With regard to the semi-supervised learning area:

  1. The model based on cooperation and competition of multiple particles is reviewed in detail [48]. This technique performs semi-supervised learning for classification tasks, as we will see in Chap. 10. A rigorous definition is provided, in which the particle competition is formally modeled from a stochastic nonlinear dynamical system. In essence, particles of the same class navigate in the network in a cooperative manner to propagate their labels, while particles of different classes compete with each other to determine class borders. Given that the model of several interacting particles corresponds to many natural and artificial systems, the study of this topic stands as an important task.

  2. Another interesting feature is that the particle competitive-cooperative model has a local label-spreading behavior. This property arises due to the competitive mechanism in which particles only visit portions of vertices potentially belonging to their teams. This can be roughly understood as a "divide-and-conquer" effect embedded in the scheme of competition and cooperation. In this way, many long-range redundant operations are avoided. As a result, the method has low computational complexity order. In contrast, traditional techniques of network-based semi-supervised learning normally rely on minimizing cost functions that ultimately lead to several matrix multiplication operations. Thus, the computational complexity of these techniques is usually of the order $\mathcal{O}(V^3)$ or higher [9, 10, 65], in which V is the number of vertices. Even though methods for enhancing matrix multiplication have been extensively studied,[2] the minimization of cost functions, which are usually based on a regularization framework, slows down the entire process [65]. It is expected that the models generated using competition of particles will be more efficient, which is important to treat large-scale databases.

  3. Since we construct the underlying network, in which the learning process is conducted, directly from the input data set, the correspondence between the input data and the processing result (the final network) is maintained. Consequently, the "black box" effect of artificial neural networks can be avoided to a large extent.

  4. The reliability of the labels is a crucial factor in a semi-supervised learning environment, because mislabeled samples may propagate wrong labels to a portion of or even to the entire data set. Here, we address the error propagation problem originated by these mislabeled samples by presenting detection and prevention processes embedded within the particle competition-cooperation model. Though this is an important topic, it has not received

---

[2]For instance, c.f. the generalized iterated matrix-vector multiplication technique, GMIV-M, proposed by [28].

much attention from researchers and there are still few works devoted to the study of semi-supervised learning from imperfect training data [4, 5, 25]. Usually, in supervised or semi-supervised learning, the input label information of the training data set is supposed to be completely reliable. However, in real situations, this is not always true and mislabeled samples are commonly found in the data sets due to instrumental errors, corruption from noise, or even human mistakes in the labeling process. For example, in a medical diagnostic system, the diagnostic results in the training set provided by doctors may be wrong. If these kinds of wrong labels are used to further classify new data (in the supervised learning case) or are propagated to unlabeled data (in the semi-supervised learning case), severe consequences may occur. This situation becomes more critical in autonomous learning, in which no external or minimal external intervention is involved. Thus, if the prior knowledge presented to the autonomous learning system contains errors, the performance of the learning system will get worse and worse because of the error propagation. Therefore, considering and designing mechanisms to prevent error propagation is important in the machine learning study and especially in the autonomous learning. Specifically, the prevention of error propagation can benefit the learning systems from two different aspects:

– Improvements of the performance of the learning system, i.e., the system can learn from errors;
– Avoidance of a system's catastrophe by limiting the spreading of wrong labels (input and generated errors).

To our knowledge, many semi-supervised learning techniques have been proposed [15], but the great majority considers that the label information of the labeled set is totally correct, i.e., there is no error prevention mechanism. In this way, such a mechanism make a clear contribution to general machine learning and especially to autonomous learning research.

• With regard to the supervised learning area:

1. A hybrid supervised learning framework composed of a convex combination of low- and high-level classifiers is studied in Chap. 8. Traditional supervised data classification considers only physical features (e.g., distance or similarity) of the input data. Here, this type of learning is called *low-level classification*. The human (animal) brain, in contrast, performs both low and high orders of learning and it has facility in identifying patterns according to the semantic meaning of the input data. Data classification that considers not only physical attributes but also the pattern formation is here referred to as *high-level classification*. The idea behind introducing the high-level term in a learning process is that data items often have patterns or organizational features that are left hidden within the numerous interrelationships among them. These, in turn, are not very well explored by traditional low-level classification techniques, as they are concerned with smoothness or cluster assumptions, which are in essence physical constraints. The high-level term precisely comes into play to fill in this gap by trying to find organizational

or structural features among different classes. Thus, it abstracts its decisions from physical constraints. As we will see, both the low- and high-level classifiers are often necessary to provide good accuracy rates, suggesting that they are complementary in the learning process.

2. Motivated by the intrinsic ability to describe topological structures among the data items, two types of high-level classification techniques have been proposed, all of which running in a networked environment. As a common goal, both realize the prediction scheme by extracting the features of the underlying network constructed from the input data. The high-level classification techniques are comprised of:

   – Three classical network measures borrowed from the complex network theory: assortativity, clustering coefficient, and network connectivity. The combination of these three measures can capture local to global structural patterns from the network topology, when utilized with reasonable low-level classifiers.
   – A weighted linear combination of tourist walks processes with different memory lengths. For this end, variations of dynamic variables generated by this deterministic process, the transient and cycle lengths, for different values of the tourist's memory length are employed. We show that, by adjusting the memory length of the tourist walker, we can systematically capture structural network features that range from local to global.

3. An interesting phenomenon uncovered using this hybrid classification framework is that, as the class complexity increases, a larger portion of the high-level term is required to get correct classification. The class complexity may be understood in terms of the mixture or overlap that exists among different classes. This feature confirms that the high-level classification has a special importance in complex situations of classification.

1.3   Organization of the Remainder of the Book

The remainder of this book is organized as follows. In Chap. 2, we review the complex network theory and dynamic processes that run in networks. In Chap. 3, we present the basic definitions of the machine learning area. These two first chapters elucidate the fundamental concepts that make way for the understanding of the developments reviewed and developed in this book.

Once the elementary theory is presented, in Chap. 4, we discuss the problem of constructing a network from unstructured data, which is a required step whenever we are dealing with data that is not yet in a network format. This task is crucial for network-based learning methods. Following that, we give a comprehensive review on network-based supervised learning, unsupervised learning, and semi-supervised learning in Chaps. 5–7, respectively.

In the second part of this book, we present the new developments of network-based machine learning in Chaps. 8–10. In Chap. 8, we delve into the supervised learning domain by showing a hybrid classification framework that derives its decision based on a convex combination of low- and high-level classifiers. This method is a pioneer across-network supervised learning technique. The model is analyzed in an empirical way and significant results are obtained. A real-world application (handwritten digits recognition) is supplied. In Chap. 9, we explore the unsupervised learning domain using a technique that is based on a particle competition model. Several experiments and mathematical investigations are performed, as well as a real-world application (handwritten digits and letters clustering). In Chap. 10, the particle competition model is extended to the semi-supervised learning domain. Likewise the previous chapter, we also display many experiments and mathematical investigations, as well as another real-world application (detection and prevention of mislabeled vertices).

References

1. Albert, R., Barabási, A.L.: Statistical mechanics of complex networks. Rev. Mod. Phys. **74**(1), 47–97 (2002)
2. Albert, R., Jeong, H., Barabási, A.L.: Diameter of the world wide web. Nature **401**, 130–131 (1999)
3. Albert, R., Albert, I., Nakarado, G.L.: Structural vulnerability of the north american power grid. Phys. Rev. E **69**, 025103 (2004)
4. Amini, M.R., Gallinari, P.: Semi-supervised learning with explicit misclassification modeling. In: IJCAI 03: Proceedings of the 18th International Joint Conference on Artificial Intelligence, pp. 555–560. Morgan Kaufmann, San Francisco, CA (2003)
5. Amini, M.R., Gallinari, P.: Semi-supervised learning with an imperfect supervisor. Knowl. Inf. Syst. **8**(4), 385–413 (2005)
6. Arenas, A., Guilera, A.D., Pérez Vicente, C.J.: Synchronization reveals topological scales in complex networks. Phys. Rev. Lett. **96**(11), 114102 (2006)
7. Barabási, A.L., Albert, R.: Emergence of scaling in random networks. Science (NY) **286**(5439), 509–512 (1999)
8. Barrat, A., Barthélemy, M., Vespignani, A.: Dynamical Processes on Complex Networks. Cambridge University Press, Cambridge (2008)
9. Belkin, M., Niyogi, P.: Laplacian eigenmaps for dimensionality reduction and data representation. Neural Comput. **15**(6), 1373–1396 (2003)
10. Belkin, M., Matveeva, I., Niyogi, P.: Regularization and semi-supervised learning on large graphs. In: Shawe-Taylor, J., Singer, Y. (eds.) Learning Theory. Lecture Notes in Computer Science, vol. 3120, pp. 624–638. Springer, Berlin, Heidelberg (2004)
11. Bishop, C.M.: Pattern Recognition and Machine Learning (Information Science and Statistics). Springer, Berlin (2007)
12. Boccaletti, S., Bianconi, G., Criado, R., del Genio, C., Gómez-Gardeñes, J., Romance, M., Sendiña-Nadal, I., Wang, Z., Zanin, M.: The structure and dynamics of multilayer networks. Phys. Rep. **544**(1), 1–122 (2014)
13. Bornholdt, S., Schuster, H.G.: Handbook of Graphs and Networks: From the Genome to the Internet. Wiley-VCH, Weinheim (2003)
14. Castro Miranda, R.C., Stancato de Souza, S.R., Silva, T.C., Tabak, B.M.: Connectivity and systemic risk in the brazilian national payments system. J. Complex Networks **2**(4), 585–613 (2014)

15. Chapelle, O., Schölkopf, B., Zien, A. (eds.): Semi-Supervised Learning. Adaptive Computation and Machine Learning. MIT, Cambridge, MA (2006)
16. da Silva, M., Ma, H., Zeng, A.P.: Centrality, network capacity, and modularity as parameters to analyze the core-periphery structure in metabolic networks. Proc. IEEE **96**(8), 1411–1420 (2008)
17. Danon, L., Díaz-Guilera, A., Duch, J., Arenas, A.: Comparing community structure identification. J. Stat. Mech. Theory Exp. **2005**(09), P09008 (2005)
18. Dorogovtsev, S.N., Mendes, J.F.F.: Evolution of Networks: From Biological Nets to the Internet and WWW (Physics). Oxford University Press, USA (2003)
19. Duda, R.O., Hart, P.E., Stork, D.G.: Pattern Classification. Wiley-Interscience (2000)
20. Erdös, P., Rényi, A.: On random graphs I. Publ. Math. Debr. **6**, 290–297 (1959)
21. Evans, T.S., Lambiotte, R.: Line graphs, link partitions, and overlapping communities. Phys. Rev. E **80**(1), 016105 (2009)
22. Faloutsos, M., Faloutsos, P., Faloutsos, C.: On power-law relationships of the internet topology. In: SIGCOMM 99: Proceedings of the Conference on Applications, Technologies, Architectures, and Protocols for Computer Communication, vol. 29, pp. 251–262. ACM, New York (1999)
23. Fortunato, S.: Community detection in graphs. Physics Reports **486**, 75–174 (2010)
24. Fortunato, S., Latora, V., Marchiori, M.: Method to find community structures based on information centrality. Phys. Rev. E **70**(5), 056104 (2004)
25. Hartono, P., Hashimoto, S.: Learning from imperfect data. Appl. Soft Comput. **7**(1), 353–363 (2007)
26. Jeong, H., Tombor, B., Albert, R., Oltvai, Z.N., Barabási, A.L.: The large-scale organization of metabolic networks. Nature **407**(6804), 651–654 (2000)
27. Jiang, Y., Jiang, J.: Understanding social networks from a multiagent perspective. IEEE Trans. Parallel Distrib. Syst. **25**(10), 2743–2759 (2014)
28. Kang, U., Tsourakakis, C.E., Faloutsos, C.: PEGASUS: mining peta-scale graphs. J. Knowl. Inf. Syst. **27**(2), 303–325 (2011)
29. Karypis, G., Han, E.H., Kumar, V.: Chameleon: hierarchical clustering using dynamic modeling. Computer **32**(8), 68–75 (1999)
30. Lambiotte, R., Delvenne, J.C., Barahona, M.: Random walks, Markov processes and the multiscale modular organization of complex networks. IEEE Trans. Netw. Sci. Eng. **1**(2), 76–90 (2014)
31. Lancichinetti, A., Fortunato, S., Kertész, J.: Detecting the overlapping and hierarchical community structure in complex networks. New J. Phys. **11**(3), 033015 (2009)
32. Milgram, S.: The small world problem. Psychol. Today **2**, 60–67 (1967)
33. Mitchell, T.M.: Machine Learning. McGraw-Hill Science/Engineering/Math, New York (1997)
34. Mizruchi, M.S.: The American corporate network. Sage **2**, 1904–1974 (1982)
35. Montoya, J.M., Solée, R.V.: Small world patterns in food webs. J. Theor. Biol. **214**, 405–412 (2002)
36. Newman, M.E.J.: The structure and function of complex networks. SIAM Rev. **45**(2), 167–256 (2003)
37. Newman, M.E.J.: Fast algorithm for detecting community structure in networks. Phys. Rev. E **69**(6), 066133 (2004)
38. Newman, M.E.J.: Modularity and community structure in networks. Proc. Natl. Acad. Sci. **103**(23), 8577–8582 (2006)
39. Newman, M.E.J.: Networks: An Introduction. Oxford University Press, Oxford (2010)
40. Newman, M.E.J., Girvan, M.: Finding and evaluating community structure in networks. Phys. Rev. Lett. **69**, 026113 (2004)
41. Nicosia, V., Mangioni, G., Carchiolo, V., Malgeri, M.: Extending the definition of modularity to directed graphs with overlapping communities. J. Stat. Mech. Theory Exp. **2009**(03), 03024 (2009)
42. Palla, G., Derenyi, I., Farkas, I., Vicsek, T.: Uncovering the overlapping community structure of complex networks in nature and society. Nature **435**(7043), 814–818 (2005)

43. Poledna, S., Molina-Borboa, J.L., Martínez-Jaramillo, S., van der Leij, M., Thurner, S.: The multi-layer network nature of systemic risk and its implications for the costs of financial crises. J. Financ. Stab. **20**, 70–81 (2015)

44. Quiles, M.G., Zhao, L., Alonso, R.L., Romero, R.A.F.: Particle competition for complex network community detection. Chaos **18**(3), 033107 (2008)

45. Reichardt, J., Bornholdt, S.: Detecting fuzzy community structures in complex networks with a potts model. Phys. Rev. Lett. **93**(21), 218701(1–4) (2004)

46. Scott, J.P.: Social Network Analysis: A Handbook. SAGE, Beverly Hills, CA (2000)

47. Shen, H., Cheng, X., Cai, K., Hu, M.B.: Detect overlapping and hierarchical community structure in networks. Physica A **388**(8), 1706–1712 (2009)

48. Silva, T.C., Zhao, L.: Network-based stochastic semisupervised learning. IEEE Trans. Neural Netw. Learn. Syst. **23**(3), 451–466 (2012)

49. Silva, T.C., Zhao, L.: Stochastic competitive learning in complex networks. IEEE Trans. Neural Netw. Learn. Syst. **23**(3), 385–398 (2012)

50. Singh, A., Nowak, R.D., Zhu, X.: Unlabeled data: now it helps, now it doesn't. In: The Conference on Neural Information Processing Systems NIPS, pp. 1513–1520 (2008)

51. Souza, S.R., Tabak, B.M., Silva, T.C., Guerra, S.M.: Insolvency and contagion in the brazilian interbank market. Physica A **431**, 140–151 (2015)

52. Sporns, O.: Networks analysis, complexity, and brain function. Complexity **8**(1), 56–60 (2002)

53. Strogatz, S.H.: Exploring complex networks. Nature **410**(6825), 268–276 (2001)

54. Sun, P.G., Gao, L., Shan Han, S.: Identification of overlapping and non-overlapping community structure by fuzzy clustering in complex networks. Inf. Sci. **181**, 1060–1071 (2011)

55. Wang, P., Yu, X., Lu, J.: Identification and evolution of structurally dominant nodes in protein-protein interaction networks. IEEE Trans. Biomed. Circuits Syst. **8**(1), 87–97 (2014)

56. Watts, D.J., Strogatz, S.H.: Collective dynamics of 'small-world' networks. Nature **393**(6684), 440–442 (1998)

57. Weng, J., Luciw, M.: Brain-inspired concept networks: Learning concepts from cluttered scenes. IEEE Intell. Syst. **29**(6), 14–22 (2014)

58. West, G.B., Brown, J.H., Enquist, B.J.: A general model for the structure, and algometry of plant vascular systems. Nature **400**, 122–126 (1999)

59. Yang, J., Leskovec, J.: Overlapping communities explain core-periphery organization of networks. Proc. IEEE **102**(12), 1892–1902 (2014)

60. Zhao, L., Park, K., Lai, Y.C.: Attack vulnerability of scale-free networks due to cascading breakdown. Phys. Rev. E **70**, 035101(1–4) (2004)

61. Zhao, L., Park, K., Lai, Y.C.: Tolerance of scale-free networks against attack-induced cascades. Phys. Rev. E (Rapid Commun.) **72**(2), 025104(R)1–4 (2005)

62. Zhang, S., Wang, R.S., Zhang, X.S.: Identification of overlapping community structure in complex networks using fuzzy C-Means clustering. Physica A **374**(1), 483–490 (2007)

63. Zhao, L., Cupertino, T.H., Park, K., Lai, Y.C., Jin, X.: Optimal structure of complex networks for minimizing traffic congestion. Chaos **17**(4), 043103(1–5) (2007)

64. Zhou, H.: Distance, dissimilarity index, and network community structure. Phys. Rev. E **67**(6), 061901 (2003)

65. Zhou, D., Bousquet, O., Lal, T.N., Weston, J., Schölkopf, B.: Learning with local and global consistency. In: Advances in Neural Information Processing Systems, vol. 16, pp. 321–328. MIT, Cambridge, MA (2004)

Chapter 2
Complex Networks

Abstract Complex network comprises an emerging interdisciplinary research area that triggers much attention from physicists, mathematicians, biologists, engineering, computer scientists, among many others. Complex network structures describe a wide variety of systems of high technological and intellectual importance, such as the Internet, World Wide Web, coupled biological and chemical systems, financial, social, neural, and communication networks. The desire to understand such interwoven systems summed with their inherent complexity are factors that explain the increasing interest in enhancing complex network tools. The data representation in complex networks permits us to unify the structural complexity and vertex and connection diversities. Several relevant questions arise when investigating dynamics in complex networks, such as learning how large ensembles of dynamical systems that interact through a complex wiring topology can behave collectively. In this way, the network topology plays an important role in that it affects the functions of the represented system. As an example, the structure of social networks affects the information and disease propagation speeds, the topology of a financial network may amplify shocks in different manners, and the disposition of power grids in networks may affect the robustness and stability of power transmission. Due to the rapid evolution and the large amount of developed theories and techniques, it becomes prohibitive to make a comprehensive review on this topic. In this chapter, we present the basic concepts and ideas of complex networks that are useful in machine learning. We start out by presenting the main concepts of networks. Since complex networks and graphs share the same definition, we first present the basic notations of graph theory. Afterwards, we explore the evolution line and milestones of the complex network research. Following that, a comprehensive list of network measurements is discussed, which enables us to capture structural features of the networks in a systematic manner. Finally, we present some well-known dynamical processes that are defined within the complex networks framework.

2.1 Basic Concepts of Graphs

In this section, we discuss fundamental concepts of the graph theory.

© Springer International Publishing Switzerland 2016

T.C. Silva, L. Zhao, *Machine Learning in Complex Networks*,

DOI 10.1007/978-3-319-17290-3_2

2.1.1   Graph Definitions

We here present the main terminology employed by the literature of graphs or networks theory. In this book, the words graphs and networks convey the same type of information and are used interchangeably. In the same spirit, the data relationships that make up a graph are termed structure, topology, or anatomy of the network.

In the following, we present the formal definition of a graph [8, 21, 35].

Definition 2.1. Graph: A graph $\mathscr{G}$ is defined as an ordered pair $\langle \mathscr{V}, \mathscr{E} \rangle$, where $\mathscr{V}$ is a finite nonempty set of vertices or nodes and $\mathscr{E}$ is the set of edges or links between the vertices $\mathscr{E} \subseteq \{(u, v) \mid u, v \in \mathscr{V}\}$. Some special graphs are defined as follows:

- **Graph with no self-loops**: When the relation $\mathscr{E}$ is irreflexive, meaning that $\forall v \in \mathscr{V}, (v, v) \notin \mathscr{E}$, the graph is said to be free of self-loops. This means that there is no way of traveling to the same vertex in a single transition.
- **Graph with self-loops**: When the relation $\mathscr{E}$ satisfies the following restriction $\exists v \in \mathscr{V}, (v, v) \in \mathscr{E}$, the graph is said to have self-loops. This means that one can travel back to the same vertex through an edge without leaving it.

Moreover, we denote by $V = |\mathscr{V}|$ and $E = |\mathscr{E}|$ the number of vertices and edges, respectively, of the graph.

For example, in the graph $\mathscr{G}$ portrayed in Fig. 2.1, the vertex set is $\mathscr{V} = \{1, \ldots, 5\}$ and the edge set is $\mathscr{E} = \{(1, 2), (1, 3), (2, 3), (3, 3), (3, 4)\}$. We often label the edges with letters or numbers. In the same example, another possible edge labeling is $\mathscr{E} = \{e_1, e_2, e_3, e_4, e_5\}$, where $e_1 = (1, 2)$, $e_2 = (1, 3)$, $e_3 = (2, 3)$, $e_4 = (3, 3)$, $e_5 = (3, 4)$. We can check from Definition 2.1 that the existence of the edge $e_4 = (v_3, v_3)$ turns $\mathscr{G}$ into a graph with self-loops.

Some well-known graph topologies are discussed in the following.

Definition 2.2. Complete graph: A complete graph is a graph in which links exist between each pair of vertices. The complete graph with V vertices is denoted by $\mathscr{K}_V$.

A complete graph can also be further classified into with or without self-loops, in accordance with Definition 2.1.

Fig. 2.1 An example of graph that is undirected, non-weighted, and with self-loops

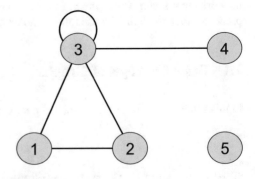

Fig. 2.2 An example of
complete graph that is
non-weighted and with no
self-loops

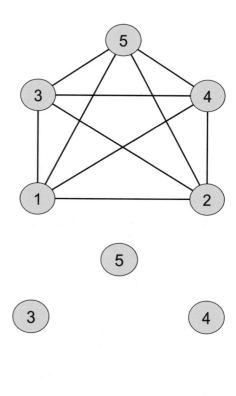

Fig. 2.3 An example of null
graph

For example, Fig. 2.2 shows a complete graph $\mathcal{K}_5$, where every pair of vertices is connected by an edge. As there are no self-loops, the graph is considered a complete graph with no self-loops.

Definition 2.3. Null graph: A null graph is a graph containing no edges, that is, $\mathcal{E} = \emptyset$.

Figure 2.3 illustrates a null graph with 5 vertices. We highlight that, even though the edge set is empty, that is, $\mathcal{E} = \emptyset$, the vertex set cannot be empty. Otherwise, we would not have a formal graph in view of Definition 2.1.

Graphs can also be classified with respect to their edge types. In the next, we discuss the main edge types encountered in the literature.

Definition 2.4. Undirected graph: When the relation $\mathcal{E}$ is symmetric, meaning that $\forall (u, v) \in \mathcal{E} \Rightarrow (v, u) \in \mathcal{E}$, it is said that the graph is undirected. In other terms, when there is an edge linking vertices u to v, so there will be a link from v to u.

Fig. 2.4 An example of a
graph that is directed
(digraph), non-weighted, and
with self-loops

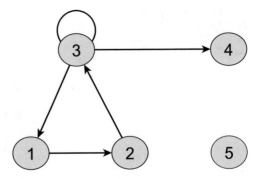

In the illustrative graph in Fig. 2.1, we note that besides the edge set $\mathscr{E}_1 =$ $\{(1, 2), (1, 3), (2, 3), (3, 3), (3, 4)\}$, the set $\mathscr{E}_2 = \{(2, 1), (3, 1), (3, 2), (3, 3), (4, 3)\}$ is also present. Therefore, the graph in Fig. 2.1 is undirected.

Commonly, when the edge $(u, v) \in \mathscr{E}$ is drawn with no arrows in its endpoints, it is assumed that the edge is undirected, implying the existence of the opposite edge $(v, u) \in \mathscr{E}$. This pictorial distinction is made clear with the definition of directed graphs in the following.

Definition 2.5. Directed graph (digraph): When the relation $\mathscr{E}$ satisfies the following restriction: $\exists (u, v) \in \mathscr{E} \mid (v, u) \notin \mathscr{E}$, it is said that the graph is directed (digraph). In other terms, this kind of graph must have at least an arbitrary edge linking u to v, with an absence of the opposite link.

Figure 2.4 gives an example of a directed graph. In this case, each edge has its direction, which is conveyed by the visual illustration of the graph itself. The directness of the graph implies that there exists at least one edge $(u, v) \in \mathscr{E}$ such that $(v, u) \notin \mathscr{E}$. This holds true in Fig. 2.4 for several cases. Among them, we can see that the edge $(1, 2) \in \mathscr{E}$, but $(2, 1) \notin \mathscr{E}$.

There is a special type of graph known as weighted graph, whose definition is given as follows. The same graph categories discussed in Definition 2.1 can be applied to it [13, 32].

Definition 2.6. Weighted graph: A weighted graph $\mathscr{G}$ is defined as a triple $\mathscr{G} = \langle \mathscr{V}, \mathscr{E}, \mathbf{W} \rangle$, where $\mathscr{V}$ and $\mathscr{E}$ are the sets of vertices and edges, respectively, and $\mathbf{W}$ is a matrix that carries the edge weights. For example, the entry $\mathbf{W}_{uv} = w$, $(u, v) \in \mathscr{E}$, fixes as $w > 0$ the weight of the edge linking vertices u to v. If $(u, v) \notin \mathscr{E} \Rightarrow \mathbf{W}_{uv} = 0$.

Figure 2.5 shows an example of a weighted graph where each edge is associated to a value. Often, when no edge weight is specified, it is assumed that the weight is unitary.

The weights can convey various types of meanings in different applications. For example, each value (weight) may represent the distance from vertex (location) i to j, or it may also represent traffic flow and so on. By setting the edge weight with large or small values, we are effectively adjusting the importance of that edge for the

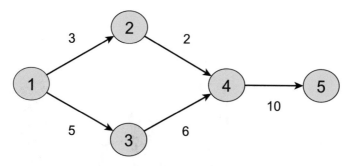

Fig. 2.5 An example of a graph that is weighted, directed (digraph), and with no self-loops

application that we are dealing with. For instance, in graph-based machine learning, each weight frequently represents the similarity degree between two vertices (data samples). As such, large values denote a close proximity of those vertices and, hence, a high importance is given to that relationship in the learning process.

Remark 2.1. When **W** is a binary matrix, then the weighted graph reduces to a non-weighted graph, which is the special graph supplied in Definition 2.1.

Definition 2.7. Bipartite graph: A bipartite graph is a graph whose set of vertices $\mathcal{V}$ can be split into two disjoint non-empty subsets $\mathcal{V}_1$ and $\mathcal{V}_2$, $\mathcal{V} = \mathcal{V}_1 \bigcup \mathcal{V}_2$, in such a way that $(u, v) \in \mathcal{E} \Rightarrow u \in \mathcal{V}_1, v \in \mathcal{V}_2$. Therefore, no edge exists between pairs of vertices in the same subsets $\mathcal{V}_1$ and $\mathcal{V}_2$.

Remark 2.2. Note that, if $\mathcal{G}$ is a bipartite graph, then $\mathcal{G}$ cannot have self-loops.

Remark 2.3. We say that $\mathcal{G}$ is a complete bipartite graph $\mathcal{K}_{M,N}$ when $|\mathcal{V}_1| = M$ and $|\mathcal{V}_2| = N$ and $\forall (v, u) \in \mathcal{V}_1 \times \mathcal{V}_2, (v, u) \in \mathcal{E}$.

When modeling relations between two different classes of objects, bipartite graphs very often arise naturally. Some examples are:

- The graph of football players and clubs, in which an edge exists between a player and a club if that player has played for that club, is a natural example of an affiliation network, a type of bipartite graph used in social network analysis.
- The graph that represents job allocation in a company. A boss must allocate in a company N open jobs for M workers. Each worker is qualified to do some of the N jobs, but not others. Links will exist between a worker and his/her specified qualified jobs.

Figure 2.6 depicts a bipartite graph with $V = 5$ vertices, where $\mathcal{V}_1 = \{1, 2, 3\}$ and $\mathcal{V}_2 = \{4, 5\}$. Note that the existence of links only occurs between vertices of $\mathcal{V}_1$ and $\mathcal{V}_2$.

Fig. 2.6 An example of a
graph that is bipartite,
non-weighted, and directed
(digraph)

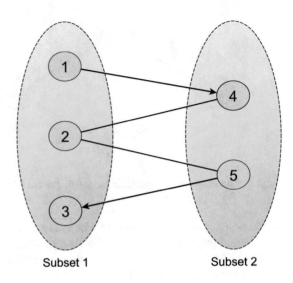

Subset 1   Subset 2

2.1.2   Connectivity

In this section, we introduce common terms related to graph connectivity, which are
used throughout this book [8, 13, 21, 32, 35].

Definition 2.8. Adjacent vertices: Two vertices $u \in \mathcal{V}$ and $v \in \mathcal{V}$ are called
adjacent if they share a common edge, in which case the common edge is said to
join the two vertices.

Remark 2.4. In undirected graphs, if u is adjacent to v, then v must be adjacent to
u as well.

Remark 2.5. In digraphs, u adjacent to v does not imply that v is adjacent to u.
Specifically, if $(u, v) \in \mathcal{E}$ and $(v, u) \notin \mathcal{E}$, then v is adjacent to u, but the opposite
does not hold.

For instance, in the undirected graph portrayed in Fig. 2.7a, vertices 1 and 3 are
adjacent to each other. In contrast, vertex 1 is not adjacent to 4. Now, in the directed
graph depicted in Fig. 2.7b, vertex 1 is adjacent to 3, but the converse is not true.

Definition 2.9. Neighborhood of a vertex: The neighborhood of a vertex $v \in \mathcal{V}$,
in a graph $\mathcal{G}$ is the set of vertices adjacent to v. The neighborhood is denoted by
$\mathcal{N}(v)$ and is formally given by $\mathcal{N}(v) = \{u : (v, u) \in \mathcal{E}\}$.

For illustrative purposes, in the undirected graph shown in Fig. 2.7a, the neighborhood of vertex 1 is $\mathcal{N}(1) = \{2, 3\}$. Now, in the directed graph exhibited in
Fig. 2.7b, $\mathcal{N}(1) = \{2\}$.

Remark 2.6. Some authors further distinguish the neighborhood of a vertex in open
and closed neighborhoods. The *open neighborhood* of v never includes v itself. The

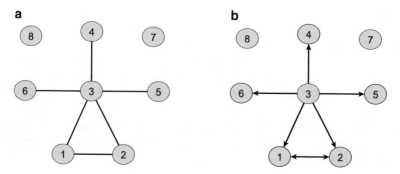

Fig. 2.7 Illustrative non-weighted graphs for exemplifying basic graph concepts. (**a**) Undirected graph. (**b**) Directed graph

closed neighborhood extends the previous one by adding v itself into $\mathcal{N}(v)$, i.e., $\mathcal{N}^{(\text{closed})}(v) = \mathcal{N}(v) \bigcup \{v\}$. In this book, we opt not to discriminate between these classes of neighborhood, because in some machine learning algorithms, self-loops are allowed to prevent transition to other vertices. Therefore, if $v \in \mathcal{N}(v) \iff (v, v) \in \mathcal{E}$. That is, the condition of v being neighbor to itself only depends on the existence of a self-loop in v. We find that this notation is more intuitive and consistent with the machine learning literature.

Definition 2.10. Degree (valency or connectivity) of a vertex: In an undirected graph, the degree of a vertex v is the total number of vertices adjacent to v. The degree of a vertex v is denoted by k_v. We can equivalently define the degree of a vertex as the cardinality of its neighborhood set and say that, for any vertex v, $k_v = |\mathcal{N}(v)|$, i.e.,

$$k_v = |\mathcal{N}(v)| = |\{u : (v, u) \in \mathcal{E}\}| = \sum_{u \in \mathcal{V}} \mathbb{1}_{[(v,u) \in \mathcal{E}]}, \qquad (2.1)$$

in which $\mathbb{1}_{[K]}$ represents the Kronecker delta or indicator function that yields 1 if the logical expression K is true; otherwise, it returns 0.

Remark 2.7. The feasible values of k_v are within the discrete-valued interval $\{0, \ldots, V - 1\}$ if self-loops are not allowed, and in $\{0, \ldots, V\}$ if self-loops are permitted.

Remark 2.8. When $k_v = 0$, then v is said to be a *singleton* or *isolated* vertex.

Remark 2.9. When k_v assumes relative large values than the remainder of the vertices in the network, we say that v is a *hub*.

In the undirected graph depicted in Fig. 2.7a, vertices 7 and 8 are singleton, for $k_7 = k_8 = 0$. In contrast, vertex 3 is considered as a hub, for its degree is relative large in relation to the remainder vertices of the network.

We have so far discussed definitions mostly suited to undirected graphs. For directed graphs, some of the previously defined connectivity measures suffer slight modifications, mainly due to the fact that distinctions in the edge endpoints must be brought into consideration. In special, a directed edge has two distinct ends: an origin and a destination. The measures use these two endpoints independently. In light of these considerations, we now extend the connectivity definitions to the case of directed networks.

Definition 2.11. In-degree and out-degree: In a directed graph, the notion of vertex degree can be further extended into the in-degree, $k_v^{(in)}$, and out-degree, $k_v^{(out)}$, as follows:

$$k_v^{(in)} = \sum_{u \in \mathcal{V}} \mathbb{1}_{[v \in \mathcal{N}(u)]} = \sum_{u \in \mathcal{V}} \mathbb{1}_{[(u,v) \in \mathcal{E}]}, \qquad (2.2)$$

$$k_v^{(out)} = \sum_{u \in \mathcal{V}} \mathbb{1}_{[u \in \mathcal{N}(v)]} = \sum_{u \in \mathcal{V}} \mathbb{1}_{[(v,u) \in \mathcal{E}]}, \qquad (2.3)$$

$$k_v = k_v^{(in)} + k_v^{(out)}. \qquad (2.4)$$

Remark 2.10. The domains of $k_v^{(out)}$ and $k_v^{(in)}$ are $\{0, \ldots, V-1\}$ if self-loops are not allowed, and $\{0, \ldots, V\}$ if self-loops are permitted. Therefore, k_v may assume the values $\{0, \ldots, 2(V-1)\}$ when no loops are present and $\{0, \ldots, 2V\}$ when loops are allowed.

Remark 2.11. Note that $k_v^{(out)} = |\mathcal{N}(v)|$.

For example, in the directed graph exhibited in Fig. 2.7b, $k_3^{(out)} = 5$, $k_3^{(in)} = 0$, and $k_3 = 5 + 0 = 5$. In addition, $k_1^{(out)} = 1$, $k_1^{(in)} = 2$, and $k_1 = 1 + 2 = 3$.

Definition 2.12. Average network degree: The average degree of the network, or network connectivity, is given by:

$$\bar{k} = \frac{1}{V} \sum_{v \in \mathcal{V}} k_v = \frac{1}{V} \sum_{(v,u) \in \mathcal{V}^2} \mathbb{1}_{[(v,u) \in \mathcal{E}]}. \qquad (2.5)$$

For instance, in the undirected graph exhibited in Fig. 2.7a, the average degree is:

$$\bar{k} = \frac{1}{8} [k_1 + \ldots + k_8] = \frac{1}{8} [2 + 2 + 5 + 1 + 1 + 1 + 0 + 0] = 1.5,$$

i.e., on average, a vertex belonging to that network has 1.5 links.

Definition 2.13. Average in-degree and out-degree: In a directed graph, the average in-degree and out-degree have the same numerical value and are evaluated as:

$$\bar{k}^{(in)} = \bar{k}^{(out)} = \frac{1}{V} \sum_{v \in \mathscr{V}} k_v^{(in)} = \frac{1}{V} \sum_{v \in \mathscr{V}} k_v^{(out)}. \tag{2.6}$$

In the example shown in Fig. 2.7b, the average in- and out-degree are given by:

$$\bar{k}^{(in)} = \frac{1}{8}\left[k_1^{(in)} + \ldots + k_8^{(in)}\right] = \frac{1}{8}[2 + 2 + 0 + 1 + 1 + 1 + 0 + 0] = \frac{7}{8},$$

$$\bar{k}^{(out)} = \frac{1}{8}\left[k_1^{(out)} + \ldots + k_8^{(out)}\right] = \frac{1}{8}[1 + 1 + 5 + 0 + 0 + 0 + 0 + 0] = \frac{7}{8}.$$

In the rest of this section, we define some connectivity measurements that are useful for weighted graphs.

Definition 2.14. Strength: In an undirected weighted graph, the strength of a vertex $v \in \mathscr{V}$, indicated by s_v, represents the total sum of weighted connections of v towards its neighbors.

$$s_v = \sum_{u \in \mathscr{V}} \mathbf{W}_{vu}, \tag{2.7}$$

in which $\mathbf{W}_{vu}$ is the edge weight of v to u, as introduced in Definition 2.6.

In the graph exhibited in Fig. 2.8a, $s_1 = 3 + 2 = 5$, and $s_2 = 3 + 5 + 10 = 18$.

Definition 2.15. In-strength and out-strength: In a directed weighted graph, the notion of vertex strength can be further extended into the in-strength, $s_v^{(in)}$, and out-strength, $s_v^{(out)}$, as follows:

$$s_v^{(in)} = \sum_{u \in \mathscr{V}} \mathbf{W}_{uv}, \tag{2.8}$$

$$s_v^{(out)} = \sum_{u \in \mathscr{V}} \mathbf{W}_{vu}, \tag{2.9}$$

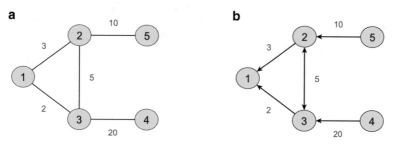

Fig. 2.8 Illustrative weighted graphs for exemplifying basic graph concepts. (**a**) Undirected graph. (**b**) Directed graph

$$s_v = s_v^{(in)} + s_v^{(out)}, \tag{2.10}$$

in which $\mathbf{W}_{vu}$ is the edge weight linking v to u.

In the example supplied in Fig. 2.8b, for vertex 1, we have $s_1^{(in)} = 3 + 2 = 5$ and $s_1^{(out)} = 0$. Similarly, for vertex 2, $s_2^{(in)} = 5 + 10 = 15$ and $s_2^{(out)} = 3 + 5 = 8$.

With the basic connectivity concepts introduced, we present another well-known graph topology in the following.

Definition 2.16. Regular graph: A graph is regular if all of the graph vertices have the same degree. In particular, if the degree of each vertex is k, $\mathscr{G}$ is said to be k-regular.

Remark 2.12. If $\mathscr{G}$ is a complete graph with V vertices, then it is $(V - 1)$-regular. An example is the complete graph in Fig. 2.2, which is 4-regular with 5 vertices.

Examples of regular graphs that are not complete are supplied in Fig. 2.9. In special, Fig. 2.9a has six vertices and is a 2-regular network, while Fig. 2.9b has ten vertices and is a 3-regular network.

2.1.3 Paths and Cycles

Definition 2.17. Walk: Let $v_1, \ldots, v_K \in \mathscr{V}$, $K \geq 2$. A walk $\mathscr{W}$ is an ordered sequence of edges: $\mathscr{W} = \{(v_1, v_2), (v_2, v_3), \ldots, (v_{K-1}, v_K)\}$, such that $\forall k \in \{2, \ldots, K\} : (v_{k-1}, v_k) \in \mathscr{E}$. In this case, v_1 and v_k are called the walk's origin and destination, respectively. Note that vertices can be revisited in the same walk.

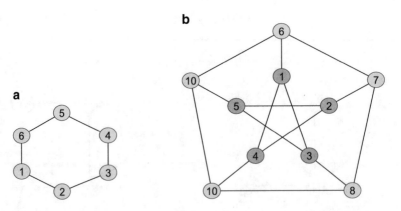

Fig. 2.9 Examples of regular graphs that are not complete. (**a**) 2-regular graph. (**b**) 3-regular graph (Petersen graph)

Remark 2.13. A walk is called *closed* if $v_1 = v_K$ and *open* otherwise.

Remark 2.14. A walk consisting of a single vertex is called a *trivial* walk.

Definition 2.18. Trail: A trail is a walk in which no edge is repeated. Trails can also be further classified into open and closed trails, according to Remark 2.13.

Definition 2.19. Tour or circuit: A tour is a closed trail.

Definition 2.20. Walk length: The length of a walk $\mathcal{W} = \{(v_1, v_2), (v_2, v_3), \dots,$ $(v_{K-1}, v_K)\}$, $K \geq 2$, is the number of edges that the walk traverses, i.e., $|\mathcal{W}| = K - 1 \geq 1$.

In the undirected graph portrayed in Fig. 2.10, $\mathcal{W}_1 = \{(1, 3), (3, 4), (4, 6), (6, 7),$ $(7, 4)\}$ is an open walk. In contrast, $\mathcal{W}_2 = \{(1, 3), (3, 4), (4, 6), (6, 7), (7, 4), (4, 3),$ $(3, 1)\}$ is a closed walk. There are no trivial walks, as the graph in Fig. 2.10 has no self-loops. $\mathcal{W}_3 = \{(5, 8), (8, 7)\}$ is an open trail and $\mathcal{W}_4 = \{(5, 8), (8, 7), (7, 5)\}$ is a closed trail or a tour. The lengths of these walks are: $|\mathcal{W}_1| = 5$, $|\mathcal{W}_2| = 7$, $|\mathcal{W}_3| = 2$, and $|\mathcal{W}_4| = 3$. There are no walks that visit vertex 10.

Definition 2.21. Path: A path $\mathcal{P}$ is a non-trivial walk in which all vertices (except possibly the first and last) are distinct.

Remark 2.15. A path is always a walk.

Definition 2.22. Cycle: A cycle is closed path.

In Fig. 2.10, $\mathcal{P}_1 = \{(1, 2), (2, 5), (5, 7)\}$ is a path and $\mathcal{P}_2 = \{(1, 2), (2, 5), (5, 7),$ $(7, 4), (4, 3), (3, 1)\}$ is a cycle. Note that $\mathcal{P}_3 = \{(5, 8), (8, 7), (7, 6), (6, 4), (4, 7),$ $(7, 5), (5, 8)\}$ is a walk and tour but not a cycle, because it is not even a path.

Definition 2.23. Walk or path distance: The distance d of the walk $\mathcal{W} = \{(v_1, v_2), (v_2, v_3), \dots, (v_{K-1}, v_K)\}$, $K \geq 2$ is given by:

Fig. 2.10 Illustrative undirected graph to introduce graph traversal measures

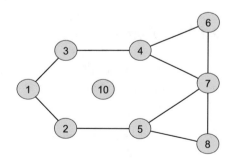

$$d(\mathscr{W}) = \sum_{k=2}^{K} |(v_{k-1}, v_k)| = \sum_{k=2}^{K} \mathbf{W}_{k-1,k}, \tag{2.11}$$

in which $|(v_{k-1}, v_k)|$ is the edge weight linking vertex v_{k-1} to v_k.

Definition 2.24. Shortest path (geodesic path) between vertices: The shortest path between $u \in \mathscr{V}$ and $v \in \mathscr{V}$, denoted here as d_{uv}, is given by the path starting from u and ending at v with the least distance. Mathematically,

$$d_{uv} = \min_{\mathscr{W}_{u \to v}} d(\mathscr{W}_{u \to v}), \tag{2.12}$$

in which $\mathscr{W}_{u \to v}$ represents walks starting from u and ending at v.

Remark 2.16. For measures that require two inputs, such as the shortest path between vertices, we use d_{uv} when the subscripts are variables and $d_{1,2}$ when they are numbers. That is, we maintain the notation as succinct as possible. The comma is employed for clarity when numbers are indexed.

Definition 2.25. Distance between vertices: The distance d_{uv} between two vertices u and v is always their shortest path distance.

Remark 2.17. Note that d_{uv} is always evaluated from a path. That is, the distance between u and v cannot be a walk that is not a path.

Remark 2.18. The distance between any vertex and itself is 0.

Remark 2.19. If there is no path from u to v, then $d_{uv} = \infty$.

In Fig. 2.10, the distance between 1 and 3 is $d_{1,3} = 1$, since the shortest path from 1 to 3 is $\{(1, 3)\}$. The distance from vertex 10 to itself is $d_{10,10} = 0$. Moreover, the distance from vertex 1 to 10 is $d_{1,10} = \infty$, as no paths nor walks exist between 1 and 10.

2.1.4   Subgraphs

Definition 2.26. Reachability: We say that $v_2 \in \mathscr{V}$ is reachable from $v_1 \in \mathscr{V}$ if $d_{v_1 v_2}$ is finite. Alternatively, v_1 reaches v_2 if there is at least a walk that starts from v_1 and ends at v_2.

Definition 2.27. Connectedness: Graph $\mathscr{G}$ is connected if, for every pair of vertices v_1 and v_2, v_2 is reachable from v_1 *or* v_1 is reachable from v_2.

Definition 2.28. Strong connectedness: Graph $\mathscr{G}$ is strongly connected if, for every pair of vertices v_1 and v_2, v_2 is reachable from v_1 *and* v_1 is reachable from v_2.

Remark 2.20. Strong connectedness implies connectedness.

Remark 2.21. In undirected graphs, connectedness implies strong connectedness. This holds true because if v_1 reaches v_2, then the converse must be true, for edges are two-way in undirected graphs.

Remark 2.22. In directed graphs, connectedness does not imply strong connectedness.

In the undirected graph depicted in Fig. 2.11a, the graph is strongly connected and, hence, each pair of vertices is mutually reachable. In contrast, in the directed graph exhibited in Fig. 2.11b, the graph is connected but not strongly connected. For instance, v_1 reaches v_6 but the converse is not true. The graphs in Fig. 2.11c, d are not strongly connected nor connected. For example, v_1 and v_8 are mutually non-reachable.

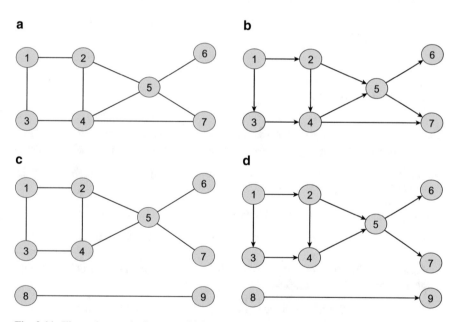

Fig. 2.11 Illustrative graphs for exemplifying subgraph concepts. (**a**) Undirected graph (1 component). (**b**) Directed graph (1 component). (**c**) Undirected graph (2 components). (**d**) Directed graph (2 components)

Definition 2.29. Graph component: The subgraph $\mathscr{G}_C$ of $\mathscr{G}$ is a component if:

- $\mathscr{G}_C$ is connected;
- All of the proper subsets of $\mathscr{G}_C$ are not connected.

Alternatively, $\mathscr{G}_C$ is a graph component if any two of its vertices are reachable at least from one to another, and if its vertex members are connected to no additional vertices in the remainder of the graph.

Remark 2.23. A connected graph has always a single component.

In Fig. 2.11a, b, there is a single component that is the graph itself. In contrast, in Fig. 2.11c, d, two components exist: $\mathscr{G}_1 = \{1, 2, 3, 4, 5, 6, 7\}$ and $\mathscr{G}_2 = \{8, 9\}$.

Definition 2.30. Clique: A clique in an undirected graph is a subset of vertices such that every two vertices in the subset are connected by an edge. Cliques therefore are subgraphs or graphs that are complete.

In Fig. 2.11a, there are two cliques: one comprises the vertices $\{4, 5, 7\}$, while the other, $\{2, 4, 5\}$.

2.1.5   Trees and Forest

Definition 2.31. Tree graph: A tree is a connected graph that has no cycles. In a tree, a leaf is a vertex of degree 1. An internal vertex is a vertex of degree at least 2.

Definition 2.32. Forest: A forest is an undirected graph in which all of its connected components are trees.

Remark 2.24. Note that a forest is a graph consisting of a disjoint union of trees.

Remark 2.25. All trees are forests, but the converse is not always true.

Remark 2.26. Special cases of forests include: a single tree and a graph with only singleton vertices (empty graph).

Figure 2.12a illustrates a tree, while Fig. 2.12b exemplifies a forest with two trees.

Definition 2.33. Spanning tree: If $\mathscr{G}$ is a connected graph, the spanning tree in $\mathscr{G}$ is a subgraph of $\mathscr{G}$ which includes every vertex of $\mathscr{G}$ and is also a tree graph.

For example, Fig. 2.13b shows a possible spanning tree from the graph exhibited in Fig. 2.13a. In this transformation process, we have removed the edges $(2, 3), (2, 4), (3, 5), (4, 5), (6, 7)$.

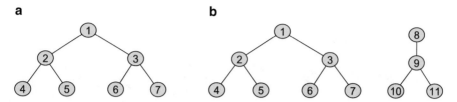

Fig. 2.12 Examples of special types of graphs: trees and forests. (**a**) A tree. (**b**) A forest with two trees

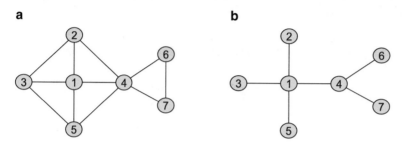

Fig. 2.13 Transformation of a graph in (**a**) into a spanning tree in (**b**)

2.1.6 Graph Representation

Mathematically, a non-weighted graph $\mathscr{G} = \langle \mathscr{V}, \mathscr{E} \rangle$ or weighted graph $\mathscr{G} = \langle \mathscr{V}, \mathscr{E}, \mathbf{W} \rangle$ are frequently represented by an adjacency matrix $\mathbf{A}$ that is constructed from the vertex and edge sets. A formal definition of the adjacency matrix is given as follows.

Definition 2.34. Adjacency matrix: Let $\mathscr{G} = \langle \mathscr{V}, \mathscr{E}, \mathbf{W} \rangle$ be an weighted graph. Then, the adjacency matrix $\mathbf{A}$ is defined as follows:

- The number of vertices $|\mathscr{V}| = V$ serves to establish the dimension of the adjacency matrix, which is always $V \times V$;
- The edge set contributes to defining the entry values of the adjacency matrix in the following manner. The (i,j)-th entry of $\mathbf{A}$ is denoted as $\mathbf{A}_{ij} = a_{ij} = \mathbf{W}_{ij}$, where $\mathbf{W}_{ij}$ is the weight of the edge linking i to j. Formally, $\forall (i,j) \in \mathscr{E} : a_{ij} \neq 0$ and $\forall (i,j) \notin \mathscr{E} : a_{ij} = 0$.

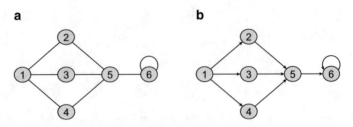

Fig. 2.14 Illustrative graphs introduced for evaluating their adjacency matrices. (**a**) Undirected graph. (**b**) Directed graph

Usually, the adjacency matrix $\mathbf{A}$ takes the following matrix form:

$$\mathbf{A} = \begin{pmatrix} a_{1,1} & a_{1,2} & \cdots & a_{1,V} \\ a_{2,1} & a_{2,2} & \cdots & a_{2,V} \\ \vdots & \vdots & \ddots & \vdots \\ a_{V,1} & a_{V,2} & \cdots & a_{V,V} \end{pmatrix}. \tag{2.13}$$

Remark 2.27. If $\mathcal{G}$ is non-weighted, then $\mathbf{A}_{ij} \in \{0, 1\}$, $\forall i, j \in \mathcal{V}$.

Remark 2.28. If the graph $\mathcal{G}$ is undirected, then $\mathbf{A}$ is symmetric. This fact implies that if $\mathbf{A}_{ij} = 1$, then $\mathbf{A}_{ji} = 1$.

Remark 2.29. Contrasting to the previous Remark, directed graphs may not have symmetric adjacency matrices, as j can be a neighbor of i and the converse may not hold.

For instance, the undirected graph shown in Fig. 2.14a has the following adjacency matrix:

$$\mathbf{A} = \mathbf{A}^T = \begin{pmatrix} 0 & 1 & 1 & 1 & 0 & 0 \\ 1 & 0 & 0 & 0 & 1 & 0 \\ 1 & 0 & 0 & 0 & 1 & 0 \\ 1 & 0 & 0 & 0 & 1 & 0 \\ 0 & 1 & 1 & 1 & 0 & 1 \\ 0 & 0 & 0 & 0 & 1 & 1 \end{pmatrix}. \tag{2.14}$$

in which the superscript T denotes the transpose operator.

In addition, note that the matrix in (2.14) is symmetric, i.e., $\mathbf{A} = \mathbf{A}^T$, as the graph in Fig. 2.14a is undirected.

In contrast, the directed graph exhibited in Fig. 2.14b has the following adjacency matrix:

Fig. 2.15 Illustrative
weighted graph introduced
for evaluating the weighted
matrix

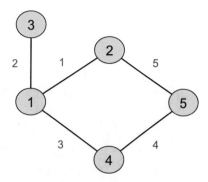

$$A = \begin{pmatrix} 0 & 1 & 1 & 1 & 0 & 0 \\ 0 & 0 & 0 & 0 & 1 & 0 \\ 0 & 0 & 0 & 0 & 1 & 0 \\ 0 & 0 & 0 & 0 & 1 & 0 \\ 0 & 0 & 0 & 0 & 0 & 1 \\ 0 & 0 & 0 & 0 & 0 & 1 \end{pmatrix}. \tag{2.15}$$

In this case, the matrix in (2.15) is not symmetric.

For weighted graphs, the entries in the adjacency matrix can assume arbitrary values. For instance, the (weighted) adjacency matrix of the weighted undirected graph portrayed in Fig. 2.15 is:

$$A = A^T = \begin{pmatrix} 0 & 1 & 2 & 3 & 0 \\ 1 & 0 & 0 & 0 & 5 \\ 2 & 0 & 0 & 0 & 0 \\ 3 & 0 & 0 & 0 & 4 \\ 0 & 5 & 0 & 4 & 0 \end{pmatrix}. \tag{2.16}$$

2.2 Complex Network Models

With the expectation of studying topological properties that are linked to real networks, several network models have been proposed. Some of these models even have inspired an extensive study due to its features of great interest. As examples of important categories of networks, one can list: random networks, small-world networks, clustered random networks, scale-free networks, and core-periphery networks. In the next sections, we review these models in detail.

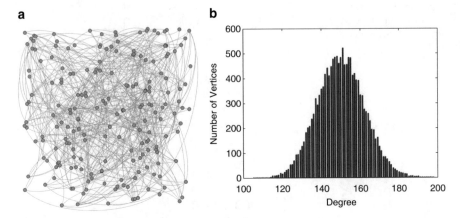

Fig. 2.16 An example of random networks of Erdös and Réyni. (**a**) A network constructed by means of the random approach proposed by Erdös e Réyni; and (**b**) the degree distribution of a network consisting of $V = 15,000$ constructed using the Erdös and Réyni methodology with $p = 0.01$

2.2.1   Random Networks

In the article dated back to 1959, Erdös and Réyni [24] developed a model that generates random networks consisting of V vertices and E edges. Starting from V vertices completely disconnected (no edges in the network), the network is built from the gradual addition of L edges randomly created, in such a way that self-looping is avoided. Another similar model sets V vertices in a network, and there is a probability $p > 0$ of connecting each possible pair of vertices. The latter model is widely recognized as the model of Erdös and Réyni. Figure 2.16a depicts an example of this type of network. Note that no spacial relation between the vertices is used. In this network formation, we merely create edges in a uniform probabilistic way, regardless of the similarity between vertices.

Since, for each vertex $i \in \mathscr{V}$ of the network (a total of V), there are $V-1$ different possibilities of connections with other vertices, it follows that the cardinality of the sample space, $|\Omega|$, which quantifies the maximum theoretical number of edges between the vertices, is given by:

$$|\Omega| = \frac{V(V-1)}{2}, \qquad (2.17)$$

in which the division by 2 comes from the fact that we are considering that the graph is undirected, i.e., the edges are always bidirectional in relation to both linked vertices. In general, the presence of these two edges represents the occurrence of the same probabilistic event, on account of the inherent coupling (bidirectionally). Having in mind that an arbitrary edge is present in a random network with probability p and is absent with probability $1 - p$, and remembering that there are

$\binom{V-1}{k}$ ways of choosing k vertices over $V - 1$ in total, and p^k denotes the joint probability of these k vertices to possess exactly k connected vertices,[1] then $\binom{V-1}{k}p^k$ provides the probability of these k vertices to have exactly k other interconnected vertices. However, in this analysis, it should be imposed that there are no more edges beyond these k, i.e., for the reminiscent quantity of vertices, $V - 1 - k$, the complementary probabilistic event of existing edges, that is, $(1 - p)^{(V-1-k)}$, must happen. In view of this reasoning, the degree distribution follows a Binomial distribution with parameters $Binomial(V - 1, p)$, whose equation is governed by the following expression:

$$P(k) = \binom{V-1}{k} p^k (1 - p)^{(V-1)-k}. \tag{2.18}$$

Given that $V \to \infty$ and $p \ll 1$, one can show that a Binomial distribution parameterized with $Binomial(V - 1, p)$ asymptotically approximates a Poisson distribution with parameter $Poisson(\lambda)$ [52], with the following linking condition:

$$(V - 1)p = \lambda. \tag{2.19}$$

Recall from the probability theory that the mean, μ, and the variance, σ^2, of a $Poisson(\lambda)$ are given by $\mu = \sigma^2 = \lambda$. If we construct an artificial random network using the discussed methodology with $V = 15,000$ e $p = 0.01$, we get the degree distribution that is displayed in Fig. 2.16b. Note that the resulting degree distribution really approximates the Poisson distribution with mean (peak) around $\lambda = (V - 1)p = (15,000 - 1)0.01 \approx 150$.

Moreover, the average shortest path $\langle d \rangle$ is small in random networks. This quantity increases proportionally to the logarithm of the network size, i.e., $\langle d \rangle \sim \frac{\ln(V)}{\ln(\langle k \rangle)}$, where $\langle k \rangle$ is given by the average value of the Poisson distribution (mean degree), meaning that $\langle k \rangle = \lambda = (V - 1)p$, [20].

The big discovery of Erdös and Réyni was that many important properties of a random network may be unveiled as one modifies the parameters of a $Binomial(V - 1, p)$. In their study, they showed that, for values of the connecting probability p larger than a critical probability p_c, almost all of the random networks present a specific property Q with probability 1. That same property is not verified whenever $p \leq p_c$. For example, if p is larger than a certain value of p_c, the random networks can present a single connected component. But, for values below this critical threshold, the random networks no longer present a single component, but instead several unconnected subgraphs. Many other interesting properties have been discussed in the literature and some of them are reviewed in [55].

---

[1]The joint probability is evaluated taking into account that the existence or absence of links are independent from each other in the random network model.

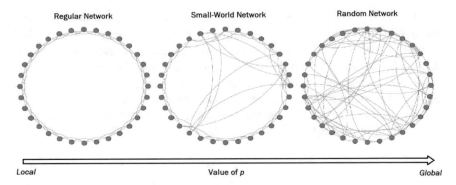

Fig. 2.17 Network behavior as we increase the parameter p, which is responsible for the relocation frequency of the edges

2.2.2   Small-World Networks

Several real-world networks exhibit the small-world property, i.e., most vertices can be reached by others by means of a small number of intermediate steps (edges). This characteristic is found, for example, in social networks, where virtually everyone in the world can be reached by a short chain of people [73, 74].

In order to build a network that presents the small-world property, one can use the following network formation process introduced in [74]:

- Initially, the network is regular, comprising V vertices, as shown in the left-most network in Fig. 2.17, in which each vertex connects to its k nearest neighbors in each direction, totalizing $2k$ connections;
- Then, each edge is randomly relocated, i.e., given an arbitrary vertex $i \in \mathcal{V}$, we randomly choose one of its original $2k$ connections. The selected edge, say linking vertices i and $j \in \mathcal{V}$, is randomly relocated, such that the destination from j is switched to another vertex $u \in \mathcal{V}, j \neq u$, with probability p.

When $p = 0$, no rearrangements are performed and, therefore, the network continues to be regular. Conversely, when $p \to 1$, all of the edges are effectively relocated [74]. Figure 2.17 brings a schematic of the behavior of the parameter p, responsible for the relocation frequency of the edges. Note that, for $p = 0$, the resulting network is virtually a regular one. As p increases (but still remains small), the property of small-world becomes apparent. When $p = 1$, the network turns out to be random. In this case, the peak of the degree distribution, following this approach, is situated close to $2k$ [73, 74].

The immediate implication for networks that have the property of small world is that the spread of any information, given that it was generated at any arbitrary vertex of the network, is very fast. For example, in viral contagion networks with the small-world property, given that a person has contracted some virus, then it is expected that, in a short time, many people will be infected by this virus due to the network topology that favors rapid propagation.

2.2.3 Scale-Free Networks

In a study conducted by Barabási and Albert [5], they noticed that some networks have a small number of vertices with large degrees, while most of them have very small degrees. With this observation in mind, in 1999, they proposed a new type of network denominated scale-free networks, in which the degree distribution obeys a power-law, as follows:

$$P(k) \sim k^{-\gamma}, \tag{2.20}$$

in which γ is a scaling exponent. Note that, by setting a fixed value for γ, as the degree k grows, the number of vertices that have degree k decreases. Thus, it is expected that $P(k)$ will have a large value for small values of k and a small value for large values of k, which is consistent with the observation found by Barabási and Albert.

The scale-free property strongly correlates with the network robustness to failure. In a scale-free network topology, it turns out that major hubs are closely followed by smaller ones. These smaller hubs, in turn, are followed by other vertices with an even smaller degree and so on until we reach peripheral or terminal vertices. This hierarchy allows for a fault tolerant behavior. If failures occur at random and the vast majority of vertices are those with small degree, the likelihood that a hub would be affected is almost negligible. Even if a hub-failure occurs, the network generally does not lose its connectedness, due to the remaining hubs. On the other hand, if we choose a few major hubs and take them out of the network, the network is turned into a set of rather isolated graphs. Thus, hubs are both a strength and a weakness of scale-free networks. In view of that, the literature often terms scale-free networks as robust to random attacks yet fragile to intentional attacks. These properties have been studied analytically using percolation theory by Cohen et al. [16, 17] and by Callaway et al. [11].

The formation of scale-free networks happens due to preferential attachment of vertices. This behavior can be understood in terms of network growth. Growth in this context means that the number of vertices in the network increases over time. Preferential attachment means that the more connected a vertex is, the more likely it is to receive new links. Vertices with larger degree have stronger ability to grab links added to the network. Intuitively, the preferential attachment can be understood if we think in terms of social networks connecting people. Here a link from A to B means that person A "knows" or "is acquainted with" person "B." Heavily linked vertices represent well-known people with lots of relations. When a newcomer enters the community, he or she is more likely to become acquainted with one of those more visible people rather than with a relative unknown. Similarly, on the web, new pages link preferentially to hubs, i.e., very well-known sites such as Google or Wikipedia, rather than to pages that hardly anyone knows. If someone selects a new page to link to by randomly choosing an existing link, the probability of selecting a particular page would be proportional to its degree. This explains the preferential attachment probability rule. Preferential attachment is an example of a positive feedback cycle where initially random variations are automatically reinforced, thus greatly magnifying differences.

Albert and Barabási [5] proposed an algorithm to generate scale-free network with this preferential attachment mechanism. The network begins with an initial connected network of V_0 vertices. New vertices are added to the network one at a time. Each new vertex is connected to $V \leq V_0$ existing vertices with a probability that is proportional to the number of links that the existing vertices already have. Formally, the probability p_i that the new vertex is connected to vertex i is:

$$p_i = \frac{k_i}{\sum_{j \in \mathcal{V}} k_j}, \tag{2.21}$$

in which k_i is the degree of vertex i. Heavily linked vertices or hubs tend to quickly accumulate even more links, while vertices with only a few links are unlikely to be chosen as the destination for a new link. The new vertices have a "preference" to attach themselves to the already heavily linked vertices.

Figure 2.18 shows an illustrative network that shares the scale-free properties. Note that there are very few vertices with large degree, while the great majority (terminal vertices) has small degree.

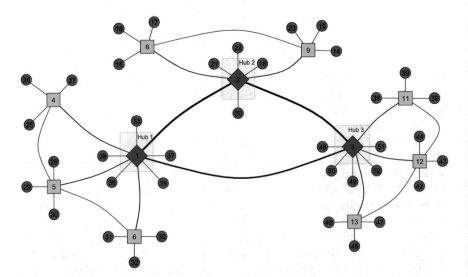

Fig. 2.18 Schematic of a scale-free network. The hubs (vertices with large degrees) have been evidenced. Note that there are very few vertices with large degrees, while the great majority (terminal vertices) has small degrees

2.2.4 Random Clustered Networks

Some real-world networks, such as social and biological ones, present modular structures called communities [31]. These communities consist of sets of vertices that satisfy a simple rule: vertices belonging to the same community have many interconnecting edges, while different communities share relatively few edges interconnecting each other. A model for generating such communities was proposed in [31]. This agglomerative method groups V initially isolated vertices into M communities. This is managed by creating a link between two vertices with probability p_{in}, if they belong to the same community, or with probability p_{out}, if they belong to distinct communities. The values for p_{in} and p_{out} can be arbitrarily chosen to control the number of intracommunity and intercommunity links, z_{in} and z_{out}, respectively, for an arbitrary average network degree $\langle k \rangle$.

High values of p_{in} and low values of p_{out} refer to networks with well-defined communities, i.e., there is a high concentration of edges confined within each community and very few edges interconnecting different communities. Conversely, low values of p_{in} and high values of p_{out} contribute to the appearance of communities highly mixed with each other. On the basis of these parameters, we can define the fraction of intracommunity links $z_{in}/\langle k \rangle$ and, likewise, the fraction of intercommunity links $z_{out}/\langle k \rangle$. The quantity $z_{out}/\langle k \rangle$ defines the mixture among different communities. Essentially, as $z_{out}/\langle k \rangle$ increases, the communities become more mixed and harder to be identified. As we will further see in Sect. 6.2.4, these quantities are usually employed to compare different competing community detection techniques using the Girvan-Newman's benchmark, which adopts the random clustered networks discussed here.

Empirically, $p_{out} \ll p_{in}$ must be satisfied in order to guarantee the presence of communities in the network. Figure 2.19 illustrates a network with four communities. Observe that the communities in this figure are well-defined, since the number of edges connecting vertices of the same community is much larger than the number of edges interconnecting those of different communities.

2.2.5 Core-Periphery Networks

Networks can be described using a combination of local, global, and intermediate-scale (mesoscale) perspectives. In this aspect, one of the key objectives of network theory is the identification of statistical summaries for large networks in order to develop frameworks that serve to analyze and compare complex structures. In such efforts, the algorithmic identification of mesoscale graph structures makes it possible to uncover features that might not be apparent neither at the local level of vertices and edges nor at the global level of statistical summaries.

In particular, several efforts have gone into the algorithmic identification and investigation of a particular type of mesoscale structure known as community

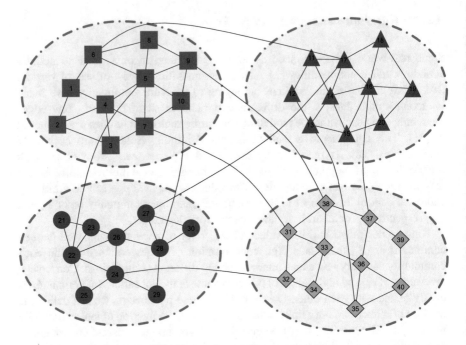

Fig. 2.19 Schematic of a random clustered network with four well-defined communities. Each community is distinguished by a unique color or format

structure, in which cohesive groups called communities consist of vertices that are densely interconnected and connections between vertices in different communities are comparatively sparse.

Although researches of community structure have been very successful [28], the investigation of other types of mesoscale structures—often in the form of different "block models" [26, 28]—have received much less attention in the literature. In this section, we deal with another kind of mesoscale network structure known as core-periphery structure. The qualitative notion that social networks can have such a structure makes intuitive sense and has a long history in subjects like sociology [22, 45], international relations [12, 70], and economics [42]. The most popular quantitative method to investigate core-periphery structure was proposed by Borgatti and Everett in 1999 [10].

By computing a network core-periphery structure, one attempts to determine which vertices are part of a densely connected core and which are part of a sparsely connected periphery. Core vertices should also be reasonably well-connected to peripheral vertices, but the latter are not well-connected to a core nor to each other. Hence, a vertex belongs to a core if and only if it is well-connected both to other core vertices and to peripheral vertices. A core structure in a network is thus not merely densely connected but also tends to be "central" to the network (e.g., in terms of short paths through the network). The goal of quantifying various notions of "centrality," which are intended to measure the importance of a vertex or other network component [58, 72], also helps in distinguishing core-periphery structure from community structure. Additionally, networks can have nested core-periphery structure as well as both core-periphery structure and community structure [46], so it is desirable to develop algorithms that allow one to simultaneously examine both types of mesoscale structure.

Hubs, which are vertices that have large degree, occur in many real-world networks and can pose a problem for community detection, as they often are connected to vertices in many parts of a network and can thus have strong ties to several different communities. For instance, such vertices might be assigned to different communities when applying different computational heuristics using the same notion of community structure [69]. Therefore, it becomes crucial to consider their strengths of membership across different communities (e.g., by using a method that allows overlapping communities) [1]. In such situations, the usual notion of a community might not be ideal for achieving an optimal understanding of the mesoscale network structure that is actually present, and considering hubs to be part of a core in a core-periphery structure might be more appropriate [46]. For example, one can consider communities as tiles that overlap to produce a network core [68, 75].

Figure 2.20 illustrates a perfect core-periphery network. We observe that core vertices are strongly interconnected to each other and also considerably connected to the remainder of the peripheral network. Peripheral vertices, in turn, are only connected to the core.

Fig. 2.20 Schematic of a core-periphery network. The core member are depicted in *square-shaped vertices (cyan color)* and the peripheral members, in *circle-shaped vertices (yellow color)*

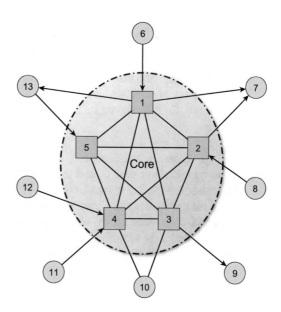

2.3 Complex Network Measures

In this section, we review network measurements that have been proposed in the complex networks literature.

2.3.1 Degree and Degree-Correlation Measures

Definition 2.35. Density: the network density D measures how strong the vertices of a graph are connected. It is defined as the fraction of actual connections over the total possible connections.

For a directed network, the density D is defined as:

$$D = \frac{E}{2\binom{V}{2}} = \frac{2E}{2V(V-1)} = \frac{E}{V(V-1)}, \tag{2.22}$$

in which $2\binom{V}{2}$ denotes the total number of possible connections in a directed graph. In special, the binomial accounts for getting the total number of pairwise combinations between two vertices in the network. We multiply by two because the ordering (start and destination vertices) of those pairwise connections matters in a directed graph.

For an undirected network, the density D is:

$$D = \frac{E}{\binom{V}{2}} = \frac{2E}{V(V-1)}, \tag{2.23}$$

in which, in this case, the ordering of the pairwise connections does not matter.

The density assumes values in the interval $[0, 1]$. When $D = 0$, we say that $\mathscr{G}$ is an empty graph. Conversely, when $D = 1$, $\mathscr{G}$ is said to be a complete or maximal clique graph.

Remark 2.30. Often in the literature, networks can also be classified as sparse, when D assumes values near 0, and dense, otherwise. As a rule of thumb, when the number of edges in the networks is of the order of the number of vertices, i.e., $E = \mathscr{O}(V)$, the network is considered sparse. As we will see, the density of networks has profound implications on the time complexity of the majority of the machine learning algorithms. As such, it is a common practice in the literature to state the time complexity of machine learning techniques both in terms of sparse and dense networks.

Definition 2.36. Network assortativity: The network assortativity captures, in a structural sense, the preference of vertices to attach to others that are similar or different in terms of the degree [54]. Assortativity is often operationalized as a degree correlation among vertices. The assortativity coefficient r is essentially the Pearson correlation coefficient of degree between pairs of linked vertices. Hence, positive values of r indicate a correlation between vertices of similar degree, while negative values indicate relationships between vertices of different degrees [53]. In general, r lies between -1 and 1. When $r = 1$, the network is said to have perfect assortative mixing patterns, while at $r = -1$ the network is completely disassortative.

Many studies have been conducted and some conclusions have been drawn on some types of real-world networks. For example, social networks often have apparent assortative mixing. On the other hand, the technological, biological, and financial networks frequently appear to be disassortative [53].

Considering that u_e and v_e are the degrees of the two vertices at the endpoints of the e-th edge of a non-empty graph $\mathscr{G}$, and that $E = |\mathscr{E}|$ is the number of edges of $\mathscr{G}$, the network assortativity r is evaluated as follows [53]:

$$r = \frac{E^{-1}\sum_{e\in\mathscr{E}} u_e v_e - \left[\frac{E^{-1}}{2}\sum_{e\in\mathscr{E}}(u_e + v_e)\right]^2}{\frac{E^{-1}}{2}\sum_{e\in\mathscr{E}}(u_e^2 + v_e^2) - \left[\frac{E^{-1}}{2}\sum_{e\in\mathscr{E}}(u_e + v_e)\right]^2}. \tag{2.24}$$

Definition 2.37. Local assortativity: Local assortativity can be used to analyze assortative or disassortative tendencies at local level [65]. Local assortativity, denoted by r_{local}, has been defined as the individual contribution of each vertex to the network assortativity. The local assortativity of a vertex u with degree $j + 1$ is given by [65, 66]:

$$r_{\text{local}}(u) = \frac{(j+1)(j\bar{k} - \mu_q^2)}{2E\sigma_q^2}, \tag{2.25}$$

in which $\bar{k}$ is the average remaining degree of the neighbors of u, E is the number of links in the network, μ_q and σ_q are the mean and standard deviation of the remaining degree distribution of the network. It follows that the network assortativity r can be retrieved using the following expression:

$$r = \sum_{u\in\mathscr{V}} r_{\text{local}}(u). \tag{2.26}$$

Definition 2.38. Non-normalized rich-club coefficient: The rich-club coefficient first appeared in the literature as an unscaled metric parametrized by vertex degree ranks [82]. More recently, this has been updated to be parameterized in terms of vertex degrees k, indicating a degree cut-off. The rich-club coefficient measures the structural property of complex networks called "rich-club" phenomenon. This property refers to the tendency of vertices with large degree (hubs) to be tightly connected to each other, thus forming clique or near-clique structures.

This phenomenon has been discussed in several instances in both social and computer sciences. Essentially, vertices with a large number of links, usually known as rich vertices, are much more likely to form dense interconnected subgraphs (clubs) than vertices with small degree. Considering that $E_{>k}$ is the number of edges among the $N_{>k}$ vertices that have degree larger than a given threshold $k \geq 0$, the scaled version of the rich-club coefficient is expressed as [19, 51, 61]:

$$\phi(k) = \frac{2E_{>k}}{N_{>k}(N_{>k} - 1)}, \tag{2.27}$$

in which the factor $N_{>k}(N_{>k}-1)/2$ represents the maximum feasible number of edges that can exist among $N_{>k}$ vertices. We note that, while the network assortativity captures how connected similar vertices are in terms of degree connectivity, the rich-club coefficient can be viewed as a more specific notation of associativity, where we are only concerned with the connectivity of vertices beyond a certain richness metric. For example, if a network consists of a collection of hub and spokes, where the hubs are well connected, such a network would be considered disassortative. However, due to the strong connectedness of the hubs in the network, the network would demonstrate a strong rich-club effect.

Definition 2.39. Normalized rich-club coefficient: A criticism of the above non-normalized rich-club coefficient is that it does not necessarily imply the existence of the rich-club effect, as it is monotonically increasing even for random networks. This is true because vertices with larger degree are naturally more likely to be more densely connected than vertices with smaller degree, simply due to the fact that they have more incident edges. In fact, for certain degree distributions, it is not possible to avoid connecting hubs with large degrees. As a result, for a proper evaluation of this phenomenon, we must normalize out this factor. This point was raised in [19], which derived an analytical expression for the rich-club coefficient of uncorrelated large-size networks at large degrees. To account for this, it is necessary to compare the above metric to the same metric on a degree distribution that preserves the randomized version of the network. This updated metric is defined as [19, 51, 61]:

$$\phi_{\mathrm{norm}}(k) = \frac{\phi(k)}{\phi_{\mathrm{rand}}(k)}, \tag{2.28}$$

in which $\phi(k)$ is the non-normalized rich-club coefficient of the network under analysis and $\phi_{\mathrm{rand}}(k)$ is the non-normalized rich-club coefficient evaluated on a maximally randomized network with the same degree distribution $P(k)$ of the network under study. This new ratio discounts unavoidable structural correlations that are a result of the degree distribution, giving a better indicator of the significance of the rich-club effect. For this metric, if for certain values of k we have $\phi_{\mathrm{norm}}(k) > 1$, this denotes the presence of the rich-club effect.

Remark 2.31. Networks with strong disassortative mixing patterns that have rich-club regions composed of vertices with large degrees point for the existence of core-periphery structures (cf. Sect. 2.2.5). The number of cores, in this case, is defined as the number of graph components that results when applying the procedure to evaluate the rich-club coefficient.

2.3.2 Distance and Path Measures

Definition 2.40. Diameter: The diameter of $\mathcal{G}$, T, is the length of the largest pairwise distance in $\mathcal{G}$. Formally, it is given by:

$$T = \max_{u,v \in \mathcal{V}} d_{uv}. \tag{2.29}$$

For a non-weighted graph, the feasible values of T are $[0, V - 1]$. The diameter can be interpreted as the largest intermediation chain in the network.

Definition 2.41. Vertex eccentricity: The eccentricity of $u \in \mathcal{V}$, e_u, is the largest distance from u to any other vertex $v \in \mathcal{V} \setminus \{u\}$, i.e.:

$$e_u = \max_{v \in \mathcal{V} \setminus \{u\}} d_{uv}. \tag{2.30}$$

Definition 2.42. Radius: The network radius, ζ, is its minimum eccentricity, i.e.:

$$\zeta = \min_{u \in \mathcal{V}} e_u. \tag{2.31}$$

Definition 2.43. Wiener index: The Wiener index, λ, is defined as the sum of geodesic distances between each pair of vertices in the graph. Mathematically, it is given by:

$$\lambda = \frac{1}{2} \sum_{\substack{u,v \in \mathcal{V} \\ u \neq v}} d_{uv}. \tag{2.32}$$

One problem of this measure is its divergence for disconnected graphs, because at least one geodesic distance is infinity. Such a problem can be avoided by computing only pairs of connected vertices. However, it introduces distortion if the graph has many pairs of disconnected vertices. The following network measures, global efficiency and average harmony, are defined in a way to solve this problem.

Definition 2.44. Global efficiency: The global efficiency, GE, considers that the efficiency of sending information between two vertices u and v is inversely proportional to the geodesic distance [2], i.e.:

$$GE = \frac{1}{V(V-1)} \sum_{\substack{u,v \in \mathscr{V} \\ u \neq v}} \frac{1}{d_{uv}}.$$
(2.33)

Definition 2.45.   Average Harmony: The average harmony, h, is the reciprocal of the overall global efficiency, i.e.:

$$h = \frac{1}{GE}.$$
(2.34)

The average harmony does not present the problem of divergence shown by Wiener index, so it is suitable for graphs with unconnected vertices [20].

2.3.3   Structural Measures

Definition 2.46.   Clustering Coefficient: The clustering coefficient measure quantifies the degree to which local vertices in a network tend to cluster together. Evidence suggests that in many real-world networks, and in particular social networks, vertices tend to create tightly knit groups characterized by a relatively high density of ties [74]. Several generalizations and adaptations of such measure have been proposed in the literature [44, 60]. Here, we define the measure originally proposed by Watts and Strogatz [74]. The local clustering coefficient of a vertex in a graph quantifies how close its neighbors are to being a clique (complete graph). Mathematically speaking, the local clustering coefficient of vertex i is given by:

$$CC_i = \frac{2|e_i|}{k_i(k_i-1)},$$
(2.35)

in which $|e_i|$ the number of links shared by the direct neighbors of vertex i (number of triangles formed by vertex i and any of its two neighbors) and k_i is the degree of vertex i. By (2.35), we see that $CC_i \in [0, 1]$.

Definition 2.47.   Network clustering coefficient: We can also evaluate the network clustering coefficient, which gives us a sense of quasi-local connectivity between vertices, as follows:

$$CC = \frac{1}{V} \sum_{i \in \mathscr{V}} CC_i,$$
(2.36)

in which V symbolizes the number of vertices and $CC \in [0, 1]$. Roughly speaking, the clustering coefficient tells how well-connected the neighborhood of the vertex is.

If the neighborhood is fully connected, the clustering coefficient is 1 and a value close to 0 means that there are hardly any triangular connections in the neighborhood.

Definition 2.48. Cyclic Coefficient: This coefficient characterizes the degree of circulation in complex networks by considering cycles of all orders from 3 up to infinity [39]. The cyclic coefficient θ_i of vertex i is the average of the inverse size of the smallest cycle that connects that vertex and any of two of its neighbor vertices. Mathematically, it is calculated as follows [39]:

$$\theta_i = \frac{2}{k_i(k_i - 1)} \sum_{j,k \in \mathcal{N}(i)} \frac{1}{S_{jk}^i}, \qquad (2.37)$$

in which S_{jk}^i is the smallest size of the closed shortest path that passes through vertex i and its two neighbor vertices j and k. Note that the sum goes over all of the neighbor pairs (j, k) of i. If vertices j and k are directly linked to each other, then vertices i, j, and k form a triangle. It is a cycle of order 3 and $S_{jk}^i = 3$, which is the smallest value of S_{jk}^i. If no paths exist that connect vertices j and k except for that one that crosses vertex i, then vertices i, j, and k form a tree structure. In this case, there is no closed loop passing through the three vertices i, j, and k, in a way that $S_{jk}^i = \infty$.

Definition 2.49. Global cyclic coefficient: The global cyclic coefficient, θ, is equal to the average of cyclic coefficients of all of the vertices, as follows [39]:

$$\theta = \frac{1}{V} \sum_{i \in \mathcal{V}} \theta_i. \qquad (2.38)$$

The global cyclic coefficient takes a value between 0 and $1/3$, where 0 means the network has a tree structure in which no cycle can be found, and the opposite case ($\theta = 1/3$) indicates that there is a connection between all pairs of vertices, in which case the clustering coefficient is 1.

Definition 2.50. Modularity: The modularity measure quantifies how good a particular division of a network is [15, 57] and is designed to measure the strength of division of a network into modules (also called groups, clusters or communities). Generally, it ranges from 0 to 1. When the modularity is near 0, it means that the network does not present community structure, suggesting that the links are disposed at random in the network. As the modularity grows, the community structure gets more and more defined, that is, the mixture between communities gets smaller and therefore the fraction of links inside communities is larger than that between different communities.

Besides the network, the modularity takes as input a hypothesis about the membership of each vertex towards a community. It then tests how those vertices inside the given network fit into well-defined communities using the aforementioned notion. Mathematically, the modularity in non-weighted networks is expressed as:

$$Q = \frac{1}{2E} \sum_{i,j \in \mathcal{V}} \left(\mathbf{A}_{ij} - \frac{k_i k_j}{2E} \right) \mathbb{1}_{[c_i = c_j]}, \tag{2.39}$$

in which E represents the total number of edges in the network; k_i stands for the degree of the vertex i; c_i is the community of vertex i; and $\mathbf{A}_{ij}$ is the edge weight linking vertex i to j. The summation term is composed of two factors, all of which are computed only for vertices of the same community due to the indicator function. That is, cross-community links do not contribute to the modularity measure. The first term, $\frac{\mathbf{A}_{ij}}{2E}$, counts the fraction of links inside pairs of vertices that are members of the same community. From that, we subtract $\frac{k_i k_j}{(2E)^2}$, the second term, which accounts for removing the fraction of edges that are expected to occur due to randomness, using a random network model (recall Sect. 2.2.1). Nonzero values of the modularity index indicate deviations from randomness and values around 0.3 or more usually indicate good divisions.

We can also define the modularity for weighted networks [56]. In this case, the terms denoting the degree k_i in (2.39) are exchanged for the strength measures s_i, as introduced in Definition 2.14, and E is given by:

$$E = \frac{1}{2} \sum_{i \in \mathcal{V}} s_i. \tag{2.40}$$

The main idea of modularity is to calculate the fraction of edges that fall within the given groups minus the expected value if edges were distributed at random. For a given division of the network vertices into some modules, modularity reflects the concentration of vertices within modules compared to a random distribution of links between all vertices, regardless of modules.

Definition 2.51. Topological overlap: The topological overlap index measures to what extent two vertices are connected to roughly the same group of other vertices in the network. In essence, the topological overlap measure evaluates how similar the direct and indirect neighborhoods of two vertices are. To calculate the topological overlap of a pair of vertices, their connections to all of the other vertices in the network are compared. If these two vertices share similar direct and indirect neighborhoods, then they have a high "topological overlap." We can adjust the depth of the neighborhood which is used in the comparison. That is, we can only compare the direct neighborhood of two vertices, up to the second order neighborhood, and so on. Specifically, the m-th order topological overlap measure is constructed by (i) counting the number of m-step neighbors that a pair of vertices share and (ii) normalizing it to assume a value between 0 and 1. The resulting vertex similarity measure is a measure of agreement between the m-step neighborhoods of two input vertices. Such a measure can be applied in a number of ways, for instance, similarity search, prediction based on k-nearest neighbors, multi-dimensional scaling and module identification by clustering.

Let $\mathcal{N}_m(i)$, $m > 0$, denote the set of vertices (excluding i itself) that is reachable from i within a shortest path of length m, i.e., $\mathcal{N}_m(i) = \{j \neq i \mid d_{ij} \leq m\}$, where d_{ij} is the geodesic distance (shortest path distance) between i and j. The m-step topological overlap is given by:

$$t_{ij}^{[m]} = \begin{cases} \frac{|\mathcal{N}_m(i) \cap \mathcal{N}_m(j)| + \mathbf{A}_{ij}}{\min[|\mathcal{N}_m(i)|, |\mathcal{N}_m(j)|] + 1 - \mathbf{A}_{ij}}, & \text{if } i \neq j \\ 1, & \text{if } i = j \end{cases}, \qquad (2.41)$$

in which $\mathbf{A}_{ij}$ denotes the (i,j)-th entry of the adjacency matrix of the graph. Thus, the m-step topological overlap measures the agreement of the m-step neighborhoods between two vertices. Note that, even in the case that two vertices have the same m-step neighborhoods, the topological overlap index only assumes its maximum value when they are directly connected, i.e., when $\mathbf{A}_{ij} = 1$.

2.3.4 Centrality Measures

Centrality measures quantify how central or how important vertices or edges are inside a network. The first centrality measure that comes to our mind may be the degree of a vertex. In this way, it is natural to assume that vertices with large degrees are central to the network, while vertices with small degrees are usually peripheral or terminal ones. In spite of its simplicity, degree is widely used as a centrality measure. In many real networks, vertices with large degree are often called *hubs*. Many centrality measures have been reported by the literature. Each one is defined according to a different heuristics that ultimately lead to different conclusions about the centrality of vertices or edges.

2.3.4.1 Distance-Based Centrality Measures

We divide these types of centrality measures in two groups that are classified according to the criterion used to calculate the centrality distance [41].

Definition 2.52. Minimax criterion: The first family consists of those problems that use a minimax criterion. As an example, consider the problem of determining the location for an emergency facility such as a hospital. The main objective of such an emergency facility location problem is to find a site that minimizes the maximum response time between the facility and the site of a possible emergency.

The aim of the first problem family is to determine a location that minimizes the maximum distance to any other location in the network. Suppose that a hospital is located at a vertex $u \in \mathcal{V}$. We denote the maximum distance from u to a random vertex v in the network, representing a possible incident, as the eccentricity e_u of u. Recall that the eccentricity is given by $e_u = \max_{v \in \mathcal{V}} d_{uv}$. The problem of finding an optimal location can be solved by determining the minimum over all e_u with $u \in \mathcal{V}$. Therefore, the centrality of vertex u based on the eccentricity is:

$$c_E(u) = \frac{1}{e_u} = \frac{1}{\max_{v \in \mathcal{V}} d_{uv}}. \tag{2.42}$$

Definition 2.53. Minisum criterion: The second family of location problems optimizes a minisum criterion that is used to determine the location of a service facility like a shopping mall. The aim here is to minimize the total travel time. We denote the sum of the distances from a vertex $u \in \mathcal{V}$ to any other vertex in a graph as the total distance $\sum_{v \in \mathcal{V}} d_{uv}$. The problem of finding an appropriate location can be solved by computing the set of vertices with minimum total distance as follows:

$$c_C(u) = \frac{1}{\sum_{v \in \mathcal{V}} d_{uv}}. \tag{2.43}$$

In social network analysis, a centrality index based on this concept is called closeness. The focus lies here, for example, on measuring the closeness of a person to all other people in the network. People with a small total distance are considered as more important as those with a high total distance.

2.3.4.2   Path- and Walk-Based Centrality Measures

Centrality measures that are based on paths do not take into consideration the distances from vertex to vertex, but they consider the flow passing through a vertex. In essence, a vertex is declared as more important if there are many shortest paths passing through it.

Definition 2.54. Betweenness: The betweenness measures the extent to which a vertex lies on the shortest paths between every pair of vertices in a network [29, 30, 58]. Suppose we have a network in which the vertices exchange messages among themselves. Let us initially make the simple assumption that every pair of vertices in the network exchanges a message with equal probability per unit time and that messages always take the shortest (geodesic) path of the network, or one of such paths, chosen at random, if there are several. Then, let us ask the following question: if we wait a suitably long time until many messages have passed between each pair of vertices, how many messages, on average, will have passed through each vertex *en route* to their destination? The answer is that, since messages are passing down each geodesic path at the same rate, the number passing through each vertex is simply proportional to the number of geodesic paths the vertex lies on [58]. This number of geodesic paths is what it is called betweenness index.

Given this definition, it follows that vertices with high betweenness may have considerable influence within a network by virtue of their control ability over information passing between others. The vertices with the highest betweenness in our message-passing scenario are the ones through which the largest number of messages pass, and if those vertices get to see the messages in question as they pass, or if they get paid for passing the messages along, they could derive a lot of power from their position within the network. The vertices with the highest betweenness are also the ones whose removal from the network will most disrupt communications between other vertices because they lie on the path of several messages. In real-world situations, of course, not all vertices exchange communications with the same frequency, and in most cases, communications do not always take the shortest path, due to, for example, political or physical reasons.

Mathematically, let η_{st}^{v} be 1 if vertex v lies on the geodesic path from s to t and 0 if it does not or if there is no such path (because s and t lie in different components of the network). Then, the betweenness centrality x_v is given by:

$$B_v = \sum_{s \neq v \in V} \sum_{t \neq v \in V} \frac{\eta_{st}^{v}}{\eta_{st}}, \tag{2.44}$$

i.e., the betweenness of v evaluates the fraction of shortest paths between all pairs of vertices s and t that passes through v over the total number of shortest paths between s and t.

Definition 2.55. Communicability [25]: Many topological and dynamical properties of complex networks are defined by assuming that most of the transport on the network flows along the shortest paths, such as the betweenness measure. However, there are different scenarios in which non-shortest paths are used to reach the network destination. For instance, in air transportation, airplanes may have to fly through more distant routes between two destinations, because in the shortest path between them there is a war or no-fly zone. Thus the consideration of only the shortest paths does not account for the global communicability of a complex network. Communicability is defined for every pair of vertices $p \in \mathcal{V}$ and $q \in \mathcal{V}$. In essence, it quantifies how easily vertex p can communicate with q by means of a combination of shortest paths and random walks with varying lengths. Mathematically, the communicability of vertex p to q is given by:

$$\mathbf{G}_{pq}(\mathbf{M}) = \frac{1}{s!} \mathbf{P}_{pq} + \sum_{k>s} \frac{1}{k!} (\mathbf{A}^k)_{pq} = (e^{\mathbf{A}})_{pq}, \tag{2.45}$$

in which $\mathbf{P}_{pq}$ denotes the number of paths with the shortest length from p to q; s is the length of such paths; and $\mathbf{A}$ is the binary adjacency matrix of the network. The term $\mathbf{A}_{pq}^{(k)}$ is the element (p, q) of the k-th power of matrix $\mathbf{A}$ that gives the number of walks of length k from p to q along the adjacency matrix $\mathbf{A}$, with k strictly greater than s steps. The communicability of $\mathbf{G}_{pq}$ and $\mathbf{G}_{qp}$ may be different for directed graphs. A large $\mathbf{G}_{pq}$ reveals that p can reach q by several routes. Conversely, when $\mathbf{G}_{pq}$ is small, there are few possibilities for p to reach q.

2.3.4.3   Vitality

Let $\mathcal{Q}$ be the set of all simple, undirected and non-weighted graphs $\mathcal{G} = \langle \mathcal{V}, \mathcal{E} \rangle$ and let $f : \mathcal{G} \to \mathbb{R}$ be any real-valued function on $\mathcal{G} \in \mathcal{Q}$. A vitality index $V(\mathcal{G}, u)$, $u \in \mathcal{V}$, is then defined as the difference of the values of f on $\mathcal{G}$ and on $\mathcal{G}$ without element or vertex u: $V(\mathcal{G}, u) = f(\mathcal{G}) - f(\mathcal{G} \backslash \{u\})$ [41].

Definition 2.56. Flow betweenness vitality: define the max-flow betweenness vitality for a vertex $u \in \mathcal{V}$ by:

$$BV(u) = \sum_{\substack{s,t \in \mathcal{V} \\ u \neq s, u \neq t}} \frac{f_{st}(u)}{f_{st}}, \qquad (2.46)$$

in which $f_{st}(u)$ is the amount of flow which must go through u. We determine $f_{st}(u)$ by $f_{st}(u) = f_{st} - \tilde{f}_{st}$ where $\tilde{f}_{st}$ is the maximal s-t-flow in $\mathcal{G} \backslash \{u\}$. That is, $\tilde{f}_{st}$ is determined by removing u from $\mathcal{G}$ and computing the maximal s-t-flow in the resulting reduced network $\mathcal{G} \backslash \{u\}$.

Definition 2.57. Closeness vitality: Let the distance between two vertices s and t represent the costs of sending a message from s to t. Then, the closeness vitality of u denotes how much the transport costs in an all-to-all communication will increase if the corresponding element u is removed from the network. That is,

$$CV(u) = I(\mathcal{G}) - I(\mathcal{G} \backslash \{u\}), \qquad (2.47)$$

in which $I(\mathcal{G}) = \sum_{v,w \in \mathcal{V}} d_{vw}$, i.e, the total distance of the network.

Definition 2.58. Dynamical vitality [67]: Consider a network as a directed graph with V vertices, $\mathbf{A}u = \lambda u$ and $v^T \mathbf{A} = \lambda v^T$, where $\mathbf{A}$ is the adjacency matrix, λ is the largest eigenvalue of $\mathbf{A}$, u and v are right and left eigenvectors of $\mathbf{A}$. The dynamic importance of edge (i, j), DI_{ij}, is defined as:

$$DI_{ij} = -\frac{\Delta \lambda_{ij}}{\lambda}, \qquad (2.48)$$

that is, it is the amount $-\Delta \lambda_{ij}$ by which λ decreases upon the removal of edge (i, j), normalized by λ. Similarly, the dynamical importance of vertex k is defined in terms of the amount $-\Delta \lambda_k$ by which λ decreases upon removal of that vertex:

$$DI_k = -\frac{\Delta \lambda_k}{\lambda}. \qquad (2.49)$$

By removing the edge (i, j), we get $(\mathbf{A} + \Delta \mathbf{A})(u + \Delta u) = (\lambda + \Delta \lambda)(u + \Delta u)$. If we multiply by v^T, expand the formula, and neglect second order terms $v^T \Delta \mathbf{A} \Delta u$ and

$\Delta\lambda v^T\Delta u$, we obtain $\Delta\lambda = \frac{v^T\Delta\mathbf{A}u}{v^Tu}$. Upon the removal of edge (i,j), the perturbation matrix is $(\Delta\mathbf{A})_{lm} = -A_{ij}\delta_{il}\delta_{jm}$, and therefore:

$$\widehat{DI}_{ij} = -\frac{A_{ij}v_iu_j}{\lambda v^Tu}. \tag{2.50}$$

By removing the vertex k, the perturbation matrix is given by $(\Delta\mathbf{A})_{lm} = -A_{ij}(\delta_{il}+\delta_{jm})$, since $\Delta u_k = -u_k$,[2] therefore, we set $\Delta u = \delta u - u_k e_k$, where e_k is the unit vector for the k-th component, and we assume that δu is small. By multiplying v^T and again neglecting the second order terms $v^T\Delta\mathbf{A}\delta u$ and $\Delta\lambda v^T\delta u$, we obtain $\Delta\lambda = \frac{(v^T\Delta\mathbf{A}u - u_k v^T\Delta\mathbf{A}e_k)}{(v^Tu - v_k u_k)}$. Using the expression of $\Delta\mathbf{A}$, we get $v^T\Delta\mathbf{A}u = -2\Delta u_k v^k$ and $u_k v^T\Delta\mathbf{A}e_k = \lambda u_k v_k$. Considering that the network is large ($V \gg 1$), we assume that $u_k v_k < v^Tu$. Thus, we obtain:

$$\widehat{DI}_k = -\frac{v_k u_k}{\lambda v^Tu}. \tag{2.51}$$

2.3.4.4   General Feedback Centrality

Now we present measures that are built on the concept of feedback centrality. In this respect, a vertex has larger feedback centrality the more central are its neighbors [41].

Definition 2.59. Bonacich's eigenvector centrality: In 1972, Phillip Bonacich [9] introduced a centrality measure that is computed using eigenvectors of adjacency matrices. In special, he presented three different approaches to evaluate the centrality measure and all three of them result in the same valuation of the vertices. The difference between these methodologies are in a constant factor. In the following, we assume that the graph $\mathscr{G}$ is undirected, connected, without self-loops, and non-weighted. As the graph is undirected and without self-loops, the adjacency matrix $\mathbf{A}$ is symmetric and all diagonal entries are zero. The three methods that score each vertex are:

1. The factor analysis approach;
2. The convergence of an infinite sequence; and
3. The solution of a simultaneous linear equation system.

Here, we only focus on the third approach. It follows the idea of calculating an eigenvector of a linear equation system. If we define the centrality of a vertex to be a weighted sum of the centralities of its adjacent vertices, where the weight is given by the network topology, we get the following equation system:

---

[2]Recall that the left and right eigenvectors have zero k-th entries after the removal of vertex k.

$$s_i = \sum_{j \in \mathcal{V}} \mathbf{A}_{ij} s_j, \tag{2.52}$$

in which s_i is the Bonacich score or centrality of vertex i. In a matrix form,

$$s = \mathbf{A}s. \tag{2.53}$$

Equation (2.53) has a single solution only if $\det(\mathbf{A} - \mathbf{I}) = 0$, where $\mathbf{I}$ is the identity matrix. We can instead solve for s using the eigenvalue problem of $\mathbf{A}$, i.e., $\lambda s = \mathbf{A}s$.

Definition 2.60. Katz index: This index first appeared in the context of social networks to determine the importance or status of an individual [38]. To take the number of intermediate individuals into account, a damping factor $\alpha > 0$ is introduced: the longer the path between two vertices i and j is, the smaller should its impact on the status of j be. The associated mathematical model is hence a non-weighted, directed graph $\mathcal{G} = \langle \mathcal{V}, \mathcal{E} \rangle$ without self-loops and associated adjacency matrix $\mathbf{A}$. Using the fact that $(\mathbf{A}^k)_{ji}$ holds the number of paths from j to i with length k, the status of vertex i is:

$$C_k(i) = \sum_{k=1}^{\infty} \sum_{j \in \mathcal{V}} \alpha^k (\mathbf{A}^k)_{ji}. \tag{2.54}$$

In matrix notation, we have:

$$C_K = \sum_{k=1}^{\infty} \alpha^k (\mathbf{A}^T)^k \mathbf{1}_V, \tag{2.55}$$

in which $\mathbf{1}_V$ is the V-dimensional vector where every entry is 1. Assuming that $\alpha |\lambda_0| < 1$, where λ_0 is the largest eigenvalue of $\mathbf{A}$, the infinite series converges. Thus, we can find a closed form expression for the status index of Katz:

$$C_K = \sum_{k=1}^{\infty} \alpha^k (\mathbf{A}^T)^k \mathbf{1}_V = (\mathbf{I} - \alpha \mathbf{A}^T)^{-1} \mathbf{1}_V \tag{2.56}$$

or in another form:

$$(\mathbf{I} - \alpha \mathbf{A}^T) C_K = \mathbf{1}_V, \tag{2.57}$$

which is an inhomogeneous system of linear equations that emphasizes the feedback nature of the centrality: the value of $C_K(i)$ depends on the centrality values of neighbors of i in the graph, i.e., $C_K(j), j \neq i$.

Definition 2.61. Web page centrality—PageRank: PageRank (PR) is a well-known measure used by Google to rank web pages. It is supposed to simulate the behavior of a user browsing the Web. Most of the time, the user visits pages just by surfing, i.e., by clicking on hyperlinks of the page he/she is on. Another manner is to jump to another page by typing its URL on the browser, or going to a bookmark, etc. In a network, this process can be modeled by a simple combination of a random walk with occasional jumps toward randomly selected vertices. This can be described by the simple set of implicit relations [64]:

$$p(i) = \frac{q}{V} + (1-q) \sum_{j \in \mathcal{V}: j \to i} \frac{p(j)}{k_j^{(out)}}. \tag{2.58}$$

Here, V is the number of vertices of the graph, $p(i)$ is the PR value of vertex i, $k_j^{(out)}$ the out-degree of vertex j, and the sum runs over the vertices pointing toward (direct connection to) i. The damping factor $q \in [0, 1]$ is a probability that weighs the mixture between the realized random walk and random jumps.

For any $q > 0$, the process reaches stationarity, as a walker has a finite (no matter how small) probability to escape from a dangling end, whenever it lands there. When $q = 0$, the process may not be stationary and PR is ill defined. When $q = 1$, instead, the jumping process dominates and all of the vertices have the same PR-value $1/v$.

PR goes beyond the concept of in-degree. In order to have a large PR for a vertex, it is important to have many neighbors pointing at that vertex, i.e., large in-degree, but it is also important that the neighbors have large PR values themselves. So, if two vertices have equal in-degree, the vertex with more "important" neighbors will have larger PR.

Definition 2.62. Eigenvector centrality: The eigenvector centrality, like the PageRank, relies on the principle that the importance of a vertex depends on the importance of its neighbors [64]. The relationship that the eigenvector centrality captures is more straightforward than that in PageRank: the prestige x_i of vertex i is simply proportional to the sum of the prestiges of the neighboring vertices pointing to it. Numerically,

$$\lambda x_i = \sum_{j \in \mathcal{V}: j \to i} x_i = \sum_{j \in \mathcal{V}} A_{ji} x_j = (\mathbf{A}^T x)_i. \tag{2.59}$$

We see that x_i is basically the i-th component of the transposed eigenvector of the adjacency matrix $\mathbf{A}$ associated to the eigenvalue λ. We observe that the trivial eigenvector with all of the components equal to zero is always a solution of (2.59). From (2.59), we also see that singleton vertices have zero centrality. In general, vertices pointed at by vertices with zero centrality also have zero centrality and this effect will propagate to other vertices, so that in many cases the eigenvector centrality would not give any information about a large fraction of vertices. To avoid this, it is useful to make the following modification: to each vertex, we assign a prestige ϵ, which is independent of its relationships with the other vertices. As a result, Eq. (2.59) becomes [64]:

$$x_i = \alpha(\mathbf{A}^T x)_i + \epsilon. \tag{2.60}$$

The role of the parameter ϵ reminds that of the damping factor q in PageRank. The parameter α weighs the relative importance of the contribution of the peers versus that of the vertex itself.

2.3.5   Classification of the Network Measurements

As it can be noticed, the complex network literature has proposed a myriad of network measurements that capture different aspects of the network structure. The provided list is far from being exhaustive. New network measurements are introduced to suit the needs of computational problems that arise in our day-to-day problems. Some of them may be domain-dependent and others may even require external information to be computed. In the previous sections, we have introduced the network measurements by dividing them into functional roles. In this section, we re-compile these network measurements using a meta-information approach. We classify them in accordance with the type of information they use in their computation. We define three classes of network measurements, as follows:

- *Strictly local measures*: these measures only employ information from the vertex itself to be computed. Strictly local measures are always vertex-level measures.
- *Mixed measures*: besides using strictly local information, these measures also use topological information from its direct and indirect neighborhoods. This additional information can vary from simply quasi-local topology, such as the number of triangles in the neighborhood, to long-range information, such as the shortest path between the two most distant pair of vertices. Mixed measures are always vertex-level measures.
- *Global measures*: these network measurements make use of the entire network structure to be computed. Global measures are always network-level measures.

Figure 2.21 portrays a schematic of the three classes of network measurements. Strictly local and mixed measures are vertex-level, while global measures must be network-level. Table 2.1 reports the classification of the network measurements we have discussed so far in this chapter.

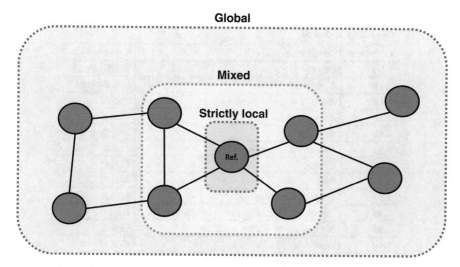

Fig. 2.21 Intuition for classifying network measurements in terms of the type of information they need to be computed

2.4 Dynamical Processes in Complex Networks

One of the fundamental differences between graph theory and complex network studies is that the latter focus not only on the static structures but also on the dynamical properties of networks under study. Therefore, in this section, we review five dynamical processes in networks: random walk, lazy random walk, self-avoiding walks, tourist walk, and epidemic spreading. Besides these ones, there are many other dynamical processes in networks, such as information transmission, percolation in regular lattices and in complex networks, and synchronization among oscillators (vertices). However, the last ones are not the focus of this book.

2.4.1 Random Walks

A random walk is a mathematical formalization of a trajectory that consists of taking successive random steps [63]. It has been used to describe many natural phenomena and it has also been applied to solve a wide range of engineering problems. Some of these include graph matching and pattern recognition [33], image segmentation [34], neural network modeling [37, 47], network centrality measure [59], network partition [81], construction and analysis of communication networks [78, 80].

Given a network $\mathscr{G} = \langle \mathscr{V}, \mathscr{E} \rangle$ and a starting vertex $v \in \mathscr{V}$, we select a neighbor of it at random, and move to this neighbor; then we select a neighbor of this new

Table 2.1 Classification of the network measurements using a meta-information approach

Definition	Description	Classification
2.10	Degree	Strictly local
2.11	In- and out-degree	Strictly local
2.12	Average degree (connectivity)	Global
2.13	Average in- and out-degree	Global
2.14	Strength	Strictly local
2.15	In- and out-strength	Strictly local
2.35	Density	Global
2.36	Assortativity	Global
2.37	Local assortativity	Mixed
2.38	Non-normalized rich-club coefficient	Global
2.39	Normalized rich-club coefficient	Global
2.40	Diameter	Global
2.41	Vertex eccentricity	Mixed
2.42	Radius	Global
2.43	Wiener index	Global
2.44	Global efficiency	Global
2.45	Average harmony	Global
2.46	Clustering coefficient	Mixed
2.47	Network clustering coefficient	Global
2.48	Cyclic coefficient	Mixed
2.49	Global cyclic coefficient	Global
2.50	Modularity	Global
2.51	Topological overlap	Mixed
2.52	Eccentricity centrality (minimax criterion)	Mixed
2.53	Total distance centrality (minisum criterion)	Mixed
2.54	Betweenness	Mixed
2.55	Communicability	Mixed
2.56	Flow betweenness vitality	Mixed
2.57	Closeness vitality	Mixed
2.58	Dynamical vitality	Mixed
2.59	Bonacich centrality	Mixed
2.60	Katz index	Mixed
2.61	PageRank	Mixed
2.62	Eigenvector centrality	Mixed

vertex again at random, and move to it, and so on. The random sequence of vertices selected this way is a random walk on the graph. A finite random walk of length $t > 0$ has the same intuition, but we stop after making $t - 1$ random transitions. If the graph is weighted, then we transition to a neighbor u with probability that is proportional to the edge weight $\mathbf{A}_{vu}$.

In essence, the theory of random walks on networks and the theory of finite discrete Markov chains are basically the same, so that every discrete Markov chain can be conceived as random walk on a graph. Discrete Markov chains are stochastic processes whose future states are conditionally independent of the past states provided that the present state is known. In graph theory, the states are denoted by the vertices in the graph. In a graph theory context, given that a walker is at vertex v, the Markovian property affirms that the probability of visiting a neighboring vertex is independent on the past trajectories of that walker. We formalize that concept in the following.

Definition 2.63. Discrete-time Markov chain: A discrete-time Markov chain is a stochastic process $\{X_t : t \in \mathbb{N}\}$, where the random variable X assumes values in a countable set $\mathcal{N}$ at any given time t. The transition probability to state $q \in \mathcal{N}$ is:

$$P[X_t = q \mid X_{t-1}, X_{t-2}, \dots, X_0] = P[X_t = q \mid X_{t-1}], \tag{2.61}$$

i.e., the probability of the next outcome only depends on the last value of the process. Therefore, past trajectories are irrelevant.

Remark 2.32. In the context of graph theory, the countable set is composed of the vertex set, i.e., $\mathcal{N} = \mathcal{V}$.

Remark 2.33. In Markovian processes, each feasible value in the countable set $\mathcal{V}$ is called a state.

Definition 2.64. Transition probability: The transition probability of going from state (vertex) q to u is denoted by $\mathbf{P}_{qu}(t)$, $q, u \in \mathcal{V}$, which is a shorthand for $\mathbf{P}_{qu}(t) = P[X_t = u \mid X_{t-1} = q]$. Mathematically, the transition probability is defined in accordance with the network topology, as follows:

$$\mathbf{P}_{qu} = \frac{\mathbf{A}_{qu}}{\sum_{i \in \mathcal{V}} \mathbf{A}_{qi}}, \tag{2.62}$$

i.e., the stronger is the edge weight linking q to u, the more likely will be that transition.

Remark 2.34. Rewrite (2.62) as:

$$\mathbf{P}_{qu} = \frac{\mathbf{A}_{qu}}{K(q)}, \tag{2.63}$$

in which $K(q) = \sum_{i \in \mathscr{V}} \mathbf{A}_{qi}$. Then,

- If the network is undirected and non-weighted, then $K(q) = k_q$, where k_q is the degree of vertex q.
- If the network is directed and non-weighted, then $K(q) = k_q^{(out)}$, where $k_q^{(out)}$ is the out-degree of vertex q.
- If the network is undirected and weighted, then $K(q) = s_q$, where s_q is the strength of vertex q.
- If the network is directed and weighted, then $K(q) = s_q^{(out)}$, where $s_q^{(out)}$ is the out-strength of vertex q.

Definition 2.65. Transition matrix: In Markovian processes, we can map all of the feasible transitions using the transition matrix $\mathbf{P}(t)$ as follows:

$$\mathbf{P}(t) = \begin{pmatrix} \mathbf{P}_{1,1}(t) & \mathbf{P}_{1,2}(t) & \cdots & \mathbf{P}_{1,V}(t) \\ \mathbf{P}_{2,1}(t) & \mathbf{P}_{2,2}(t) & \cdots & \mathbf{P}_{2,V}(t) \\ \vdots & \vdots & \ddots & \vdots \\ \mathbf{P}_{V,1}(t) & \mathbf{P}_{V,2}(t) & \cdots & \mathbf{P}_{V,V}(t) \end{pmatrix}. \tag{2.64}$$

Note that the transition matrix completely characterizes the Markovian process because, the immediate future state $X(t + 1)$ is only determined by the current state $X(t)$, regardless of the past trajectories.

Remark 2.35. If $\mathbf{P}(t)$ is immutable for all $t \in \mathbb{N}$, then the Markov process is said to be *time-homogenous*. In a graph theory perspective, this is equivalent to saying that the graph topology does not change during the walk. For clarity, if the Markov process (or random walk) is time-homogeneous, we drop the time indexing of the transition matrix.

Definition 2.66. m-step transition matrix: For a time-homogeneous Markovian process, we can define the m-step transition matrix, $m > 0$, as $\mathbf{P}^m$. Essentially, the entry $\mathbf{P}_{qu}^m$ encodes the transition probability of starting from state or vertex q and arriving at state or vertex u after exactly m transitions.

Remark 2.36. The original transition matrix defined in (2.64) is a 1-step transition matrix.

For each realization of the Markovian process $\omega \in \Omega$, let $\mathrm{pt}(j)$ be the number of times j appears in the random walk that visits the states $X_0(\omega), X_1(\omega), X_2(\omega), \ldots$. Then, $\mathrm{pt}(j)$ is the total number of times the state j is visited by the stochastic process X in realization ω. If $\mathrm{pt}(j)$ is finite, then X eventually leaves state j never to return. Mathematically, there must be an integer n such that $X_n(\omega) = j$ and $X_m(\omega) \neq j$, $\forall m > n$. In contrast, if $\mathrm{pt}(j) = \infty$ for a realization ω, then X keeps on visiting j again and again. These two classes that state j can assume are important from a practical point-of-view [14]. We now turn our attention in providing formal tools to classify states according to those perspectives.

The passage time function counts the number of times a given vertex has been visited during a random walk. We formalize this notion in the following.

Definition 2.67. Passage Time: The passage time is a function pt : $\mathcal{V} \to \mathbb{N}$ such that pt(q) is the number of times the Markovian process reaches the state q. Mathematically,

$$\text{pt}(q) = |\{t \in \mathbb{N} \mid X_t = q\}|$$

$$= \sum_{t=0}^{\infty} \mathbb{1}_{[X_t(\omega)=q]}. \tag{2.65}$$

Recall that $\mathbb{1}_{[A]}$ is the indicator function that yields 1 whenever the logical expression A is true, and returns 0, otherwise. Basically, we increment pt(q) by one each time the stochastic process X visits state or vertex q.

We now define the so-called potential matrix of the Markovian process X.

Definition 2.68. Potential or fundamental matrix: The potential matrix **R** encodes the expected number of times each vertex is visited when we start from any given other vertex. Mathematically, its (i, j)-th entry is expressed as:

$$\mathbf{R}_{ij} = \mathbb{E}\left[\text{pt}(j) \mid X(0) = i\right], \tag{2.66}$$

which can be seem as the mean passage time to reach j conditioned that the walker starts at vertex i.

Plugging (2.65) into (2.66) and using the monotone convergence theorem, we get:

$$\mathbf{R}_{ij} = \mathbb{E}\left[\sum_{n=0}^{\infty} \mathbb{1}_{[X_n=j]} \,\middle|\, X(0) = i\right]$$

$$= \sum_{n=0}^{\infty} \mathbb{E}\left[\mathbb{1}_{[X_n=j]} \,\middle|\, X(0) = i\right]$$

$$= \sum_{n=0}^{\infty} P\left(X_n = j \,\middle|\, X(0) = i\right)$$

$$= \sum_{n=0}^{\infty} \mathbf{P}_{ij}^m. \tag{2.67}$$

Let T be the time that state or vertex j is first visited by a realization of the Markovian process.

Definition 2.69. Recurrent state: State j is recurrent if:

$$P(T < \infty \mid X(0) = j) = 1. \tag{2.68}$$

As a consequence, the number of returns of a recurrent state is always infinite, that is:

$$\mathbf{R}_{jj} = \mathbb{E}\left[\text{pt}(j) \mid X(0) = j\right] = \infty. \tag{2.69}$$

Definition 2.70. Transient state: State j is transient if:

$$P(T = +\infty \mid X(0) = j) > 0. \tag{2.70}$$

As a consequence, the number of returns of a transient state is always finite, that is:

$$\mathbf{R}_{jj} = \mathbb{E}\left[\text{pt}(j) \mid X(0) = j\right] < \infty. \tag{2.71}$$

Remark 2.37. There are only two states: recurrent or transient. In this way, if j is not recurrent, then it must be a transient state, and vice versa.

Remark 2.38. Let j be a recurrent state. Then, we sub-classify it as *null recurrent* if:

$$\mathbb{E}[T \mid X(0) = j] = \infty, \tag{2.72}$$

otherwise, we call it *non-null recurrent*.

Remark 2.39. Let j be a recurrent state. Then, we sub-classify it as *periodic with period δ* if $\delta \geq 2$ is the largest integer for which:

$$P(T = n\delta \text{ for some } n \geq 1) = 1. \tag{2.73}$$

otherwise, we call it *aperiodic*.

Definition 2.71. Closed set of states: A set of states is said to be closed if no state outside it can be reached from any state inside it.

Definition 2.72. Absorbing state: A state forming a closed set by itself is called an absorbing state. We say that state q is absorbing if there is a probability 1 to go from q to itself. In other words, once an absorbing state has been reached in a random walk, the walker stays in this state forever.

Definition 2.73. Irreducible closed set: A closed set is irreducible if no proper subset of it is closed.

Definition 2.74. Irreducible Markov chain: A Markov chain is called irreducible if its only closed set is the set of all states. Therefore, a Markov chain is irreducible if and only if all states can be reached from each other.

The state set of the Markov chain process can be divided into the absorbing state set $\mathcal{V}_A$ and its complementary set, the transient state set $\mathcal{V}_T = \mathcal{V} \setminus \mathcal{V}_A$.

Remark 2.40. The mean passage time for transient states can be obtained by computing the fundamental matrix only for the transient states $\mathbf{R}^{(\text{transient})}$:

$$\mathbf{R}^{(\text{transient})} = (\mathbf{I} - \mathbf{P}_T)^{-1}, \tag{2.74}$$

in which $\mathbf{I}$ is the $|\mathcal{V}_T| \times |\mathcal{V}_T|$ identity matrix and $\mathbf{P}_T$ is the transition probability matrix restricted to the transient states. The entry $\mathbf{R}^{(\text{transient})}_{q'q}$ contains the mean passage time in state $q \in \mathcal{V}_T$ during random walks starting in state q'. Hence,

$$\mathbb{E}\left[\text{pt}(q)\right] = [p'^{(\text{transient})} \mathbf{R}^{(\text{transient})}]_q, \tag{2.75}$$

in which $p'^{(\text{transient})}$ is the transpose of the initial probability vector when we only consider transient states. Note that the expectation operation is taken over random walks with arbitrary lengths.

Given a distribution $p(t)$, $\dim(p(t)) = 1 \times V$, where the v-th entry denotes the probability that the system will be at vertex $v \in \mathcal{V}$, the evolution of $p_v(t)$ is:

$$p_v(t+1) = \sum_{(u,v) \in \mathcal{E}} \mathbf{P}(t)_{uv} p_u(t). \tag{2.76}$$

Analogously, the evolution of the distribution $p(t)$ is:

$$p(t+1) = p(t)\mathbf{P}(t). \tag{2.77}$$

Intuitively, the evolution of the probability distribution $p(t)$ as a function of t can be seen as describing a diffusion process in the underlying graph. The diffusion is completely characterized once we know the initial distribution $p(0)$ and the transition matrices $\mathbf{P}(t)$.

Definition 2.75. Stationary distribution: If the network $\mathcal{G}$ is a finite, irreducible, time-homogenous, and aperiodic Markov chain, then it has a unique stationary distribution $\pi = [\pi_1, \ldots, \pi_V]$ that can be reached from any initial distribution $p(0)$. In the dynamic equation, the stationarity is reached when the following holds:

$$\pi = \pi\mathbf{P}. \tag{2.78}$$

Each entry of the stationary distribution is of the form:

$$\pi_i = \frac{1}{\mathbb{E}\left[T \mid X(0) = i\right]}, \tag{2.79}$$

in which recall that $\mathbb{E}\left[T \mid X(0) = i\right]$ is the expected time to regress to vertex i starting from i.

For an undirected network, we have that:

$$\mathbb{E}\left[T \mid X(0) = i\right] = \frac{\sum_{j \in \mathscr{V}} k_j}{k_i} == \frac{2E}{k_i}, \tag{2.80}$$

in which E is the number of edges in the network and k_i is the degree of vertex i.

Substituting (2.80) in (2.79), we get:

$$\pi_i = \frac{k_i}{2E}. \tag{2.81}$$

2.4.2   Lazy Random Walks

The unique stationary distribution in Definition 2.75 only holds true, among other things, for aperiodic networks. However, if the network is periodic, there is an easy way to fix the periodicity problem by introducing the lazy random walk. In a lazy random walk at time t, the walker may decide upon two different actions:

1. It can transition to a neighboring vertex in accordance with the transition matrix with probability $1/2$; or
2. It can stay at the current vertex[3] with probability $1/2$.

Remark 2.41. The lazy random walk can be viewed as a vanilla version of the classical random walk in a network in which we add k_u self-loops to every vertex u in the original graph $\mathscr{G}$.

Formally, the evolution of the probability distribution $p(t)$ of a lazy random walk is given by:

$$\begin{aligned}
p(t + 1) &= \frac{1}{2}p(t) + \frac{1}{2}p(t)\mathbf{P}(t) \\
&= p(t)\frac{1}{2}\left[\mathbf{I} + \mathbf{P}(t)\right] \\
&= p(t)\mathbf{P}'(t), \tag{2.82}
\end{aligned}$$

---

[3] Hence, the terminology "lazy" random walk.

in which $\mathbf{P}'(t)$ is the modified transition matrix for the lazy random walk:

$$\mathbf{P}'(t) = \frac{1}{2}[\mathbf{I} + \mathbf{P}(t)]. \tag{2.83}$$

Note also that the stationary distribution of a lazy random walk is identical to that of the classical random walks portrayed in Definition 2.75. To see that, it suffices to see that $\mathbf{P}'(t)$ is also a valid transition matrix, just like the original $\mathbf{P}(t)$. As long as the graph $\mathscr{G}$ is finite, irreducible, time-homogenous, and aperiodic, the unique stationary distribution always exists.

2.4.3 Self-Avoiding Walks

A self-avoiding walk on a network $\mathscr{G}$ is a path that visits no vertex more than once. Self-avoiding walks were first introduced in the chemical theory of polymerization [27], and since then their critical behavior has attracted attention of mathematicians and physicists [49].

Broadly speaking, self-avoiding walks are usually considered in infinite lattices, so that steps are only allowed in a discrete number of directions and of certain lengths. Self-avoiding walks cannot be Markovian, because we need to check the past trajectory in order to list the possible futures states that the process can assume. The research in [49] provides a comprehensive review on self-avoiding walks.

2.4.4 Tourist Walks

A tourist walk can be conceptualized as a walker (tourist) aiming at visiting sites (data items) in a P-dimensional map, representing the data set. At each discrete timestep, the tourist follows a simple deterministic rule: it visits the nearest site that has not been visited in the previous μ steps. In other words, the walker performs partially self-avoiding deterministic walks over the data set, where this self-avoiding factor is limited to the memory window $\mu - 1$. This quantity can be understood as a repulsive force emanating from the sites in this memory window, which prevents the walker from visiting them in this interval (refractory time). Therefore, it is prohibited that a trajectory intersects itself inside this memory window. In spite of being a simple rule, it has been shown that this kind of movement possesses complex behavior when $\mu > 1$ [48]. Note that tourist walks differ from self-avoiding random walks in that the former is a deterministic process, while the latter is a random process.

The tourist's behavior heavily depends on the data set's configuration and the starting site. In computational terms, the tourist's movements are entirely realized

Fig. 2.22 Illustration of a
tourist walk with $\mu = 1$. The
red (*dark gray*) and *green*
(*light gray*) dots represent
visited and unvisited sites,
respectively. The *dashed lines*
indicate the transient part of
the walk, whereas the
continuous lines, the attractor
of the walk

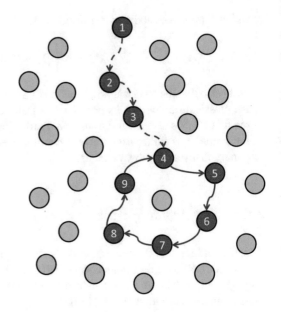

by means of a neighborhood table. This table is constructed by ordering all the data
items in relation to a specific site. This procedure is performed for every site of the
data set.

Each tourist walk can be decomposed in two terms: (1) the initial *transient part*
of length t and (2) a *cycle* (attractor) with period c. Figure 2.22 shows an illustration
of a tourist walk with $\mu = 1$. In this case, one can see that the transient length is
$t = 3$ and the cycle length, $c = 6$.

Considering the attractor or cycle period as a walk section that begins and ends
at the same site of the data set may lead one to think that, once the tourist visits a
specific site, a new visit to it would configure an attractor. Nevertheless, this is a very
simple, and likely to fail, approach for attractors' detection. In fact, during a walk,
a site may be re-visited without configuring an attractor. Besides, the tourist's finite
memory μ allows some steps of the walk to be repeated without configuring an
attractor. For instance, if we had chosen a $\mu = 6$ for the walk in Fig. 2.22, the re-visit
performed by the tourist on the site 4 would have not configured an attractor, since
the site 5 would still be forbidden to be visited again; hence, the tourist would be
compelled to visit another site. This characteristic enables sophisticated trajectories
over the data set, at cost of also increasing the difficulty in detecting an attractor.

In the majority of the works related to tourist walk [40, 48, 71], the tourist
may visit any other site other than the ones contained in its memory window. As
μ increases, there is a significant chance that the walker will begin performing
large jumps in the data set, since the neighborhood is most likely to be already
visited in its entirety within the time frame μ. In the context of data classification,
this is an undesirable characteristic that can be simply avoided by using a graph
representation of the input data. In this way, the walker is only permitted to visit

vertices, represented now by the sites, that are in its connected neighborhood (link). With this modified mechanism, for large values of μ, it is very likely that the walker will get trapped within a vertex, not being able to further visit other vertices of the neighborhood. In this scenario, we say that the walk only had a transient part and the cycle period is null ($c = 0$).

2.4.5 Epidemic Spreading

Epidemic spreading in complex networks has triggered much attention to many researchers. It is a dynamic process within a network and the main concern is how the network structure attenuates or amplifies disease breakouts or immunization. Since epidemic spreading processes can be considered as information transmission, it is useful for machine learning. For example, epidemic spreading may be directly related to data label propagation in semi-supervised learning. Although we have not found works connecting epidemic spreading and machine learning in literature yet, we would like to share such a prediction with the readers. The readers who are interested in this topic are invited to develop their new techniques in this direction. For the above-mentioned purpose, we here review two basic models of epidemic spreading in complex networks. For a comprehensive review, see [23, 62, 83]. For some development related to information transmission in complex networks, see [18, 50, 76, 77, 79].

The most extensively studies of epidemic models are about the susceptible-infected-recovered and susceptible-infected-susceptible models [3, 4, 36]. We review these models in the following.

2.4.5.1 Susceptible-Infected-Recovered (SIR) Model

In the SIR model, each individual is at one of the three states: susceptible (does not infect others but may be infected), infected, or recovered (will not be affected again). At each time step, assume that a susceptible individual may be infected by another infected person with probability α and that the recovering rate of infected individuals is β. Then, the epidemic process in the SIR model can be described by the following dynamic equations:

$$\frac{dx}{dt} = -\alpha yx, \tag{2.84}$$

$$\frac{dy}{dt} = -\alpha yx - \beta y, \tag{2.85}$$

$$\frac{dz}{dt} = -\beta y, \tag{2.86}$$

in which x, y and z are the ratios of susceptible, infected, and removed individuals to the entire population, respectively. In a network setting, each individual is represented by a vertex and links exist when two individuals have some kind of contact. In this network, a susceptible vertex will be infected only if it has at least one infected neighbor.

2.4.5.2   Susceptible-Infected-Susceptible (SIS) Model

For some diseases, such as influenza and pulmonary tuberculosis, the recovered individual can be infected again. This situation is not considered by the SIR model. For this reason, the SIS model was introduced. The only difference between them is that in SIS model, the infected individuals will return to the susceptible state after recovering. The SIS model is defined by the following equations:

$$\frac{dx}{dt} = -\alpha yx + \beta y, \tag{2.87}$$

$$\frac{dy}{dt} = \alpha yx - \beta y. \tag{2.88}$$

2.4.5.3   Epidemic Spreading in Complex Networks

In [43], the authors studied the SIS model on small-world networks of Watts and Strogatz, which have been presented in Sect. 2.2.2. They found that even when the rewiring probability p is very small (for instance, $p = 0.01$), the disease can permanently exist with very small infection ratios and without fluctuations in the population ratios. In contrast, when p gets large enough (for example, $p = 0.9$), periodic oscillations of the number of infected individuals start to appear.

Consider the SIS model in random networks and assume that λ denotes the spreading rate. In [6, 7], the authors uncovered a spreading threshold λ_c. If the value of λ is above the threshold, i.e., $\lambda > \lambda_c$, the infection spreads and becomes persistent. Below it, the infection disappears. Such a result implies that the disease can persist only if it infects a sufficiently large amount of individuals. However, in real situations, many diseases can persistently exist with just a small fraction of the population being infected, such as computer viruses and measles. In [6, 7], the authors obtained the epidemic threshold of the SIS dynamics in general networks as follows:

$$\lambda_c = \frac{\langle k \rangle}{\langle k^2 \rangle}, \tag{2.89}$$

in which $\langle . \rangle$ represents an averaging operator over all of the network vertices, and k denotes the degree. Note that $\langle k \rangle = \bar{k}$, which is the network connectivity. In scale-free networks, when the network size goes to infinite, we have that $\lambda_c = 0$. The absence of epidemic threshold in scale-free networks provides a good explanation for the empirical data [6, 7].

2.5 Chapter Remarks

In this chapter, we have introduced the basic notion of graphs and some of the network topologies that are well-known by the complex network community. We have also explored a comprehensive list of network measurements, which are able to extract structural information of the data relationships in a systematic manner. Finally, we have reviewed classical dynamic processes, such as the random walk, self-avoiding walk, tourist walk, and epidemic spreading with a focus on networked environments.

References

1. Ahn, Y.Y., Bagrow, J.P., Lehmann, S.: Link communities reveal multiscale complexity in networks. Nature **466**, 761–764 (2010)
2. Albert, R., Jeong, H., Barabási, A.L.: Diameter of the world wide web. Nature **401**, 130–131 (1999)
3. Anderson, R.M., May, R.M.: Infectious Diseases of Humans: Dynamics and Control. Oxford University Press, Oxford, NY (1992)
4. Bailey, N.: The Mathematical Theory of Infectious Diseases and Its Applications. Griffin, London (1975)
5. Barabási, A.L., Albert, R.: Emergence of scaling in random networks. Science - New York **286**(5439), 509–512 (1999)
6. Boguñá, M., Pastor-Satorras, R.: Epidemic spreading in correlated complex networks. Phys. Rev. E **66**, 047104 (2002)
7. Boguñá, M., Pastor-Satorras, R., Vespignani, A.: Absence of epidemic threshold in scale-free networks with connectivity correlations. Phys. Rev. Lett. **90**, 028701 (2003)
8. Bollobas, B.: Modern Graph Theory. Springer, Berlin (1998)
9. Bonacich, P.: Factoring and weighting approaches to status scores and clique identification. J. Math. Sociol. **2**(1), 113–120 (1972)
10. Borgatti, S.P., Everett, M.G.: Models of core/periphery structures. Soc. Netw. **21**(4), 375–395 (2000)
11. Callaway, D.S., Newman, M.E.J., Strogatz, S.H., Watts, D.J.: Network robustness and fragility: Percolation on random graphs. Phys. Rev. Lett. **85**, 5468–5471 (2000)
12. Chase-Dunn, C.K.: Global Formation: Structures of the World-Economy. Blackwell, Oxford (1989)
13. Chung, F.R.K.: Spectral Graph Theory. CBMS Regional Conference Series in Mathematics, vol. 92. American Mathematical Society, Philadelphia (1997)
14. Çinlar, E.: Introduction to Stochastic Processes. Prentice-Hall, Englewood Cliffs, NJ (1975)
15. Clauset, A., Newman, M.E.J., Moore, C.: Finding community structure in very large networks. Phys. Rev. E **70**(6), 066111+ (2004)
16. Cohen, R., Erez, K., ben Avraham, D., Havlin, S.: Resilience of the internet to random breakdowns. Phys. Rev. Lett. **85**, 4626–4628 (2000)
17. Cohen, R., Erez, K., ben Avraham, D., Havlin, S.: Breakdown of the internet under intentional attack. Phys. Rev. Lett. **86**, 3682–3685 (2001)
18. Cohen, R., Havlin, S., ben Avraham, D.: Efficient immunization strategies for computer networks and populations. Phys. Rev. Lett. **91**, 247901 (2003)
19. Colizza, V., Flammini, A., Serrano, M.A., Vespignani, A.: Detecting rich-club ordering in complex networks. Nat. Phys. **2**(2), 110–115 (2006)

20. Costa, L.F., Rodrigues, F.A., Travieso, G., Villas Boas, P.R.: Characterization of complex networks: a survey of measurements. Adv. Phys. **56**(1), 167–242 (2007)
21. Diestel, R.: Graph Theory. Graduate Texts in Mathematics. Springer, Berlin (2006)
22. Doreian, P.: Structural equivalence in a psychology journal network. J. Am. Soc. Inf. Sci. **36**(6), 411–417 (1985)
23. Draief, M., Massouli, L.: Epidemics and Rumours in Complex Networks. Cambridge University Press, New York, NY (2010)
24. Erdös, P., Rényi, A.: On random graphs I. Publ. Math. (Debrecen) **6**, 290–297 (1959)
25. Estrada, E., Hatano, N.: Communicability in complex networks. Phys. Rev. E **77**, 036111 (2008)
26. Everett, M., Borgatti, S.: Regular equivalence: general theory. J. Math. Sociol. **18**(1), 29–52 (1994)
27. Flory, P.: Principles of Polymer Chemistry. Cornell University Press, Ithaca (1953)
28. Fortunato, S.: Community detection in graphs. Phys. Rep. **486**, 75–174 (2010)
29. Freeman, L.C.: A set of measures of centrality based upon betweenness. Sociometry **40**, 35–41 (1977)
30. Freeman, L.C. (ed.): The Development of Social Network Analysis. Adaptive Computation and Machine Learning. Empirical Press, Vancouver (2004)
31. Girvan, M., Newman, M.E.J.: Community structure in social and biological networks. Proc. Natl. Acad. Sci. USA **99**(12), 7821–7826 (2002)
32. Godsil, C.D., Royle, G.: Algebraic Graph Theory. Graduate Texts in Mathematics. Springer, Berlin (2001)
33. Gori, M., Maggini, M., Sarti, L.: Exact and approximate graph matching using random walks. IEEE Trans. Pattern Anal. Mach. Intell. **27**(7), 167–256 (2005)
34. Grady, L.: Random walks for image segmentation. IEEE Trans. Pattern Anal. Mach. Intell. **28**(11), 1768–1783 (2006)
35. Gross, J., Yellen, J.: Graph Theory and Its Applications. CRC Press Inc., Boca Raton, FL (1999)
36. Hethcote, H.W.: The mathematics of infectious diseases. SIAM Rev. **42**(4), 599–653 (2000)
37. Jiang, D., Wang, J.: On-line learning of dynamical systems in the presence of model mismatch and disturbances. IEEE Trans. Neural Netw. **11**(6), 1272–1283 (2000)
38. Katz, L.: A new status index derived from sociometric analysis. Psychometrika **18**(1), 39–43 (1953)
39. Kim, H.J., Kim, J.M.: Cyclic topology in complex networks. Phys. Rev. E **72**(3), 036109+ (2005)
40. Kinouchi, O., Martinez, A.S., Lima, G.F., Lourenço, G.M., Risau-Gusman, S.: Deterministic walks in random networks: an application to thesaurus graphs. Physica A **315**, 665–676 (2002)
41. Koschützki, D., Lehmann, K.A., Peeters, L., Richter, S., Tenfelde-Podehl, D., Zlotowski, O.: Centrality indices. In: Brandes, U., Erlebach, T. (eds.) Network Analysis: Methodological Foundations. Lecture Notes in Computer Science, vol. 3418, pp. 16–61. Springer, Berlin (2005)
42. Krugman, P.: The Self-Organizing Economy. Oxford University Press, Oxford (1996)
43. Kuperman, M., Abramson, G.: Small world effect in an epidemiological model. Phys. Rev. Lett. **86**(13), 2909–2912 (2001)
44. Latapy, M., Magnien, C., Vecchio, N.D.: Basic notions for the analysis of large two-mode networks. Soc.Netw. **30**(1), 31–48 (2008)
45. Laumann, E.O., Pappi, F.U.: Networks of Collective Action: A Perspective on Community Influence Systems. Academic Press, New York, NY (1976)
46. Leskovec, J., Lang, K.J., Dasgupta, A., Mahoney, M.W.: Community structure in large networks: natural cluster sizes and the absence of large well-defined clusters. Internet Math. **6**(1), 29–123 (2009)
47. Liang, J., Wang, Z., Liu, X.: State estimation for coupled uncertain stochastic networks with missing measurements and time-varying delays: the discrete-time case. IEEE Trans. Neural Netw. **20**(5), 781–793 (2009)

48. Lima, G.F., Martinez, A.S., Kinouchi, O.: Deterministic walks in random media. Phys. Rev. Lett. **87**, 010603 (2001)
49. Madras, N., Slade, G.: The Self-Avoiding Walk. Birkhäuser, Boston (1996)
50. May, R.M., Lloyd, A.L.: Infection dynamics on scale-free networks. Phys. Rev. E **64**, 066112 (2001)
51. McAuley, J.J., Costa, Caetano, T.S.: Rich-club phenomenon across complex network hierarchies. Appl. Phys. Lett. **91**, 084103 (2007)
52. Meyn, S.P., Tweedie, R.L.: Markov Chains and Stochastic Stability. Cambridge University Press, Cambridge (2009)
53. Newman, M.E.J.: Assortative mixing in networks. Phys. Rev. Lett. **89**(20), 208701 (2002)
54. Newman, M.E.J.: Mixing patterns in networks. Phys. Rev. E **67**(2), 026126 (2003)
55. Newman, M.E.J.: The structure and function of complex networks. SIAM Rev. **45**(2), 167–256 (2003)
56. Newman, M.E.J.: Analysis of weighted networks. Phys. Rev. E **70**, 056131 (2004)
57. Newman, M.E.J.: Modularity and community structure in networks. Proc. Natl. Acad. Sci. **103**(23), 8577–8582 (2006)
58. Newman, M.E.J.: Networks: An Introduction. Oxford University Press, Oxford (2010)
59. Noh, J.D., Rieger, H.: Random walks on complex networks. Phys. Rev. Lett. **92**, 118701 (2004)
60. Opsahl, T., Panzarasa, P.: Clustering in weighted networks. Soc. Netw. **31**(2), 155–163 (2009)
61. Opsahl, T., Colizza, V., Panzarasa, P., Ramasco, J.J.: Prominence and control: The weighted rich-club effect. Phys. Rev. Lett. **101**, 168702 (2008)
62. Pastor-Satorras, R., Castellano, C., Mieghem, P.V., Vespignani, A.: Epidemic processes in complex networks. Rev. Mod. Phys. **87**, 925 (2015)
63. Pearson, K.: The problem of the random walk. Nature **72**(1867), 294 (1905)
64. Perra, N., Fortunato, S.: Spectral centrality measures in complex networks. Phys. Rev. E **78**, 036107 (2008)
65. Piraveenan, M., Prokopenko, M., Zomaya, A.Y.: Local assortativeness in scale-free networks. Europhys. Lett. **84**(2), 28002 (2008)
66. Piraveenan, M., Prokopenko, M., Zomaya, A.: Local assortativity and growth of Internet. Eur. Phys. J. B **70**, 275–285 (2009)
67. Restrepo, J.G., Ott, E., Hunt, B.R.: Characterizing the dynamical importance of network nodes and links. Phys. Rev. Lett. **97**, 094102 (2006)
68. Rombach, M.P., Porter, M.A., Fowler, J.H., Mucha, P.J.: Core-periphery structure in networks. SIAM J. Appl. Math. **74**(1), 167–190 (2014)
69. Silva, T.C., Zhao, L.: Uncovering overlapping cluster structures via stochastic competitive learning. Inf. Sci. **247**, 40–61 (2013)
70. Smith, D.A., White, D.R.: Structure and dynamics of the global economy: network analysis of international trade. Soc. Forces **70**(4), 857–893 (1992)
71. Stanley, H.E., Buldyrev, S.V.: Statistical physics – the salesman and the tourist. Nature **413**, 373–374 (2001)
72. Wasserman, S., Faust, K.: Social Network Analysis: Methods and Applications, vol. 8. Cambridge University Press, Cambridge (1994)
73. Watts, D.J.: Small Worlds: The Dynamics of Networks Between Order and Randomness. Princeton Studies in Complexity. Princeton University Press, Princeton (2003)
74. Watts, D.J., Strogatz, S.H.: Collective dynamics of 'small-world' networks. Nature **393**(6684), 440–442 (1998)
75. Yang, J., Leskovec, J.: Overlapping communities explain core-periphery organization of networks. Proc. IEEE **102**(12), 1892–1902 (2014)
76. Yang, H.X., Wang, W.X., Lai, Y.C., Xie, Y.B., Wang, B.H.: Control of epidemic spreading on complex networks by local traffic dynamics. Phys. Rev. E **84**, 045101+ (2011)
77. Yang, H.X., Tang, M., Lai, Y.C.: Traffic-driven epidemic spreading in correlated networks. Phys. Rev. E **91**, 062817 (2015)
78. Zeng, Y., Cao, J., Zhang, S., Guo, S., Xie, L.: Random-walk based approach to detect clone attacks in wireless sensor networks. IEEE J. Sel. Areas Commun. **28**(5), 677–691 (2010)

79. Zhang, H.F., Xie, J.R., Tang, M., Lai, Y.C.: Suppression of epidemic spreading in complex networks by local information based behavioral responses. Chaos **24**(4), 043106 (2014)
80. Zhong, M., Shen, K., Seiferas, J.: The convergence-guaranteed random walk and its applications in peer-to-peer networks. IEEE Trans. Comput. **57**(5), 619–633 (2008)
81. Zhou, H.: Distance, dissimilarity index, and network community structure. Phys. Rev. E **67**(6), 061901 (2003)
82. Zhou, S., Mondragon, R.J.: The rich-club phenomenon in the Internet topology. IEEE Commun. Lett. **8**(3), 180–182 (2004)
83. Zhou, T., Fu, Z.Q., Wang, B.H.: Epidemic dynamics on complex networks. Progress Nat. Sci. **16**(5), 452–457 (2006)

Chapter 3
Machine Learning

Abstract Machine learning relates to the study, design and development of algorithms that give computers the capability to learn without being explicitly programmed. Machine learning techniques are fairly generic and can be applied in various settings. To utilize such kinds of algorithms, one has to translate the problem to the domain of machine learning, which usually expects a set of features and a desirable output or grouping criterion. In this chapter, we introduce the three machine learning paradigms often employed by the literature: supervised, unsupervised, and semi-supervised. We show that supervised algorithms exclusively utilize external information to induce or to train their hypotheses. In contrast, unsupervised learning methods are guided exclusively by the intrinsic structure of the data items throughout the learning process, i.e., without any sort of external knowledge. In-between these two learning paradigms lies the semi-supervised learning, which employs both the labeled and unlabeled data in the learning process. Here, we focus on supplying the shortcomings and potentialities of traditional and representative techniques that are well-known by the machine learning community. We will not go into technical details of traditional machine learning techniques in this chapter, because these are not the focus of this book.

3.1 Overview of Machine Learning

Machine learning aims at developing computational methods that are capable of "learning" with accumulated experiences [9, 19, 36, 43–45]. Traditionally, there are two fundamental types of learning in machine learning. The first is entitled *unsupervised learning*, whose main task consists in revealing intrinsic structures that are embedded within the data relationships. The learning process, in this case, is solely guided by the provided data, for no prior knowledge about the data is supplied [38, 44, 46]. Essentially, a typical problem in unsupervised learning consists in estimating the underlying density function that generated the data distribution under analysis [10]. Among the main tasks of unsupervised learning, one can highlight: clustering [23, 33, 48, 49], outlier detection [40, 41], dimensionality reduction [39], and association [53]. In a clustering task, we expect to find groups in which data items in the same group are very similar to each other, while data items that correspond to different groups are expected to be dissimilar. In this case, the

© Springer International Publishing Switzerland 2016
T.C. Silva, L. Zhao, *Machine Learning in Complex Networks*,
DOI 10.1007/978-3-319-17290-3_3

resemblance or similarity of different data items is judged according to an adopted similarity function [44]. In outlier detection, the goal is to find data items that differ, to a large extent, from the majority of the other data items, i.e., from the original data distribution [40]. In dimensionality reduction, the objective is to dispose the data items over a lower dimensional space in relation to its original distribution, so as to simplify the relationships among the data items [39]. In association, one seeks to generate rules that relate subsets of predictive attributes [53].

The second type of learning is referred to as *supervised learning*, whose objective is to deduce concepts regarding the data. This inference is performed using the presented labeled instances, which are commonly denoted as the training set. In this regard, the learning process tries to construct a mapping function conditioned to the provided training set [1, 26, 31, 36]. Often, the learners are tested using unseen data items that compose the so-called test set. When the labels comprise discrete values, then the problem is denominated *classification*, whereas when the values are continuous, *regression* [9].

The main difference between supervised and unsupervised learning paradigms is as follows. In the first, the learner explores external information of the training set, which is available at the training stage, in order to induce the hypothesis of the classifier. In a classification task, for example, this external information is brought to the learning process in the form of classes or labels. The goal of classification, in this case, is to create a predictive function that can generalize from this training set, when applied to unknown data (test set). The performance of the classifier in this test is often termed as the classifier's generalization power. In contrast, the unsupervised learning paradigm seeks behaviors or trends in the data, trying to group them in a way that similar data tend to be agglomerated together while dissimilar data are segregated into distinct groups. Note that, in this case, no external information or labels are employed during the learning process.

Besides these two well-defined areas, a new machine learning paradigm has received attention from the related community, which is denominated *semi-supervised learning*. This paradigm has been proposed in order to combine the strengths of the supervised and unsupervised learning paradigms [10, 64, 65]. In a typical semi-supervised classification, few data are labeled, while most of them are unlabeled. Observe that this corresponds to the typical scenario nowadays, as thousands of data are generated very quickly, while only a few of them can effectively be processed and labeled. This fact is true because, in general, the labeling task is expensive, time-consuming, and prone to errors. In the semi-supervised learning, the goal is to propagate these few labels from the labeled examples to the large amount of unlabeled examples. A semi-supervised task, therefore, utilizes both information from the training and the test sets to make predictions in a simultaneous manner.

For didactic purposes, Fig. 3.1a shows a clustering task in unsupervised learning. Note that no external information is available *a priori*, and the clustering is performed by using similarity or topological information of the data. In Fig. 3.1b, it is illustrated a semi-supervised classification in semi-supervised learning. Note now that some data already possess labels (external information) beforehand, while most

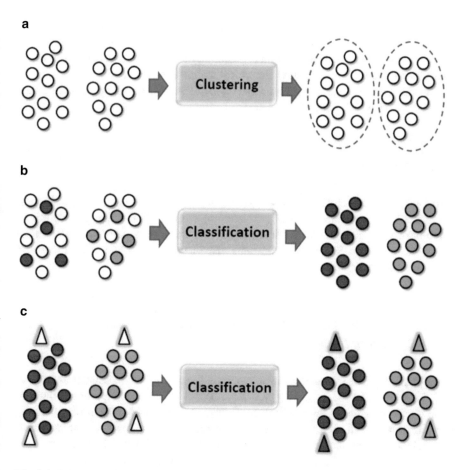

Fig. 3.1 Schematic of the three paradigms of the machine learning area. (**a**) Unsupervised learning; (**b**) Semi-supervised learning; (**c**) Supervised learning

of them are unlabeled. The learning process propagates these labels to the remaining unlabeled data. Finally, in Fig. 3.1c, a classification process in supervised learning is displayed. Initially, the classifier is trained by solely using information from the training set, which is always fully labeled. In the next phase, called classification phase, the classifier is used to predict class labels of unseen data items.

For completeness, it is worth mentioning a novel machine learning area that is not subject of study in this book. Deep learning is a parallel branch of machine learning that relies on sets of algorithms that attempt to model high-level abstractions in data by using model architectures, with complex structures composed of multiple non-linear transformations [18, 57]. Deep learning is part of a broader family of machine learning methods based on learning representations of data. An observation (e.g., an image) can be represented in many ways such as a vector of intensity values per pixel, or in a more abstract way as a set of edges, regions of particular shape, etc.

Some representations make it easier to learn tasks (e.g., face recognition or facial expression recognition) from examples. One of the promises of deep learning is replacing handcrafted features with efficient algorithms for unsupervised or semi-supervised feature learning and hierarchical feature extraction.

In the next sections, we give an overview of the supervised, unsupervised, and semi-supervised learning paradigms.

3.2 Supervised Learning

Algorithms that exclusively utilize external information to induce or to train their hypotheses are considered supervised learning methods. In this chapter, we supply definitions and reviews on traditional supervised learning methods presented in the literature. In Chap. 5, we revisit the supervised learning paradigm with a focus on network-based methods.

3.2.1 Mathematical Formalization and Fundamental Assumptions

The mathematical formulation of a supervised learning task is defined as follows [9, 38, 44, 62]. Let $\mathscr{X}_{training} = \{(x_1, y_1), \ldots, (x_L, y_L)\}$ denote the training set, where the first component of the i-th tuple $x_i = (x_{i1}, \ldots, x_{iP})$ denotes the attributes of the P-dimensional i-th training instance. The second component $y_i \in \mathscr{Y}$ characterizes the class label or target associated to that training instance. The training set is composed of $L = |\mathscr{X}_{training}|$ data items. The goal here is to learn a mapping $x \mapsto y$ using only $\mathscr{X}_{training}$, i.e., the training data distribution and the associated labels. To check the generalization performance of the trained model, the constructed classifier is checked against a test set $\mathscr{X}_{test} = \{x_{L+1}, \ldots, x_{L+U}\}$, for which labels are not provided. The test set is composed of U data items. Each data item in that set is termed as test instance. For an unbiased learning, the training and test sets must be disjoint, i.e., $\mathscr{X}_{training} \cap \mathscr{X}_{test} = \emptyset$. Usually, $N = L + U$ denotes the total number of data items in the learning process.

In the supervised learning scheme, there are two learning phases: the *training phase* and the *classification phase* [9, 26, 31, 38]. In the training phase, the classifier is induced or trained by using the training instances (labeled data) in $\mathscr{X}_{training}$ that are always fully labeled. In the classification phase, the labels of the test instances in $\mathscr{X}_{test}$ are predicted using the induced classifier.

Once trained, one must verify how well the trained supervised learner can generalize. Without any additional assumptions over the data properties, this problem cannot be solved exactly since unseen situations might have arbitrary output values. The necessary assumptions about the nature of the target function constitute the *inductive bias* of the learner. The following is a list of common inductive biases in supervised learning algorithms:

- *Maximum conditional independence*: if the hypothesis can be cast in a Bayesian framework, usually an objective function that aims at maximizing the conditional independence is employed. This is the bias used, for example, in the Naïve Bayes classifier [45, 47].
- *Maximum margin*: when drawing a boundary between two classes, attempt to maximize the width of the boundary, i.e., the distance from the data items. This is the bias used, for instance, in support vector machines. The assumption is that distinct classes tend to be separated by wide low-density boundaries.
- *Minimum description length*: when forming a hypothesis, attempt to minimize the length of the description of the hypothesis. The assumption is that simpler hypotheses are more likely to be true. The rationale here is that complex hypotheses are likely to incorporate noise coming from the training data. Hence, the model becomes overfit to the training data, in a way that its generalization power is jeopardized. A classical principle of this type of inductive bias is the Occam's Razor, which conveys the idea that the simplest consistent hypothesis about the target function is actually the best. Here, consistent means that the hypothesis of the learner yields correct outputs for all of the training data examples that have been given to the algorithm.
- *Minimum number of features*: unless there is good evidence that a feature is useful, it should be deleted. This is the assumption behind feature selection algorithms. Note that if correlation exists between different features, it correspondingly increases the variance of the overall model. In this way, it is only useful to add new features if they explain orthogonal parcels of the output or target variable that have not been explained by other features.
- *Nearest neighbors*: relying on the smoothness or continuity assumption, this inductive bias assumes that, in most of the cases, data items in small neighborhoods tend to be quite similar. Given a test instance, for which the class is unknown, we infer that its class is the same as of the majority of data items in the local neighborhood. This is the bias used, for example, in the k-nearest neighbor algorithm. The assumption is that data items that are near each other tend to belong to the same class.

In machine learning, we can form different models or hypotheses for the same data set, depending on the nature and inductive bias of the selected algorithm. A natural problem that arises is of how well the classifier can perform on unseen data. For that, we need to employ error estimation techniques. Instead of employing the entire training set to train the supervised learner, the basic idea consists in further partitioning the training set into two sub-sets, where the first one is used to train the classifier and the second one is purposed to verify its performance. The performance can be estimated because we now can compare the output value of the algorithm with the ground truth, as we are dealing with "in-sample" or training data items, for which labels are known. The method of k-fold cross-validation is the most employed error estimation procedure in the literature.[1] When trying to choose among different models, we select the hypothesis with the lowest cross-validation error. Although cross-validation may seem to be free of bias, the "no free lunch" theorem states that cross-validation must be biased. In cross-validation, we partition a data set into k equally sized non-overlapping subsets $\mathscr{F}$. For each fold $\mathscr{F}_i$, a model is trained on $\mathscr{F} \setminus \mathscr{F}_i$ (reduced training set), which is then evaluated on $\mathscr{F}_i$ ("in-sample" test instances). Another error estimation procedure is the leave-one-out, which is a special case of the k-fold cross-validation when $k = L - 1$, where L is the number of labeled data items in the training set. The cross-validation estimator of the prediction error is defined as the average of the prediction errors obtained on each fold. While there is no overlap between the test sets on which the models are evaluated, there is overlap between the training sets whenever $k > 2$. The overlap is largest for leave-one-out cross-validation. This means that the learned models are correlated, i.e., dependent, and the variance of the sum of correlated variables increases with the amount of covariance. Therefore, leave-one-out cross-validation has large variance in comparison to smaller k. However, remark that while twofold cross validation does not have the problem of overlapping training sets, it often also has large variance because the training sets are only half the size of the original sample. A good compromise often accepted by the literature is to use ten-fold cross-validation. For a thorough review on this topic, c.f. [3, 7, 31].

With regard to the training and test set characteristics, supervised learning algorithms often assume that [9, 44]:

- For a valid estimation process, the test set must not be biased toward the training set. It must, however, be sampled from the same data distribution process that generated the training set data. This assumption makes clear that, since the classifier has been trained in accordance with the training set distribution, it is fair enough that it is only capable of efficiently inferring unseen examples of that same data distribution. In practice, however, this assumption is often violated to a certain degree. Strong violations will clearly result in poor classification rates.
- The training set must be a representative sample of the distribution or population that generated the analyzed data. Since the hypothesis that the classifier induces is based upon the training set, if the available data is not a representative sample of the true underlying data distribution process, the classifier has a great chance of being mistrained. As such, it predicts in accordance with another distribution, that of the non-representative training set.

---

[1] If model selection is embedded within the error estimation procedure, then the nested k-fold cross-validation is frequently indicated. The nested cross-validation works as follows:

- An inner loop is used as part of model fitting procedure. Typically, we conduct a grid search over the parameters of the technique. For instance, in the k-nearest neighbors technique, we may run through several values of the parameter k.
- An outer loop is employed to measure the performance of the model that had the best performance on the inner loop on a separate external fold (but still in the training set).

3.2.2 Overview of the Techniques

Supervised learning techniques are divided into the following groups:

- *Decision trees*: A decision tree consists of vertices and branches that serve the purpose of breaking a set of samples into a set of covering decision rules. In each vertex, in its original inception, a single test or decision is made to obtain a partition. The starting vertex is usually referred to as the root vertex. In the terminal vertices or leaves, a decision is made on the class assignment. In each vertex, the main task is to select an attribute that makes the best partition between the classes of the samples in the training set [55].
- *Rule-based induction*: One of the most expressive and human readable representations for learned hypotheses are sets of IF-THEN rules. In these kinds of rules, the IF part contains conjunctions and disjunctions of conditions composed by the predictive attributes of the learning task, and the THEN part contains the predicted class for the samples that satisfy the IF clause [54].
- *Artificial neural networks*: Neural networks are interconnected groups of neurons that use mathematical or computational models for information processing based on a connectionist approach. In most cases, an artificial neural network is an adaptive system that changes its structure, usually represented by the connection weights between pairs of neurons, based on external or internal information that flows through the network. In more practical terms, neural networks are nonlinear statistical data modeling or decision making tools. They can be used to model complex relationships between inputs and outputs or to find patterns in data. The neural network, ignorant at the start, through a repetitive "learning" process, becomes a model of the dependencies between the descriptive variables and the target behavior. The key part in developing neural networks is to choose a suitable architecture (how many layers, thresholds utilized by the neurons, etc.) and a corrective learning algorithm (back-propagation, etc.) [27].
- *Bayesian networks*: Bayesian networks constitute a probabilistic framework for reasoning under uncertainty. From an informal perspective, Bayesian networks are directed acyclic graphs, where the vertices are random variables and the edges specify the dependence assumptions that must be held between different random variables. Bayesian networks are based upon the concept of conditional independence among variables. Once the network is constructed, it is used as an efficient device to perform probabilistic inference. This probabilistic reasoning inside the network can be carried out by exact methods, as well as by approximate methods [37]. A special case of Bayesian networks is when no dependencies on the predictive variables exist. In this case, the classifier is known as Naïve Bayes [47].
- *Statistical learning theory*: Maybe the most well-known technique of this type of learning is the support vector machines (SVM), which is based on the principle of structural risk minimization. Originally, it was worked out for linear two-class classification with margins, where margin means the minimal distance from the separating hyperplane to the closest data points. SVM seeks for an optimal

separating hyperplane, where the margin is maximal. An important and unique feature of this approach is that the solution is based only on those data points that are at the margin. These points are called support vectors. The linear SVM can be extended to a nonlinear one when first the problem is transformed into a feature space using a set of nonlinear basis functions. In the feature space—which can be very high dimensional—the data points can be separated linearly. An important advantage of the SVM is that it is not necessary to implement this transformation and to determine the separating hyperplane in the possibly very-high dimensional feature space. Instead, a kernel representation can be used, where the solution is written as a weighted sum of the values of a certain kernel function evaluated at the support vectors [59].

- *Instance-based learning*: Instance-based learning has its root in the study of the nearest neighbor algorithm. The simplest form of nearest neighbor or, more generally, k-nearest neighbors (k-NN) algorithms, simply stores the training instances and classifies a new instance by predicting that it has the same class as its nearest stored instance or the majority class of its k nearest stored instances, according to some similarity measure. The essence of this learning method resides in the form of the similarity function that computes the distances from the new test instance to the training instances [15].
- *Network-based methods*: The inference is done by means of a network constructed from the training set. Up to now, there are still few network-based supervised learning techniques [25]. In Chap. 5, we work on some representative network-based supervised learning techniques.

3.3 Unsupervised Learning

Unsupervised learning methods are guided exclusively by the intrinsic structure of the data items throughout the learning process, i.e., without any sort of external knowledge. In this chapter, we provide definitions and reviews on traditional methods presented in the literature. In Chap. 6, we revisit the unsupervised learning paradigm with a focus on network-based methods.

3.3.1 Mathematical Formalization and Fundamental Assumptions

The unsupervised learning can be defined as follows [2, 9, 21, 30, 44, 45]. Let $\mathcal{X} = \{x_1, x_2, \ldots, x_N\}$ be a data set, where N is the total number of data items involved in the learning process. Then, x_i is a P-dimensional vector, where each of the entries is called a feature or descriptor, which has the role to qualitatively or quantitatively describe the data item. Typically, it is assumed that the points are independently and identically distributed in accordance with a common distribution. In the unsupervised learning case, no professors or external sources are used, i.e., no labels are provided throughout the learning process. Therefore, it is valid to say that no training phase is involved in this type of learning. As a consequence, the algorithm must be guided by the data distribution itself to infer useful knowledge or trends. For example, in data clustering or community detection tasks, the objective is to find subgroups of data items, such that the constituents or members of the same subgroup are more similar than those of different groups. The grouping is complete over the data, i.e., one seeks subgroups $\{\mathscr{X}_1, \ldots, \mathscr{X}_k\}$, in such a way that $\bigcup_{i=1}^{k} \mathscr{X}_i = \mathscr{X}$.

Clustering is a discovery process in data mining. It groups a set of data in a way that maximizes the similarity within clusters and minimizes the similarity between two different clusters. These discovered clusters can help in explaining the characteristics of the underlying data distribution and serve as the foundation for other data mining and analysis techniques. Clustering is useful in characterizing customer groups based on purchasing patterns, categorizing Web documents, grouping genes and proteins that have similar functionality, grouping spatial locations prone to earthquakes based on seismological data, and so on.

Most existing clustering algorithms find clusters that fit some static model. Although effective in some cases, these algorithms can break down if the analyst does not choose appropriate static-model parameters. Or, sometimes, the model simply cannot adequately capture the clusters' characteristics. Most of these algorithms break down when the data contains clusters of diverse shapes, densities, and sizes. For example, Fig. 3.2 shows some illustrative cluster shapes.

Clustering results differ according to the assumptions made about the data. Some of these forms include [9, 21]:

- *The underlying process that generated the observed data has some analytical form*: the methods are biased towards finding clusters with pre-defined forms (similar with estimation). The well-known algorithm *K*-Means, for instance, seeks circular-shaped clusters [42]. This bias has a strong impact on the final results of the unsupervised learner and thus limits its power of detection. These data assumptions, on one hand, may enhance the detection power of the learner if they really reflect the data properties. On the other hand, they severely hamper the learner's detection capabilities if they are invalid.
- *Neighborhood shares the same data characteristics*: the algorithms rely on local information to infer their decisions. For a given data item, for example, the *k*-NN clustering decides in accordance with some function on the neighborhood of that data item [26, 31].
- *Data are produced in some "hierarchical" organization*: the techniques that fall into this category are the hierarchical techniques, including the divisive and agglomerative ones. For practical methods of the former, the community detection based on continuously removing edges with the maximum betweenness [50] and the bisecting *K*-Means [34] are representative examples. As good examples

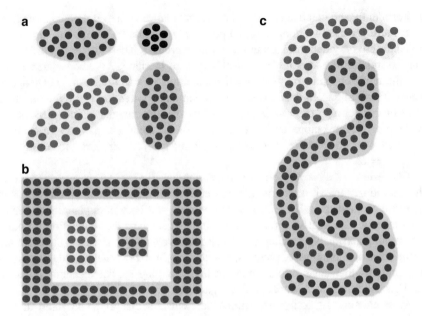

Fig. 3.2 Examples of clusters with different shapes. (**a**) Ellipsoidal clusters. (**b**) Quadrangular (hollow and opaque) clusters. (**c**) Concave (banana-shaped) clusters

of the latter, one can cite the simple-link [30, 44], complete-link [2, 30], and modularity greedy optimization [13].

- *Data fall into clusters according to some preferred directions*: the methods of this type deal with the transformation of the original data into a modified space, usually with a different number of dimensions. Examples include the principal component analysis (PCA) [32], independent component analysis (ICA) [32], and spectral graph algorithms [11].

3.3.2   Overview of the Techniques

One of the most common tasks of unsupervised learning is data clustering. Formally, data clustering aims at discovering natural groups. The groups may be defined as sets of patterns, points, or objects, all of which characterized by suitable similarity measures [9, 21, 29, 30, 63]. Each cluster is a collection of data items that are similar between them and are dissimilar to the objects belonging to other clusters. Data clustering is vital in several exploratory pattern-analysis, grouping, decision-making, and machine-learning situations. Some of them include data mining, document retrieval, image segmentation, bioinformatics, and pattern classification [12, 20, 28–30]. Unfortunately, in the majority of such tasks, only a little prior information is available about the data. In this way, advances in the methodology to automatically understand, process, and summarize the data are required. Nowadays, this becomes even more critical by virtue of the exponential increase in both the volume and the variety of data [29, 63]. In this scenario, the decision-maker must perform as few assumptions about the data as possible. It is under these practical restrictions that the clustering procedure is specially appropriate for the exploration of inter-relationships among the data points to make assessments (perhaps preliminary) of their structure [29, 30]. Data clustering algorithms are generally divided into two types: hierarchical or partitional [8, 20, 29]. The former finds successive clusters using previously established clusters, whereas the latter determines all clusters at once. Hierarchical algorithms can be agglomerative ("bottom-up") or divisive ("top-down"). Agglomerative algorithms begin with each element as a separate cluster and merge them into successively larger clusters. Divisive algorithms begin with the whole set and proceed to divide it into successively smaller clusters. Two-way clustering, co-clustering, or bi-clustering are the names for clusterings where not only the objects are clustered but also the features of the objects, i.e., if the data is represented in a data matrix, the row and columns are clustered simultaneously [9, 21, 63]. In both approaches, the algorithms may be further categorized in network-based algorithms or non network-based algorithms.

In relation to partitional methods, several techniques have been studied in the literature [1, 26, 35, 63]. The most well-known one and the pioneer in the field is the K-Means method [42]. Even though it suffers from several caveats, such as the strong dependence on the system's initial conditions and the inherent bias to find only circular-shaped clusters, it has been further enhanced and studied by the community until today. Some methods, which have improved on the basic idea of K-Means, have been proposed, such as the K-Medoids and fuzzy C-Means [2, 21, 30]. Using a similar strategy, Clarans [51] also attempts to break a data set into K clusters such that the partition optimizes a given criterion, but it also assumes that clusters are hyper-ellipsoidal and of similar sizes. Hence, it also cannot find clusters that vary in size.

DBScan (Density-Based Spatial Clustering of Applications with Noise) [56], a well-known spatial and partitional clustering algorithm, can find clusters of arbitrary shapes. DBScan defines a cluster to be a maximum set of density-connected points, which means that every core point in a cluster must have at least a minimum number of points within a given radius. DBScan assumes that all points within genuine clusters can be reached from one another by traversing a path of density-connected points and points across different clusters cannot. DBScan can find arbitrarily shaped clusters if the cluster density can be determined beforehand and the cluster density is uniform [33].

With regard to hierarchical methods, the most well-known techniques are the single-link and the complete-link [1, 2]. Among other traditional techniques, we can also include average-link and the Ward methods [44]. These algorithms differ in the way the groups similarity is evaluated. In the single-link, for instance, the similarity is established by finding the minimum distance between pairwise data items of any two given groups. The complete-link technique, in contrast, finds the similarity based on the maximum distance of two examples of different clusters, rather than the minimum [21].

CURE (Clustering Using Representatives) [24] represents a cluster by selecting a constant number of well-scattered points and shrinking them toward the cluster's centroid, according to a shrinking factor. CURE measures the similarity between two clusters by the similarity of the closest pair of points belonging to different clusters. Unlike centroid/medoid-based methods, CURE can find clusters of arbitrary shapes and sizes, as it represents each cluster via multiple representative points. Shrinking the representative points toward the centroid allows CURE to avoid some of the problems associated with noise and outliers. However, this technique fails to account for special characteristics of individual clusters. As such, it can make incorrect merge decisions when the underlying data does not follow the assumed model or when noise is present [33].

In general, hierarchical algorithms usually provide more information than partitional algorithms [30]. For instance, though susceptible to outliers, the single-link method is able to find groups very apart from each other, concentric groups, and chain-like groups in which the K-Means technique would have trouble with. However, hierarchical algorithms often take more time to process data items, which may invalidate their employment in large-scale data sets.

3.4 Semi-Supervised Learning

Algorithms that are able to learn using only a few labeled examples have aroused the interest of the artificial intelligence community. Among other features, semi-supervised learning aims at reducing the work of human experts in the labeling process. This feature is quite interesting especially when the labeling process is expensive and time consuming. This is often the case, for example, in video indexing, classification of audio signals, text categorization, medical diagnostics, genome data, among others [10, 65]. In this chapter, we introduce definitions and reviews on traditional methods presented in the literature. In Chap. 7, we revisit the semi-supervised learning paradigm with a focus on network-based methods.

3.4.1 Motivations

Semi-supervised learning is a new paradigm in relation to unsupervised and supervised learning. In this way, we first try to understand the reason behind its creation. From an engineering standpoint, it is clear that the collection of labeled data is much more intensive and costly in relation to that of unlabeled data. The main purpose of semi-supervised learning, nonetheless, goes way beyond a purely utilitarian tool. Most natural learning (human and animal) occurs in a semi-supervised regime basis. In the world in which we live, living beings are in a constant exposure to a flow of natural stimuli. Such stimuli include unlabeled data that are easily noticeable. In the context of recognition and phonological acquisition, for example, a child is exposed to many perceptible acoustic sounds. Many of these sounds are not familiar to the child. A positive feedback by part of another person is the main source of labeled data. In many cases, a small amount of feedback is sufficient to allow the child to master the acoustic-phonetic mapping of any languages [5, 6].

Humans have ability to learn unsupervised concepts (for example, clusters and categories of objects), suggesting that unlabeled data could be satisfactorily used for learning natural invariance to form categories and to construct classifiers. In many pattern recognition tasks, humans only have access to small amounts of labeled patterns. Hence, the success of human learning in environments with little knowledge indubitably happens with effective use and manipulation of large sets of unlabeled data to extract information that is useful for generalization purposes. Consequently, if the goal is to know how natural learning is processed, there is a need to think in terms of semi-supervised learning [4, 6].

Another motivation for studying semi-supervised learning is intrinsically linked to improving the performance of computational models. It has been shown that, by means of a finite sample analysis, if the complexity of the underlying data distribution is too high to be learned by L labeled data, but it is small enough to be learned by $U \gg L$ unlabeled data, then the semi-supervised learning is really able to improve the performance of a typical fully supervised learning task [58]. As an example, consider Fig. 3.3, where the numbered circles denote labeled data, while unnumbered circles represent unlabeled data. Applying a fully supervised algorithm to this problem, the decision boundary would most likely be established in the surroundings of the dotted vertical line.[2] Applying a semi-supervised learning algorithm, in contrast, the decision boundary would probably be fixed around the continuous line, because it is a low-density region that is away both from unlabeled and labeled data. In this example, supervised algorithms would not be able to efficiently classify the unlabeled examples. In contrast, semi-supervised methods, with the aid of the unlabeled data used in the training process, would perform better.

3.4.2 Mathematical Formalization and Fundamental Assumptions

Semi-supervised learning can be defined as follows [10, 65]. Let $\mathscr{X} = \{x_1, x_2, \ldots, x_{L+U}\}$ be a data set, divided into two parts: $\mathscr{X}_L = \{x_1, x_2, \ldots, x_L\}$ and $\mathscr{X}_U = \{x_{L+1}, \ldots, x_{L+U}\}$. There are L and U labeled and unlabeled data items, respectively, in a total of $N = L + U$ data items. Consider that $\mathscr{Y}$ is the label or class

---

[2]This is the case for algorithms with a maximum margin inductive bias. As one labeled item is positioned at -1 and the other at 1, the decision boundary that maximizes the margin between these two data items must cross the zero mark, as Fig. 3.3 shows.

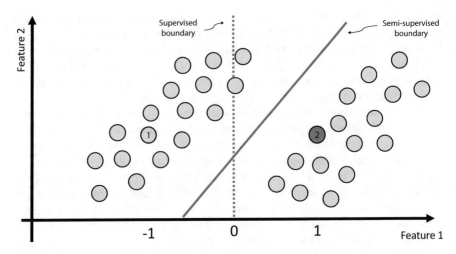

Fig. 3.3 An example of data set where semi-supervised learning techniques would lead to more robust results than supervised learning techniques. Data items are described by two numerical attributes, whose values are plotted in the horizontal and vertical axis. The numbered *circles* represent labeled data instances, while the *circles* without a number represent unlabeled ones. The *dotted line* denotes the decision boundary that would be probably output by a supervised learning method. The *continuous line* displays the same information for a semi-supervised learning algorithm

set. Supposing that the label set is discrete, this task is referred to as semi-supervised classification. (The same reasoning can be applied to regression.) The labels of the subset $\mathcal{X}_U$ are not known *a priori*. Normally, $L \ll U$, i.e., the great majority of data items does not possess labels. As we have already stressed, this often happens because the task of manual labeling is cumbersome and often is performed by human experts. The goal is to propagate labels from labeled instances to unlabeled instances in accordance to some diffusion rule. Based on these definitions, semi-supervised learning can be used in both data classification and clustering tasks. In the former case, the labeled examples are used in the process of labeling unlabeled examples. In the latter case, the labeled samples are responsible for imposing restrictions on the formation process of clusters [10].

It is worth mentioning that, for the proper functioning of semi-supervised learning techniques, some assumptions about the data consistency are essential [61]. Typically, a semi-supervised learning method relies on one or more of the following assumptions [10]:

- *Cluster assumption*: data points that belong to the same high-density region, i.e., are located in the same group, are plausible candidates for belonging to the same class.
- *Smoothness assumption*: data points that are near in the attribute space are probable candidates of being members of the same class. This assumption forces the decision function yielded by the classifier to be smoother in high-density

regions than in locations with low density. This analysis is in line with the cluster assumption and hence they complement each other.

- *Manifold assumption*: this idea is based upon the premise that a set of data points in a high dimensional space may be, approximately, reduced to a smaller space (manifold data) via a nonlinear mapping function. This hypothesis is usually employed to soften the curse of dimensionality problem, which is related to the fact that the volume of the space increases exponentially with the number of dimensions, and an exponentially larger number of examples would be needed for constructing induction classifiers with the same accuracy power.

The way that the semi-supervised learning algorithms treat these assumptions represents one of the fundamental differences among them.

Semi-supervised learning may refer to either *transductive learning* or *inductive learning*. The goal of transductive learning is to only infer the correct labels of the unlabeled data set $\mathscr{X}_U$. In contrast, the goal of inductive learning is to estimate a mapping from the available data to the output variable. Therefore, while in transductive learning the model is only used to predict the labels of $\mathscr{X}_U$, inductive learning goes beyond by also allowing other unlabeled data out of $\mathscr{X}_U$ to be estimated.

3.4.3 Overview of the Techniques

Traditionally, the semi-supervised learning methods are divided into the following categories:

- *Generative models*: The inference via generative models involves the estimation of a conditional density. The Expectation Maximization technique is the most known technique pertaining to this approach [52]. Besides that, a myriad of techniques proposed so far in the literature can be encountered in [2, 10, 22, 65].
- *Cluster-and-label models*: The inference is done based on the results obtained by a clustering task that is subjected to restrictions on the labeled data set. Some representative methods are given in [16, 17].
- *Low-density region separation models*: The inference is based on the development of decision boundary functions, such that each decision boundary is created as far as possible from high-density regions of labeled and unlabeled data items. The most well-known method of this category is the Transductive SVM [14, 60]. More related techniques can be found in [2, 10, 14, 65].
- *Network-based models*: The inference and label propagation are performed in a networked environment. In Chap. 7, we explore these techniques in detail.

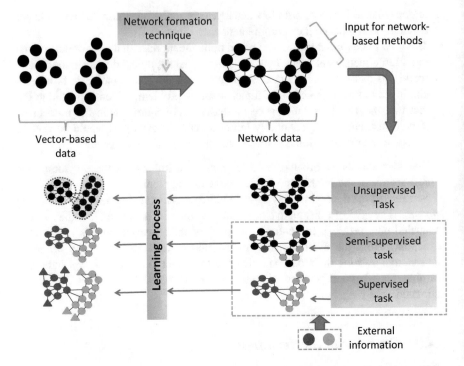

Fig. 3.4 Steps for applying machine learning methods in vector-based data. First, we transform data into a representative network using some network formation technique. That network is employed as input for network-based machine learning tasks, which in turn output the desired output

3.5 Overview of Network-Based Machine Learning

In the upcoming chapters of this book, we provide a comprehensive review on machine learning methods based on complex networks, including detailed case studies on the three learning paradigms.

When dealing with network-based methodologies, we need to follow a set of steps to complete machine learning tasks. Figure 3.4 illustrates the main steps involved to apply network-based methods in vector-based data, which we detail as follows:

1. *Gather the vector-based data set and preprocess it accordingly.* The preprocess may involve attribute transformation (scaling, normalization, standardization, demeaning, combination, decomposition, aggregation etc) or deletion, cleaning (removal of outliers, imputation of missing attributes etc), sampling, among many other preprocessing operations. The machine learning tools that we use influence the type of preprocessing we need to perform. For instance, if we have a large data set and an algorithm with high time complexity order, we need to

perform a sampling process on that data set so as to become viable the learning process.

2. *Transform the vector-based into network-based data.* This step is performed using a network formation technique. This is a crucial problem for network-based problems, because the resulting network must reliably represent the data relationships. Chapter 4 deals with this problem in a comprehensive manner.

3. *Apply the network-based machine learning task in the network constructed from the vector-based data.* The generated network in the previous step serves as input for network-based machine learning tasks. Three types of tasks can be performed when we are at the machine learning domain:

- *Network-based unsupervised learning*: only the generated network is employed in the learning process. Chapter 6 reviews the state-of-the-art unsupervised learning techniques presented in the literature, while Chap. 9 presents a case study of unsupervised learning in data clustering and in community detection that is based on competition of multiple interacting particles.
- *Network-based supervised learning*: besides the generated network, we are also given external information in the form of labels. Chapter 5 surveys representative supervised learning algorithms that learn in a networked environment. In addition, Chap. 8 explores a case study of supervised learning in data classification and in handwritten digits recognition that is based on a high-level classification framework.
- *Network-based semi-supervised learning*: besides the generated network, we are also given external information for some of the training items in the form of labels. Chapter 7 compiles a set of semi-supervised techniques that relies on networks to learn. Moreover, Chap. 10 examines a case study of semi-supervised learning in data classification and in imperfect learning that is based on a competitive-cooperative scheme of multiple interacting particles.

3.6 Chapter Remarks

In this section, we have introduced the machine learning area and its three main learning paradigms: supervised, unsupervised, and semi-supervised. Machine learning is a field of study that gives computers the ability to learn without being explicitly programmed.

We have seen that supervised algorithms utilize external information in the form of labels or classes to induce or to train their hypotheses. In the supervised learning scheme, there are two learning phases: the training and the classification phases. In the training phase, the classifier is induced or trained by using the training instances (labeled data) that are always fully labeled. In the classification phase, the labels of the test instances are predicted using the induced classifier. We have seen that, once trained, one must verify how well the trained supervised learner can generalize.

Without any additional assumptions over the data properties, this problem cannot be solved exactly since unseen situations might have an arbitrary output values. The necessary assumptions about the nature of the target function constitute the inductive bias of the learner.

Unsupervised learning methods are guided exclusively by the intrinsic structure of the data items throughout the learning process, i.e., without any sort of external knowledge. One of the most common tasks of unsupervised learning is in data clustering. The clusters may be defined as sets of patterns, points, or objects, all of which are characterized by suitable similarity measures. Each cluster is a collection of data items that are similar between them and are dissimilar to objects belonging to other clusters. Likewise supervised learning, we have seen that unsupervised algorithms also make assumptions about the underlying data generation process and that fact leads to shortcomings and potentialities of each available algorithm.

Between supervised and unsupervised learning lies semi-supervised learning. In this paradigm, algorithms are able to learn using only a few labeled examples. Among other features, semi-supervised learning aims at reducing the work of human experts in the labeling process. The perceptive difference between semi-supervised and supervised learning is that the latter only uses labeled data in the training phase, while the former employs both the labeled and unlabeled data. Under some conditions, we have seen that the introduction of unlabeled data in the training phase can really improve the overall classification performance of the classifier.

The goal in this chapter is to provide an overview on the usual hypotheses and limitations that we encounter in machine learning tasks. In the next four chapters, we first review network formation techniques and then revisit each of the learning paradigms with a focus on network-based approaches.

References

1. Aggarwal, C.C., Reddy, C.K.: Data Clustering: Algorithms and Applications. CRC, Boca Raton (2014)
2. Alpaydin, E.: Introduction to Machine Learning (Adaptive Computation and Machine Learning). MIT, Cambridge (2004)
3. Arlot, S., Celisse, A.: A survey of cross-validation procedures for model selection. Stat. Surv. **4**, 40–79 (2010)
4. Belkin, M., Matveeva, I., Niyogi, P.: Regularization and semi-supervised learning on large graphs. In: Shawe-Taylor, J., Singer, Y. (eds.) Learning Theory. Lecture Notes in Computer Science, vol. 3120, pp. 624–638. Springer, Berlin/Heidelberg (2004)
5. Belkin, M., Niyogi, P., Sindhwani, V.: On manifold regularization. In: Proceedings of the Tenth International Workshop on Artificial Intelligence and Statistics (AISTAT 2005), pp. 17–24. Society for Artificial Intelligence and Statistics, New Jersey (2005)
6. Belkin, M., Niyogi, P., Sindhwani, V.: Manifold regularization: a geometric framework for learning from labeled and unlabeled examples. J. Mach. Learn. Res. **7**, 2399–2434 (2006)
7. Bengio, Y., Grandvalet, Y.: No unbiased estimator of the variance of k-fold cross-validation. J. Mach. Learn. Res. **5**, 1089–1105 (2004)
8. Berkhin, P.: Survey of clustering data mining techniques. Technical Report, Accrue Software (2002)

9. Bishop, C.M.: Pattern Recognition and Machine Learning (Information Science and Statistics). Springer, New York (2007)
10. Chapelle, O., Schölkopf, B., Zien, A. (eds.): Semi-supervised learning. Adaptive Computation and Machine Learning. MIT, Cambridge (2006)
11. Chung, F.R.K.: Spectral graph theory. CBMS Regional Conference Series in Mathematics, vol. 92. American Mathematical Society, Providence (1997)
12. Cinque, L., Foresti, G.L., Lombardi, L.: A clustering fuzzy approach for image segmentation. Pattern Recogn. **37**, 1797–1807 (2004)
13. Clauset, A., Newman, M.E.J., Moore, C.: Finding community structure in very large networks. Phys. Rev. E **70**(6), 066111+ (2004)
14. Cortes, C., Vapnik, V.: Support-vector networks. Mach. Learn. **20**, 273–297 (1995)
15. Cover, T.M., Hart, P.: Nearest neighbor pattern classification. IEEE Trans. Inf. Theory **13**, 21–27 (1967)
16. Dara, R., Kremer, S.C., Stacey, D.A.: Clustering unlabeled data with SOMs improves classification of labeled real-world data. In: Proceedings of the World Congress on Computational Intelligence (WCCI), pp. 2237–2242 (2002)
17. Demiriz, A., Bennett, K.P., Embrechts, M.J.: Semi-supervised clustering using genetic algorithms. In: Proceedings of Artificial Neural Networks in Engineering (ANNIE-99), pp. 809–814. ASME (1999)
18. Deng, L., Yu, D.: Deep learning: Methods and applications. Founda. Trends Signal Process. **7**(3), 197–387 (2014)
19. Duda, R.O., Hart, P.E., Stork, D.G.: Pattern Classification. Wiley-Interscience, Chichester (2000)
20. Duda, R.O., Hart, P.E., Stork, D.G.: Pattern Classification. Wiley, New York (2001)
21. Gan, G.: Data Clustering: Theory, Algorithms, and Applications. Society for Industrial and Applied Mathematics, Philadelphia (2007)
22. Gärtner, T.: Kernels for Structured Data, vol. 72. World Scientific Publishing, Singapore (2008)
23. Girvan, M., Newman, M.E.J.: Community structure in social and biological networks. Proc. Natl. Acad. Sci. U.S.A. **99**(12), 7821–7826 (2002)
24. Guha, S., Rastogi, R., Shim, K.: CURE: an efficient clustering algorithm for large databases. Inf. Syst. **26**(1), 35–58 (2001)
25. Hasan, M.A., Chaoji, V., Salem, S., Zaki, M.: Link prediction using supervised learning. In: Proceedings of SDM 06 workshop on Link Analysis, Counterterrorism and Security (2006)
26. Hastie, T., Tibshirani, R., Friedman, J.: The Elements of Statistical Learning: Data Mining, Inference, and Prediction. Springer, New York (2011)
27. Haykin, S.S.: Neural Networks and Learning Machines. Prentice Hall, Englewood Cliffs (2008)
28. Husek, D., Pokorny, J., Rezanková, H., Snášel, V.: Data clustering: from documents to the web. In: Web Data Management Practices: Emerging Techniques and Technologies, pp. 1–33. IGI Global, Hershey, PA (2006)
29. Jain, A.K.: Data clustering: 50 years beyond K-means. Pattern Recogn. Lett. **31**, 651–666 (2010)
30. Jain, A.K., Murty, M.N., Flynn, P.J.: Data clustering: A review. ACM Comput. Surv. **31**(3), 264–323 (1999)
31. James, G., Witten, D., Hastie, T., Tibshirani, R.: An Introduction to Statistical Learning: with Applications in R. Springer, New York (2013)
32. Jolliffe, I.T.: Principal Component Analysis. Springer Series in Statistics. Springer, New York (2002)
33. Karypis, G., Han, E.H., Kumar, V.: Chameleon: hierarchical clustering using dynamic modeling. Computer **32**(8), 68–75 (1999)
34. Kashef, R., Kamel, M.S.: Enhanced bisecting K-Means clustering using intermediate cooperation. Pattern Recogn. **42**(11), 2557–2569 (2009)
35. Kaufman, L., Rousseeuw, P.J.: Finding Groups in Data: An Introduction to Cluster Analysis. Wiley, New York (2005)

36. Kodratoff, Y., Michalski, R.S.: Machine Learning: An Artificial Intelligence Approach, vol. 3. Morgan Kaufmann, San Mateo (2014)
37. Korb, K.B., Nicholson, A.E.: Bayesian Artificial Intelligence. Chapman and Hall, Boca Raton (2010)
38. Kuhn, M., Johnson, K.: Applied Predictive Modeling. Springer, New York (2013)
39. Lim, G., Park, C.H.: Semi-supervised dimension reduction using graph-based discriminant analysis. In: Computer and Information Technology (CIT), vol. 1, pp. 9–13. IEEE Computer Society, Xiamen (2009)
40. Liu, H., Shah, S., Jiang, W.: On-line outlier detection and data cleaning. Comput. Chem. Eng. **28**, 1635–1647 (2004)
41. Lu, C.T., Chen, D., Kou, Y.: Algorithms for spatial outlier detection. In: Proceedings of the 3rd IEEE International Conference on Data Mining (ICDM 2003). IEEE Computer Society (2003)
42. MacQueen, J.B.: Some methods for classification and analysis of multivariate observations. In: Proceedings of the fifth Berkeley Symposium on Mathematical Statistics and Probability, vol. 1, pp. 281–297. University of California Press (1967)
43. Marsland, S.: Machine Learning: An Algorithmic Perspective. CRC, Boca Raton (2014)
44. Mitchell, T.M.: Machine Learning. McGraw-Hill Science/Engineering/Math, New York, NY (1997)
45. Müller, P., Quintana, F.A., Jara, A., Hanson, T.: Bayesian Nonparametric Data Analysis. Springer, New York (2015)
46. Murphy, K.P.: Machine Learning: A Probabilistic Perspective. MIT, Cambridge (2012)
47. Neapolitan, R.E.: Learning Bayesian Networks. Prentice Hall, Upper Saddle River (2003)
48. Newman, M.E.J.: Finding community structure in networks using the eigenvectors of matrices. Phys. Rev. E **74**(3), 036104 (2006)
49. Newman, M.E.J.: Modularity and community structure in networks. Proc. Natl. Acad. Sci. **103**(23), 8577–8582 (2006)
50. Newman, M.E.J., Girvan, M.: Finding and evaluating community structure in networks. Phys. Rev. Lett. **69**, 026113 (2004)
51. Ng, R.T., Han, J.: CLARANS: A method for clustering objects for spatial data mining. IEEE Trans. Knowl. Data Eng. **14**(5), 1003–1016 (2002)
52. Nigam, K., McCallum, A.K., Thrun, S., Mitchell, T.: Text classification from labeled and unlabeled documents using EM. Mach. Learn. **39**(2–3), 103–134 (2000)
53. Piatetsky-Shapiro, G.: Discovery, Analysis, and Presentation of Strong Rules, chap. 12. AAAI/MIT, Cambridge (1991)
54. Quinlan, J.R.: Generating production rules from decision trees. In: Proceedings of the 10th International Joint Conference on Artificial Intelligence (IJCAI'87), vol. 1, pp. 304–307. Morgan Kaufmann, San Mateo (1987)
55. Quinlan, J.R.: C4.5: Programs for Machine Learning. Morgan Kaufmann Series in Machine Learning. Morgan Kaufmann, San Mateo (1992)
56. Sander, J., Ester, M., Kriegel, H.P., Xu, X.: Density-based clustering in spatial databases: the algorithm GDBSCAN and its applications. Data Min. Knowl. Disc. **2**(2), 169–194 (1998)
57. Schmidhuber, J.: Deep learning in neural networks: an overview. Neural Netw. **61**, 85–117 (2015)
58. Singh, A., Nowak, R.D., Zhu, X.: Unlabeled data: Now it helps, now it doesn't. In: The Conference on Neural Information Processing Systems NIPS, pp. 1513–1520 (2008)
59. Vapnik, V.N.: The Nature of Statistical Learning Theory. Springer, New York (1995)
60. Vapnik, V.N.: Statistical Learning Theory. Wiley-Interscience, New York (1998)
61. Wang, F., Li, T., Wang, G., Zhang, C.: Semi-supervised classification using local and global regularization. In: AAAI'08: Proceedings of the 23rd National Conference on Artificial Intelligence, pp. 726–731. AAAI (2008)
62. Witten, I.H., Frank, E.: Data Mining: Practical Machine Learning Tools and Techniques. Morgan Kauffman, San Mateo (2005)
63. Xu, R., II, D.W.: Survey of clustering algorithms. IEEE Trans. Neural Netw. **16**(3), 645–678 (2005)

64. Zhu, X.: Semi-supervised learning literature survey. Technical Report 1530, Computer Sciences, University of Wisconsin-Madison (2005)
65. Zhu, X., Goldberg, A.B.: Introduction to Semi-Supervised Learning. Synthesis Lectures on Artificial Intelligence and Machine Learning. Morgan and Claypool Publishers, San Rafael (2009)

Chapter 4
Network Construction Techniques

Abstract In many areas of machine learning, networks are used to model local relationships between data points and to build global structures from local information. Building networks is often a necessary step when dealing with problems arising from applications in machine learning or data mining. This fact becomes crucial when we want to apply network-based learning methods to vector-based data sets, in which a network must be constructed from the input data set using some convenient network formation criteria. In this chapter, we review the main ingredients that are needed to construct a graph from non-networked data. In special, we discuss transformation of vector-based and time series data. Several similarity functions are also discussed.

4.1 Introduction

Networks are essential for encoding information, and data in network format is increasingly abundant in fields ranging from computational biology to computer vision. The transformation from unstructured data to a network data representation can always be performed in a lossless manner. The inverse transformation, however, is often a lossy one. Let us give an example. Consider the WWW that is inherently represented by a network format. In such a network, pages are vertices and links exist between different pages if one page references another one. Now suppose we desire to extract a vector-based data out of this network. A very difficult task would be to model the recursiveness of cycles in the network topology when we now go to a vector-based format. Moreover, the local and global topologies of the pages relationships would probably be distorted by the transformation. In addition, considering that there are more than one network component, then some shortest paths between members of different components would be infinity in the network. To model this extreme dissimilarity in a vector-based format would be difficult, because the information of whether or not vertices are in the same component is structural and depends on the topology of the data relationships, which in turn is not easily modeled in a vector-based format.

From that example, it is clear that networks embed more information than vector-based data sets. This additional information is made up of several ingredients, among which the most important one is the structural or topological information

© Springer International Publishing Switzerland 2016

T.C. Silva, L. Zhao, *Machine Learning in Complex Networks*,

DOI 10.1007/978-3-319-17290-3_4

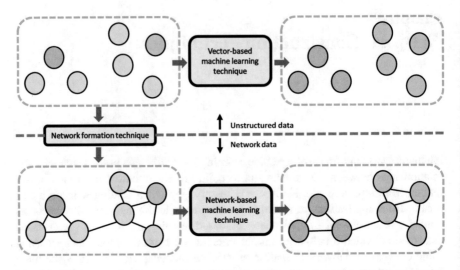

Fig. 4.1 Differences of vector- and network-based machine learning tasks. We illustrate using a semi-supervised learning classification. *Colored* vertices denote labeled data, while *gray* data symbolize unlabeled data. Network formation methods interface between unstructured and structured network data

of the data relationships. In this way, the network topology is able to encode in an elegant manner interactions of the data items in a systematic manner, going from local to global structural information. Thus, a natural question that arises is how can we build networks from unstructured or vector-based data, such that the resulting network encodes as much information as possible? The structure, in principle, must be estimated using the network formation technique based on some heuristics. In this chapter, we discuss the problem of network formation, a task that serves as interface between unstructured and structured network data.

Figure 4.1 illustrates where network formation techniques stand in an overall machine learning scheme. We illustrate using a semi-supervised learning task.[1] First, we see that there is a one-to-one correspondence of $\mathscr{X}$ and $\mathscr{V}$, i.e., each data item in the data set is a vertex in the resulting network. Edges are created using some heuristics that capture similarity among data items. Note these edges that naturally encode similarity are only explicitly modeled in a network environment. Thus, they are estimated by the network formation procedure that interfaces between unstructured and structured network data.

Following our previous notation on general machine learning tasks, given a set of N data points $\mathscr{X} = \{x_1, \ldots, x_N\}$ that is not in a networked format, we can transform it into a network $\mathscr{G}$ that consists of (1) the vertex set $\mathscr{V} = \{v_1, \ldots, v_V\}$ and (2) the

---

[1] Recall that, in a semi-supervised setting, the goal is to propagate the labels from the labeled to the unlabeled set.

edge set $\mathscr{E}$, which is a subset of $\mathscr{V} \times \mathscr{V}$. The transformation is performed a mapping procedure $g : \mathscr{X} \to \mathscr{G} = \langle \mathscr{V}, \mathscr{E} \rangle$. We now discuss how the sets $\mathscr{V}$ and $\mathscr{E}$ that compose the network are obtained.

In relation to the vertex set, for the majority of machine learning applications, $\mathscr{V} = \mathscr{X}$ holds, i.e., each data item exactly corresponds to a vertex in the resulting network. To illustrate, in a handwritten digits recognition, each digit in $\mathscr{X}$ would correspond to a vertex in $\mathscr{V}$. Some learning techniques, however, may use reduced or expanded sets of the data items. For example, we can compact several very similar date items in $\mathscr{X}$ into a super-vertex that essentially represents in a summarized manner all of those data items.

We have been using N to denote the cardinality or number of data items in $\mathscr{X}$. In a network setting, the number of vertices in $\mathscr{V}$ is symbolized as $V = |\mathscr{V}|$, which is not necessarily (but often is) equal to N, as we have previously discussed.

Now we discuss how to obtain the set of edges $\mathscr{E}$. We process the decision of establishing or not edges in $\mathscr{E}$ in accordance with two factors:

- *A proper similarity function s*: the similarity function $s : \mathscr{V} \times \mathscr{V} \mapsto \mathbb{R}$ enables us to quantify how different or similar two data items are with respect to their attributes. That is, the similarity function transforms two data items[2] into a scalar value. Applying the similarity function to all of the pairs of vertices, we are able to construct (1) the similarity matrix $\mathbf{S}$, in which $\mathbf{S}_{ij} = s(v_i, v_j)$, where $v_i, v_j \in \mathscr{V}$, or (2) equivalently the dissimilarity matrix $\mathbf{D}_{ij} = d(v_i, v_j)$.
- *A network formation technique*: we decide whether or not to add a link between v_i and v_j by using some rules applied on the similarity matrix $\mathbf{S}$ or on the dissimilarity matrix $\mathbf{D}$.

In this chapter, we discuss these two ingredients that are needed to build up a network from non-networked data. First, we review the theory behind the definition of similarity functions as well as some well-known examples. Following that we show network formation techniques that are applied to vector-based data, in which each data item is represented by a feature vector.[3] After, we deal with the issue of constructing networks from time series. In this case, there is an additional caveat of temporal data dependency that introduces some complexity in the network formation process.

---

[2] The term data item is used in a wide sense. It may denote feature vectors, time series, graphical objects, among many other types of objects.

[3] Essentially, the representation of data in the format of feature vector covers a wide spectrum of applications in the real world. In image-based applications, for instance, where each data item is symbolized by an image, we can always extract some features from that image and construct a feature vector. Normally, face recognition systems and image processing tools do this for robustness (due to relative uniqueness and invariance between different images) and computational efficiency reasons.

4.2   Similarity and Dissimilarity Functions

The concepts of similarity and dissimilarity are widely employed in the artificial intelligence domain. Among the several fields that they appear, we highlight applications in data mining, information retrieval, pattern matching, genetics, drug discovery, and fuzzy logic [1, 2, 21, 30, 33, 48]. In a general sense, similarity and dissimilarity express a comparison between two elements. Though intuitive, several different formalizations of similarity and dissimilarity exist in the literature. Another prominent characteristic is the duality linking the similarity and dissimilarity concepts, which are opposite terms yet somehow interrelated. This duality also extends to properties of the data items, which could be very useful if properly exploited. Thus, every property of a similarity should have a correspondence with one property of a dissimilarity and vice versa.

Several researches have tried to formalize these concepts but the main properties of similarity or dissimilarity are still under discussion [13, 40]. The lack of basic common theory underlying these two functions leads to incompatible definitions or results. Duality is often neglected and there are few studies about how transformations of similarity to dissimilarity functions can alter their properties [40]. If the similarity function s is well-behaved, one way to calculate its dual, the dissimilarity function d, is:

$$d(x_i, x_j) = \sqrt{s(x_i, x_i) + s(x_j, x_j) - 2s(x_i, x_j)}, \qquad (4.1)$$

in which x_i and x_j are two arbitrary data items.

Similarity measures are peculiar kinds of indicators that are mainly descriptive coefficients and not estimators of some statistical parameter. We note that it is difficult to give reliable confidence intervals for most measures of similarity and probable errors can be estimated only by certain types of randomization procedures [28, 49].

4.2.1   Formal Definitions

Similarity and dissimilarity express the degree of coincidence or divergence between two elements of a given domain. Thus, it is reasonable to treat them as functions since the objective is to measure or calculate this value between any two elements of the domain. We first define a similarity function in the following [38].

Definition 4.1. Similarity function: Let $\mathscr{X}$ be a non-empty set where an equality relation is defined. If s is a similarity function, then s is upper bounded, exhaustive, and total, whose domain and range are as follows:

$$s : \mathscr{X} \times \mathscr{X} \longmapsto I_s \subset \mathbb{R}, \qquad (4.2)$$

in which I_s is upper bounded by construction, since we assumed s is upper bounded, where $\max_{\mathbb{R}} I_s = s_{max}$. Moreover, the similarity function s satisfies the following properties:

1. *Reflexivity*: $s(x, x) = s_{max}$.
2. *Strong reflexivity*: $s(x, y) = s_{max} \iff x = y$.
3. *Symmetry*: $s(x, y) = s(y, x)$.
4. *Boundedness*[4]: s is lower bounded when $\exists a \in \mathbb{R} : s(x, y) \geq a, \forall x, y \in \mathscr{X}$. This statement is equivalent to affirm that $s = s_{min} = \min_{\mathbb{R}} I_s$ exists.
5. *Closedness*[5]: This property assures the existence of a lower bound. In special, the closedness property asks for the existence of $x, y \in \mathscr{X} : s(x, y) = s_{min}$.
6. *Transitivity*: If τ_s is a transitivity operator, then the following must hold: $s(x, y) \geq \tau_s(s(x, z), s(z, y)), \forall x, y, z \in \mathscr{X}$.

We see that s takes as input two data items from $\mathscr{X}$ and outputs a bounded real-valued scalar value. The more similar two objects are, the greater is the similarity value between them. In the next paragraph, we define the concept of dissimilarity or distance function [38].

Definition 4.2. Dissimilarity function: Let $\mathscr{X}$ be a non-empty set where an equality relation is defined. If d is a dissimilarity function, then d is lower bounded, exhaustive, and total, whose domain and range are as follows:

$$d : \mathscr{X} \times \mathscr{X} \longmapsto I_d \subset \mathbb{R}, \tag{4.3}$$

in which I_d is lower bounded by construction, since we assumed d is lower bounded, where $\min_{\mathbb{R}} I_d = d_{min}$. Moreover, the dissimilarity or distance function d satisfies the following properties:

1. *Reflexivity*: $d(x, x) = d_{min}$.
2. *Strong reflexivity*: $d(x, y) = d_{min} \iff x = y$.
3. *Symmetry*: $d(x, y) = d(y, x)$.
4. *Boundedness*: d is upper bounded when $\exists a \in \mathbb{R} : d(x, y) \leq a, \forall x, y \in \mathscr{X}$. This statement is equivalent to affirm that $d = d_{max} = \max_{\mathbb{R}} I_d$ exists.
5. *Closedness*: This property assures the existence of an upper bound. The closedness property asks for the existence of $x, y \in \mathscr{X} : d(x, y) = d_{max}$.
6. *Transitivity*: If τ_d is a transitivity operator, then the following must hold: $d(x, y) \leq \tau_d(d(x, z), d(z, y)), \forall x, y, z \in \mathscr{X}$.

---

[4]Recall that a set $\mathscr{S} \in \mathbb{R}^m, m > 0$, is *bounded* if there exists a number B such that $\|x\| \leq B, \forall x \in \mathscr{S}$, that is, if $\mathscr{S}$ is contained in some ball in $\mathbb{R}^m$.

[5]Recall that a set $\mathscr{S} \in \mathbb{R}^m, m > 0$, is *closed* if, whenever $\{x_n\}_{n=1}^{\infty}$ is convergent sequence completely contained in $\mathscr{S}$, its limit is also contained in $\mathscr{S}$. For example, the sets $\mathbb{R}^m$ and $\{(x, y) \in \mathbb{R}^2 : xy = 1\}$ are closed but not bounded.

We see that d takes as input two data items from $\mathscr{X}$ and outputs a bounded real-valued scalar value. The more dissimilar are two objects, the greater is the dissimilarity value between them.

Besides those mathematical properties, there are two desirable attributes of all similarity measures. First, the measure should be independent of sample size and of the number of classes in the population [49]. Second, the measure should increase smoothly from some fixed minimum to a fixed maximum, as the samples become more similar. We refer to the researches in [11, 12, 14, 40, 49, 50] for an extensive analysis of the properties of similarity functions.

4.2.2   Examples of Vector-Based Similarity Functions

In this section, we show some traditional examples of similarity/dissimilarity functions. Assume we have a data set $\mathscr{X} = \{x_1, \ldots, x_N\}$ with $N > 1$ data items. Moreover, we characterize each data item with a feature vector $x_i = [x_{i1}, \ldots, x_{iP}]$ with $P > 0$ features or attributes. Data items x_i and x_j are both members of $\mathscr{X}$.

Before we start to explore examples of similarity functions, we give an overall intuition of the types of features or attributes that we can face.

A feature or attribute can be classified as one of the following types:

- *Categorical or nominal attribute*: a categorical feature is one that has two or more categories with no intrinsic ordering. For example, gender is a categorical variable having two categories (male and female) and there is no intrinsic ordering to the categories. Hair color is also a categorical variable having a number of categories (blonde, brown, brunette, red, etc.) and there is no agreed way to order these from highest to lowest. A purely categorical variable is one that simply allows you to assign categories but you cannot clearly order the variables.
- *Ordinal attribute*: An ordinal variable is similar to a categorical variable. The difference between the two is that there is a clear ordering scheme for ordinal variables. For example, suppose you have a variable, economic status, with three categories: low, medium, and high. In addition to being able to classify people into these three categories, you can order the categories as low, medium, and high. Now consider a variable like educational experience with values such as elementary school graduate, high school graduate, some college, and college graduate. These also can be ordered as elementary school, high school, some college, and college graduate.
- *Numerical or quantitative attribute*: The values of a quantitative variable can be ordered and measured. Height and weight are examples of numerical attributes.

Categorical and ordinal attributes are also termed as qualitative attributes, as we cannot numerically operate on them (multiplication and division, for instance, are not defined). Numerical attributes, in contrast, are classified as quantitative attributes, since mathematical operations can be performed on these types of features.

In the next section, we provide some representative similarity and distance measures. For a comprehensive review, see [33].

4.2.2.1 Numerical Data

In this section, we suppose that the attributes in x_i and x_j are all numerical. They are called feature vectors and have an arbitrary dimension of $P > 0$. The notation $x_i(k)$, $k \in \{1, \ldots, P\}$, indexes the k-th component of the attribute vector x_i. There are a total of N data items.

Definition 4.3. Euclidean distance: The Euclidean distance between x_i and x_j is:

$$d_{\text{Euclidean}}(x_i, x_j) \triangleq \sqrt{\sum_{k=1}^{P} \left[x_i(k) - x_j(k) \right]^2}. \tag{4.4}$$

Definition 4.4. Weighted Euclidean distance: The weighted Euclidean distance between x_i and x_j is:

$$d_{\text{WEuclidean}}(x_i, x_j) \triangleq \sqrt{\sum_{k=1}^{P} W_k \left[x_i(k) - x_j(k) \right]^2}, \tag{4.5}$$

in which W_k denotes the weight given for the k-th attribute.

Remark 4.1. If we give unitary weight for all of the attributes in the feature vector, then the weighted Euclidean distance reduces to the traditional Euclidean distance.

Definition 4.5. Manhattan or city-block distance: The Manhattan or city-block distance between x_i and x_j is:

$$d_{\text{Manhattan}}(x_i, x_j) \triangleq \sum_{k=1}^{P} \left| x_i(k) - x_j(k) \right|. \tag{4.6}$$

Definition 4.6. Chebyshev or supremum distance: The Chebyshev or supremum distance between x_i and x_j is:

$$d_{\text{Supremum}}(x_i, x_j) \triangleq \max \left(\left| x_i(1) - x_j(1) \right|, \ldots, \left| x_i(P) - x_j(P) \right| \right). \tag{4.7}$$

Definition 4.7. Minkowski distance (L_λ metric): The Minkowski distance or L_λ metric, $\lambda \geq 1$, between x_i and x_j is:

$$d_{\text{Minkowski}}(x_i, x_j) \triangleq \left[\sum_{k=1}^{P} |x_i(k) - x_j(k)|^{\lambda} \right]^{\frac{1}{\lambda}}. \tag{4.8}$$

Remark 4.2. The family of Minkowski functions is obtained by varying λ over 1 to ∞. The Minkowski distance is a generalization of the previous discussed metrics. We list them in the following:

- L_1 metric: Manhattan or city-block distance as in Definition 4.5.
- L_2 metric: Euclidean distance as in Definition 4.3.
- L_∞ metric: Chebyshev or supremum distance as in Definition 4.6.

Definition 4.8. Mahalanobis distance: The Mahalanobis distance between x_i and x_j is:

$$d_{\text{Mahalanobis}}(x_i, x_j) \triangleq \sqrt{\sum_{k=1}^{P} (x_i - x_j)^T \, \Sigma^{-1} \, (x_i - x_j)}, \tag{4.9}$$

in which Σ is the $P \times P$ sample covariance matrix, whose (i,j)-th entry, Σ_{ij}, is given by:

$$\Sigma_{ij} \triangleq \frac{1}{N-1} \sum_{k=1}^{N} (x_i(k) - \bar{x}_i) (x_j(k) - \bar{x}_j), \tag{4.10}$$

in which $\bar{x}_i$ and $\bar{x}_j$ are the sample means that in turn are expressed as:

$$\bar{x}_i \triangleq \frac{1}{N} \sum_{k=1}^{N} x_i(k). \tag{4.11}$$

In the next example, we provide the intuition behind the definition of the Mahalanobis distance.

Example 4.1. Consider the problem of estimating the probability that a test point in an Euclidean space belongs to a set of training data points. A natural first step would be to find the average or center of mass of these training data points. Intuitively, the closer the test point in question is to this center of mass, the more likely it is to belong to the set.

A simple refinement would be to quantify if the set is spread out over a large or a small range, so that we can decide whether a given distance from the center is noteworthy or not. The simplistic approach is to estimate the standard deviation of the distances of the sample points from the center of mass. If the distance between the test point and the center of mass of the training set is less than one standard deviation, then we might conclude that it is highly probable that the test point belongs to the set of training data points. The further away it is, the more likely that the test point should not be classified as belonging to the set.

This intuitive approach can be made quantitative by defining the normalized distance between the test point and the set of training data points as $\frac{x-\mu}{\sigma}$, where μ is the sample average and σ, the sample standard deviation of the set. By plugging these values into the normal distribution, we can derive the probability of the test point belonging to the set.

The drawback of the above approach is that we assume that the points in the training data set are distributed around the center of mass in a spherical manner. If we were dealing with a non-spherical distribution, such as the ellipsoidal distribution, then we would expect the membership probability of that test point to depend not only on the distance from the center of mass, but also on the direction. In those directions in which the ellipsoid has a short axis, the test point is expected to be closer if it is really a member of that set, while in those directions where the axis has large amplitude, the test point can be further away from the center.

Putting this on a mathematical basis, the ellipsoid that best represents the probability distribution of the training data set can be estimated by building the covariance matrix of the samples. The Mahalanobis distance is simply the distance of the test point from the center of mass divided by the width of the ellipsoid in the direction of the test point. This behavior is mathematically represented by (4.9).

Definition 4.9. Gaussian kernel similarity (radial basis function or heat kernel): The Gaussian kernel similarity between x_i and x_j is:

$$s_{\text{Gaussian}}(x_i, x_j) \triangleq a \, \exp\left(-\frac{\| x_i - x_j \|^2}{2\sigma^2}\right), \tag{4.12}$$

in which $\sigma > 0$ is the variance of bandwidth of the Gaussian function and a is scaling constant. The term $\| x_i - x_j \|$ is the Euclidean norm between x_i and x_j.

Definition 4.10. Harmonic mean similarity: The harmonic mean similarity between x_i and x_j is:

$$s_{\text{Harmonic}}(x_i, x_j) \triangleq 2 \sum_{k=1}^{P} \frac{x_i(k)x_j(k)}{x_i(k) + x_j(k)}. \tag{4.13}$$

Definition 4.11. Cosine similarity: The cosine similarity between x_i and x_j is:

$$s_{\text{Cosine}}(x_i, x_j) \triangleq \frac{\sum_{k=1}^{P} x_i(k)x_j(k)}{\|x_i\|\|x_j\|} = \frac{\langle x_i, x_j \rangle}{\|x_i\|\|x_j\|}, \tag{4.14}$$

in which $\langle .,. \rangle$ denotes the inner product operator and $\|.\|$ is the Euclidean norm.

Definition 4.12. Pearson correlation similarity: The Pearson correlation
similarity between x_i and x_j is:

$$s_{\text{Pearson}}(x_i, x_j) \triangleq \frac{\sum_{k=1}^{P} (x_i(k) - \bar{x}_i)(x_j(k) - \bar{x}_j)}{\|x_i - \bar{x}_i\| \|x_j - \bar{x}_j\|}$$

$$= \frac{\langle x_i - \bar{x}_i, x_j - \bar{x}_j \rangle}{\|x_i - \bar{x}_i\| \|x_j - \bar{x}_j\|}$$

$$= s_{\text{Cosine}}(x_i - \bar{x}_i, x_j - \bar{x}_j). \tag{4.15}$$

Remark 4.3. Cosine similarity can be applied to deal with document similarity and
image similarity. It should be noted that cosine similarity is affected by vector
translation. The Pearson correlation similarity, however, addresses this problem by
being translation-invariant due to the demeaning process. In addition, cosine and
Pearson correlations are scale-invariant due to the normalization process.

Definition 4.13. Dice similarity [36]: The Dice similarity between x_i and x_j is:

$$s_{\text{Dice}}(x_i, x_j) \triangleq \frac{2 \sum_{k=1}^{P} x_i(k)x_j(k)}{\sum_{k=1}^{P} x_i(k)^2 + x_j(k)^2}. \tag{4.16}$$

Definition 4.14. Kumar-Hassebrook similarity [29]: The Kumar-Hassebrook
similarity between x_i and x_j is:

$$s_{\text{KH}}(x_i, x_j) \triangleq \frac{\sum_{k=1}^{P} x_i(k)x_j(k)}{\sum_{k=1}^{P} x_i(k)^2 + x_j(k)^2 - x_i(k)x_j(k)}. \tag{4.17}$$

4.2.2.2 Categorical Data

In this section, we deal with categorical data. We consider that each entry of the
P-dimensional feature vectors x_i and x_j can assume either a present or an absent
value (dichotomous feature). If there are multiple categories, we define the category
of interest as present, while all of the others are considered to be in the absent class.
Therefore, when comparing the vectors x_i and x_j, we can face four different scenarios
that are delineated in Table 4.1. We see that:

- M_{11} denotes the number of occurrences of coincident present values in x_i and x_j.
- M_{01} is the number of occurrences in which there is an absence in x_i and a presence
  in x_j.
- M_{10} represents the number of occurrences in which there is a presence in x_i and
  an absence in x_j.
- M_{00} symbolizes the number of occurrences of coincident absent values in x_i
  and x_j.

Table 4.1 Possible outcomes when comparing two entries of categorical data

		x_j	
		Present	Absent
x_i	Present	M_{11}	M_{10}
	Absent	M_{01}	M_{00}

Definition 4.15. Hamming distance: The Hamming distance between x_i and x_j is

$$d_{\text{Hamming}}(x_i, x_j) \triangleq \sum_{k=1}^{P} \mathbb{1}_{[x_i(k) \neq x_j(k)]} = M_{01} + M_{10}, \tag{4.18}$$

i.e., the Hamming distance is defined as the minimum number of replacements that are needed to transform x_i into x_j.

Definition 4.16. Jaccard similarity [26]: The Jaccard similarity between x_i and x_j is

$$s_{\text{Jaccard}}(x_i, x_j) \triangleq \frac{M_{11}}{M_{11} + M_{01} + M_{10}}. \tag{4.19}$$

Remark 4.4. In the Hamming distance, each value is equally important. In some applications, however, it may be interesting to give more importance to some classes in detriment to others. In light of that, suppose we have a problem where there is a feature vector of P movies and we want to compute the similarity of movie taste of two persons. In this case, if there are several movies that they have not watched, we can not say two persons are similar simply because none of them watched any movies in the feature vector. In contrast, if these two persons have seen a significant quantity of common watched movies, we can say they are similar to some extent. That is, we give more weight to those entries in which both persons have mutually watched the film in detriment to other configurations. This is a kind of weighted Hamming distance that is known as Jaccard similarity.

Definition 4.17. Sørensen similarity [45]: The Sørensen similarity between x_i and x_j is

$$s_{\text{Sørensen}}(x_i, x_j) \triangleq \frac{2M_{11}}{2M_{11} + M_{01} + M_{10}}. \tag{4.20}$$

Remark 4.5. As opposed to Jaccard similarity, the matches in the Sørensen similarity are given more importance than mismatches. Deciding between the two is a matter of the type of data we have at hand. If many entries are present in the population but are not present in the sample, it may be useful to use Sørensen coefficient rather than Jaccard.

Definition 4.18. Simple matching similarity: The simple matching similarity between x_i and x_j is

$$s_{\mathrm{SM}}(x_i, x_j) \triangleq \frac{M_{11} + M_{00}}{M_{11} + M_{00} + M_{01} + M_{10}}. \tag{4.21}$$

Remark 4.6. Simple matching similarity is a good option to choose when absent and present values are equally valuable in the data.

Definition 4.19. Baroni-Urbani and Buser similarity [4]: The Baroni-Urbani and Buser similarity between x_i and x_j is

$$s_{\mathrm{BUB}}(x_i, x_j) \triangleq \frac{\sqrt{M_{11} M_{00}} + M_{11}}{\sqrt{M_{11} M_{00}} + M_{11} + M_{01} + M_{10}}. \tag{4.22}$$

Remark 4.7. The square root term in (4.22) is introduced to help in removing the size bias common in other similarity measures, such as the Jaccard.

4.3   Transforming Vector-Based Data into Networks

Given the similarity matrix **S** or the dissimilarity matrix **D**, one direct approach of building a network would be to establish links between pairs of vertices with weights according to S_{ij} or, equivalently, to some function on the reciprocal of D_{ij}. This approach would frequently lead to the emergence of almost complete networks. Generally speaking, a good network satisfies the following criteria: (1) it should possess a giant component to mantain the vertices connected; (2) it should be as sparse as possible in order to better reveal the relationships between the vertices. The existence of links with very small weights, however, may lead to poor results if used by network-based algorithms. Sparsification, hence, is important because it leads to improved efficiency in the learning stage, better accuracy, and robustness to noise. We can think of small-valued links as noises that would just jeopardize the learning process, providing misleading information to the machine learning algorithm by connecting two distant vertices. Therefore, the resulting network topology would be largely distorted by these noisy links. The removal of these links stands as an important pre-processing step for enhancing the efficiency of network-based learning algorithms.

Following that reasoning, the two most traditional types of nearest neighbor networks that sparsify the similarity or dissimilarity matrices are [5]:

- *k-nearest neighbors network* (*k*-NN): this is, in general, a directed network. An edge from v_i to v_j exists if and only if v_j is among the k most similar elements to v_i. In computational terms, we have to sort, in an independent manner, all of the rows of **D** in ascending order. Once sorted, given a row i, links are established among vertex i and the first k entries that are in the sorted list corresponding to the elements in the i-th row of $\mathbf{D}$.

- ϵ-*radius network*: this is an undirected network whose edge set consists of pairs (v_i, v_j) such that $\mathbf{D}_{ij} \leq \epsilon$, where $\epsilon \in \mathbb{R}_+$.

The k-nearest neighbors network is, in general, a directed network because v_j can be one of the k nearest neighbors of v_i, but the converse may not be true. In contrast, the ϵ-radius network is by construction an undirected network, because $\mathbf{D}_{ij} = \mathbf{D}_{ji}$, as we evaluate each entry of the distance matrix using a distance function. In this way, if $\mathbf{D}_{ij} < \epsilon$, it must be the case that $\mathbf{D}_{ji} < \epsilon$. Hence, the existence of links always occurs in both directions.

The k-nearest neighbors and the ϵ-radius techniques are considered as static network formation methods. This is because they treat in a uniform manner data items that are in dense and sparse regions. We now list a set of network formation techniques that employ adaptive or dynamical information:

- *Network formation using combinations of the k-nearest neighbors and ϵ-radius techniques* [43, 44]: we can devise a network formation technique that employs both heuristics based on one or more criteria. For instance, we can activate the k-nearest neighbors network when we are at sparse regions. Conversely, we can employ the ϵ-radius technique in dense regions.
- *b-matching network* [27]: As opposed to the k-NN network, the b-matching network ensures that each vertex in the graph has the same number of edges and therefore produces a balanced or regular graph. It relies on an optimization process.
- *Linear neighborhood network* [47]: the idea is to approximate the entire network by a series of overlapped linear neighborhood patches, and the edge weights in each patch are determined by a standard quadratic programming procedure. The initial neighborhoods of the vertices are set in a static way. Then, the linear embedding makes dynamical adjustments in the edge weights.
- *Relaxed linear neighborhood network* [10]: it approximates the entire network by a series of overlapped linear neighborhood patches, where the neighborhood of any vertex is captured dynamically based on the density/sparsity of its surrounding. Moreover, the relaxed linear neighborhood technique explores the degree of neighborhood during the reconstruction method rather than using fixed assignments. As a consequence, it does not get affected by outliers, producing networks that are more robust.
- *Network formation using clustering heuristics* [15]: this method uses data clustering heuristics to perform the network formation process. Specifically, it employs the single-link method, which is a clustering heuristic that is capable of constructing connected and sparse networks, while also maintaining the cluster structure of the original data set.
- *Network formation using overlapping histogram segments* [42]: this technique uses overlapping histogram segments to perform the network construction. The resulting network always produces a connected graph with vertices of the same community densely interconnected and with vertices of different commu-

nities sparsely interconnected. In essence, it is based on the k-NN technique, but with adaptive k values that are learned from the data distribution.

In the following section, we discuss in detail each of the mentioned techniques.

4.3.1 Analysis of k-Nearest Neighbors and ε-Radius Networks

We supply in Fig. 4.2a, b the geometrical intuition of the k-nearest neighbors and the ϵ-radius networks, respectively. In the k-NN, once set a reference vertex, we simply sort all of the remainder vertices in accordance with the selected distance function, thus ranking those vertices. With the selected parameter k, we only establish links with those vertices that are ranked below that threshold k. In Fig. 4.2a, note that $k = 2$. Now, in the ϵ-radius network formation technique, we only establish links to those vertices whose similarities to the reference vertex are within the threshold ϵ. A noticeable difference between these two techniques is in the number of links that emerges from a reference vertex. In the ϵ-radius technique, the number of links is not pre-determined: the network formation process keeps establishing links as long as there are vertices within the range ϵ. Both techniques have their advantages and shortcomings that we discuss further.

Parameters k and ϵ play an important role in transforming the raw data into a corresponding network, since these parameters are sensitive to the local structure of the data. Thus, depending on the choices of k and ϵ, the resulting network topology may not reliably maintain the properties of the underlying data distribution.

We first discuss that caveat for k-NN. When k is large enough, it forces the creation of links between pairs of vertices that are not similar at all. To give an example, consider the data in Fig. 4.3a, in which we build a network with $k = 3$. Even though it is apparent that two well-behaved clusters exist in the data, the reference vertex is forced to connect to distant members of the other cluster. Conversely, if we choose k too small, we may sparsify the link structure in such a way that clusters are not formed in regions where the underlying data distribution seems to have real clusters. As an example, consider Fig. 4.3b, where we want to

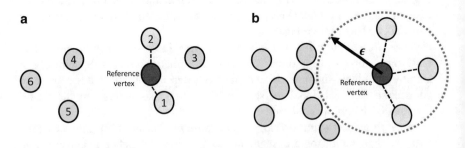

Fig. 4.2 Geometrical intuition of the k-nearest neighbor and ϵ-radius techniques, (**a**) 2-nearest neighbors, (**b**) ϵ-radius

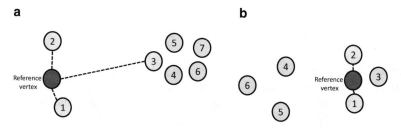

Fig. 4.3 Limitations of the k-nearest neighbors network formation technique. (**a**) Too large k ($k = 3$). (**b**) Too small k ($k = 2$)

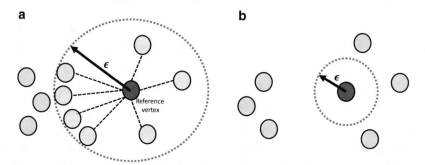

Fig. 4.4 Limitations of the ϵ-radius network formation technique. (**a**) Too large ϵ. (**b**) Too small ϵ

build a network with $k = 2$. Now, we have the opposite picture: the k-NN breaks into two apparent well-behaved clusters due to the small value of k. Note also that, provided that $k > 0$, no singletons[6] will arise in the network.

As we have pointed out, parameter ϵ also plays a major role in correctly translating the raw data into a proper network topology. Depending on the data distribution, a small increase on the value of ϵ may increase the network density to a large extent. Therefore, the network topology is very sensitive to the selected value of ϵ. Consider Fig. 4.4a, in which we employ the ϵ-radius to construct a network with a large ϵ. If we slightly reduce the value of ϵ, we get the network portrayed in Fig. 4.2b. We see here that a small increase on ϵ is responsible for an explosive increase in the number of connections established by the reference vertex. Conversely, if we choose a small value for ϵ, we may end up getting singletons in the network, as we can see in Fig. 4.4b. In both extremes, the resulting network may not represent well the true data distribution.

Some studies point that the network constructed using k-nearest neighbors and ϵ-radius techniques have dramatic influences on clustering techniques [34]. The k-NN network remains the more common approach since it is more adaptive to scale and data density.

---

[6]Singletons are those vertices that do not have connections to other vertices.

4.3.2 Combination of k-Nearest Neighbors and ε-Radius Network Formation Techniques

Recall that the ε-radius technique creates a link between two vertices if they are within a distance ε, while the k-NN sets up a link between vertices i and j if i is one of the k nearest neighbors of j. Both approaches have their limitations when sparsity or density is a concern. For sparse regions, the k-NN forces a vertex to connect to its k nearest vertices, even if they are far apart. In this scenario, one can say that the neighborhood of this vertex would contain dissimilar points. Equivalently, improper ε values could easily result in disconnected components, subgraphs, or isolated singleton vertices.

In order to combine the strengths of both approaches, a suitable combination (among many others) is given as follows. If $\mathcal{N}(v_i)$ denotes the neighborhood of v_i, then:

$$\mathcal{N}(v_i) = \begin{cases} \epsilon\text{-radius}(v_i), & \text{if } |\epsilon\text{-radius}(v_i)| > k \\ k\text{-NN}(v_i), & \text{otherwise} \end{cases} \quad (4.23)$$

in which $\epsilon-radius(v_i)$ returns the set $\{v_j, j \in \mathcal{V} : \mathbf{D}_{ij} <= \epsilon\}$, and k-NN$(v_i)$ returns the set containing the k nearest of vertex v_i. Note that the ε-radius technique is used for dense regions ($|\epsilon\text{-radius}(v_i)| > k$), while the k-NN is employed for sparse regions.

Figure 4.5 portrays how the combination of k-NN and ε-radius can transform vector-based data into networked data. In the example, we use $k = 3$ and desire to evaluate the neighbors of the colored vertices. Recall that k is employed as a threshold when defining whether vertices are located in sparse or dense regions. Whenever the number of neighbors within a radius ε of a given reference vertex is

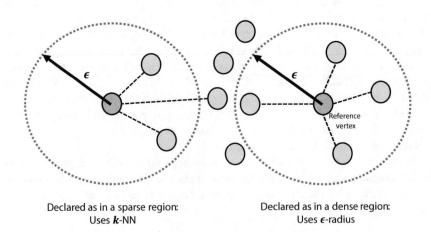

Declared as in a sparse region: Declared as in a dense region:
Uses **k**-NN   Uses ε-radius

Fig. 4.5 Combination of the k-NN and ε-radius technique as a network formation technique. In the example, we use $k = 3$ and the ε is geometrically depicted within the figure

smaller than k, we use the k-NN, which effectively forces the reference vertex to find more distant neighbors. When the number of neighbors is larger than k, then we declare the reference vertex as in a dense region. In this case, it connects to all of the neighbors within the radius ϵ, as governed by the ϵ-radius technique. One nice property of this network formation technique is that it adapts the heuristics of network formation depending on local density of data items. As it uses the k-NN in sparse regions, the combination of k-NN and ϵ-radius methods as a network formation technique tend to prevent the emergence of many network components.[7]

4.3.3   b-Matching Networks

The b-matching method is introduced in [27]. As opposed to the k-NN network, the b-matching network ensures that each vertex in the network has the same number of edges and therefore produces a balanced or regular network.

For a non-weighted adjacency matrix $\mathbf{A}$, the b-matching network formation technique relies on an optimization framework as follows:

$$\min_{\mathbf{A}} \sum_{i,j \in \mathcal{V}} \mathbf{A}_{ij} \mathbf{D}_{ij}$$

$$s.t. \sum_{j \in \mathcal{V}} \mathbf{A}_{ij} = b,$$

$$\mathbf{A}_{ii} = 0,$$

$$\mathbf{A}_{ij} = \mathbf{A}_{ji}, \tag{4.24}$$

$\forall i, j \in \mathcal{V}$. $\mathbf{D}$ denotes the dissimilarity matrix in which $\mathbf{D}_{ij}$ represents the dissimilarity between the i-th and j-th feature vectors. We can use any of the discussed metrics in Sect. 4.2 to construct $\mathbf{D}$. Note that the search space is on all binary matrices with the same dimension as of $\mathbf{A}$. The first restriction forces every vertex to link to exactly b other vertices. The second constraint prevents the emergence of self-loops. The third restriction guarantees that the resulting network is symmetric.

The k-NN network formation technique can also be stated as an optimization framework similar to that in (4.24), except for the third restriction. At the end of the network formation, the k-NN technique may not necessarily construct symmetric matrices. Though the out-degree of each of the vertices matches parameter k due to the first restriction, the in-degree varies (is at least k). This arises due to the asymmetrical behavior of the k-NN technique, in the sense that we can have vertex j as one of the k-nearest neighbors of vertex i and the converse may not hold. By introducing the third restriction, we are effectively forcing symmetry in the

---

[7]Note, however, that more than one network component may arise when k is small.

adjacency matrix. Consequently, the out- and in-degree are exactly equal to b (regular network).

The optimization framework described in (4.24) can be efficiently implemented using the algorithm of loopy belief propagation [25].

4.3.4 Linear Neighborhood Networks

The linear neighborhood network is introduced in [47]. The building block of the linear neighborhood network is based on the locally linear embedding technique, which we first explore before going to the referred network formation technique.

Locally linear embedding (LLE) is a dimensionality reduction technique [39]. The motivation behind the introduction of this class of methods is as follows. Our mental representations of the world are formed by processing large numbers of sensory inputs. Pixel intensities of images, power spectra of sounds, and the joint angles of articulated bodies are clear examples of these inputs. While complex stimuli can be represented by points in a high-dimensional vector space, they typically have a much more compact description. Coherent structure in the world leads to strong correlations between inputs (such as between neighboring pixels in images), generating observations that lie on or close to a smooth low-dimensional manifold. The dimensionality reduction problem involves mapping high-dimensional inputs into a low-dimensional summarizing space with as many coordinates as observed modes of variability. Dimensionality reduction techniques are widely employed in real-world applications, such as in the pre-processing of images of faces, spectrograms of speech, and other multidimensional signals in general. In essence, they are able to compress the signals in size and to discover compact representations of their variability.

The LLE algorithm is based on simple geometric intuitions. Suppose the data consist of V real-valued vectors x_i,[8] each with dimension P, sampled from some smooth underlying manifold. Provided there are sufficient data (such that the manifold is well-sampled), we expect that each data point and its neighbors will lie on or close to a locally linear patch of the manifold.

Instead of using pairwise relations between data items such as the k-NN and the ϵ-radius network formation techniques, the linear neighborhood network technique [47] uses locally linear embedding in the network formation step. Thus, the algorithm employs neighborhood information of the data items when establishing links. Therefore, each data point is optimally reconstructed using a linear combination of its neighbors [39]. With this simplification, we define the network formation process in terms of a constrained optimization process, whose goal is to minimize the following objective function:

---

[8]Each data item here exactly corresponds to a vertex, so that $N = V$.

$$C(\mathbf{A}) = \sum_{i \in \mathcal{V}} \left\| x_i - \sum_{j \in \mathcal{N}(x_i)} \mathbf{A}_{ij} x_j \right\|^2, \tag{4.25}$$

in which $\mathbf{A}_{ij}$ is the contribution or edge weight of x_j to x_i in the weighted adjacency matrix $\mathbf{A}$ of the network, $\mathcal{N}(x_i)$ represents the neighborhood of x_i, and $\|.\|$ is the Euclidean norm. The initial neighborhoods of all of the vertices are set in a static way. For instance, we can use k-nearest neighbors or ϵ-radius techniques.

Note that we can rewrite (4.25) in terms of the individual contributions of each vertex to the objective function, as follows:

$$C(\mathbf{A}) = \sum_{i \in \mathcal{V}} C_i(\mathbf{A}), \tag{4.26}$$

in which:

$$C_i(\mathbf{A}) = \left\| x_i - \sum_{j \in \mathcal{N}(x_i)} \mathbf{A}_{ij} x_j \right\|^2, \tag{4.27}$$

In the optimization process, we apply the following constraints:

$$\sum_{j \in \mathcal{N}(x_i)} \mathbf{A}_{ij} = 1, \forall i \in \mathcal{V}$$
$$\mathbf{A}_{ij} \geq 0, \quad i, j \in \mathcal{V} \tag{4.28}$$

The weight $\mathbf{A}_{ij}$ increases as x_j and x_i become more similar. In the extreme case, when $x_i = x_k \in \mathcal{N}(x_i)$, then $\mathbf{A}_{ik} = 1$ and $\mathbf{A}_{ij} = 0, j \neq k, x_j \in \mathcal{N}(x_i)$, is the optimal solution. Thus, we can employ $\mathbf{A}_{ij}$ to measure the similarity of x_j to x_i. As the neighborhoods of x_j and x_i may differ, then $\mathbf{A}_{ij} \neq \mathbf{A}_{ji}$ may hold in the general case. Applying some algebraic manipulations on (4.27), we see that:

$$\begin{aligned} C_i(\mathbf{A}) &= \left\| x_i - \sum_{j \in \mathcal{N}(x_i)} \mathbf{A}_{ij} x_j \right\|^2 \\ &= \left\| \sum_{j \in \mathcal{N}(x_i)} \mathbf{A}_{ij}(x_i - x_j) \right\|^2 \\ &= \sum_{j,k \in \mathcal{N}(x_i)} \mathbf{A}_{ij}\mathbf{A}_{ik}(x_i - x_j)^T(x_i - x_k) \\ &= \sum_{j,k \in \mathcal{N}(x_i)} \mathbf{A}_{ij}\mathbf{G}_{jk}^i\mathbf{A}_{ik}, \end{aligned} \tag{4.29}$$

in which $\mathbf{G}_{jk}^i$ represents the (j, k)-th entry of the local Gram matrix

$$\mathbf{G}_{jk}^i = (x_i - x_j)^T (x_i - x_k) \tag{4.30}$$

at point x_i. Thus, the reconstruction weights of each data item can be solved by the following standard quadratic programming problems:

$$\min_{\mathbf{A}} \sum_{j,k \in \mathcal{N}(x_i)} \mathbf{A}_{ij} \mathbf{G}_{jk}^i \mathbf{A}_{ik} \tag{4.31}$$

$$\text{s.t.} \sum_{j \in \mathcal{N}(x_i)} \mathbf{A}_{ij} = 1, \mathbf{A}_{ij} \geq 0, \forall i \in \mathcal{V}$$

Intuitively, the way we construct the entire network $\mathbf{A}$ is to first shear the whole network into a series of overlapped linear patches, and then paste them together.

4.3.5   Relaxed Linear Neighborhood Networks

This method is introduced in [9]. It is an extension of the linear neighborhood network discussed in the previous section.

The noticeable advantage of this method is that it uses dynamic neighborhood information, as opposed to fixed k neighbors of [47]. In summary, the technique approximates the entire network by a series of overlapped linear neighborhood patches, where the neighborhood $\mathcal{N}(x_i)$, $\forall x_i \in \mathcal{V}$, is captured dynamically via the data density in the vicinities.

Instead of finding fixed k neighbors of each vertex x_i, the relaxed linear neighborhood method captures the boundary of each vertex $\mathcal{B}(x_i)$ based on neighborhood information and declares as neighbors vertices within this boundary. We can capture this dynamic feature by using a combination of the k-NN and ϵ-radius approaches to define the neighborhood between any x_i and x_j as:

$$\mathcal{N}_{x_i;k,\epsilon}(x_j) = \begin{cases} 1, & |\mathcal{N}_\epsilon(x_i)| > k \\ \mathcal{N}_{x_i;k}(x_j), & \text{otherwise} \end{cases}, \tag{4.32}$$

in which $|\mathcal{N}_\epsilon(x_i)|$ denotes the number of neighbors if we apply the ϵ-radius technique in x_i, and $\mathcal{N}_{x_i;k}(x_j) \in \{0, 1\}$ returns 1 if x_j in is the k-nearest neighborhood of x_i, and 0 otherwise. Thus if there is a large enough number of vertices in the ϵ-vicinity ($> k$), then the boundary is identified. Otherwise, we employ k-NN. We define the boundary set of any x_i as:

$$\mathcal{B}(x_i) = \{j \in \mathcal{V} : \mathcal{N}_{x_i;k,\epsilon}(x_j) = 1\}. \tag{4.33}$$

It must be noted that we may run into problems if we only consider the established radius and density of the neighborhood. For instance, if we fix large values for k and ϵ, vertices located at dense regions would include more vertices than necessary. Conversely, for small values of k and ϵ, weak neighborhood bonds would be established in sparse regions. An adaptive algorithm that can handle a wide range of changing interval would be advantageous. It should also include information provided by neighboring vertices closest to the corresponding vertex, which can take neighborhood relations into consideration in a more intelligent way. One way to accomplish that is to extend the neighborhood definitions in (4.32) and (4.33) and account for the data sensitivity with varying distances to neighbor points based on the parameter $k > 0$:

$$N_{x_i}(x_j) = \max\left[1 - k\frac{d(x_i, x_j)}{d_{\max}}, 0\right], \tag{4.34}$$

in which $d_{\max}$ is the network diameter whose definition we recall:

$$d_{\max} = \max_{x_i, x_j \in \mathcal{V}} d(x_i, x_j), \tag{4.35}$$

and $d(x_i, x_j)$ is a suitable dissimilarity or distance function, such as those defined in Sect. 4.2. Parameter k plays the role in determining the neighborhood radius and is adjusted as follows:

$$1 - k\frac{\epsilon}{d_{\max}} = 0 \Rightarrow k = \frac{d_{\max}}{\epsilon}. \tag{4.36}$$

The new boundary set of any given x_i includes:

$$\mathcal{B}(x_i) = \left\{x_j \in \mathcal{V} : N_{x_i}(x_j) \in (0, 1]\right\}. \tag{4.37}$$

Instead of measuring pairwise relations, neighborhood information to represent the network is employed. Similarly to the studies in [39, 47], each vertex is reconstructed using a linear combination of its dynamic neighbors:

$$\min_{\mathbf{A}} \sum_{i \in \mathcal{V}} \left\| x_i - \sum_{j\,:\,x_j \in \mathcal{N}(x_i)} N_{x_i}(x_j)\mathbf{A}_{ij}x_j \right\|^2 \tag{4.38}$$

$$\text{s.t.} \sum_{j \in \mathcal{V}} \mathbf{A}_{ij} = 1, \mathbf{A}_{ij} \geq 0, \forall i \in \mathcal{V},$$

in which $N_{x_i}(x_j) \in [0, 1]$ is the degree of neighborhood to the boundary set $\mathcal{B}(x_i)$ and $\mathbf{A}_{ij}$ is the degree of contribution of x_j to x_i. When $N_{x_i}(x_j) = 0$, no links exist.

4.3.6 Network Formation Using Clustering Heuristics

This network formation technique is introduced in [15]. Recall that k-NN and ϵ-radius network formation methods generate either disconnected or densely connected networks. Neither of the two situations is desirable for the majority of data mining and machine learning tasks. For example, Fig. 4.6 shows a data set with 300 data samples and the resulting networks generated using the k-NN method with various values of k. Note that the generated networks become connected only when $k \geq 33$. This means that sometimes we need a very large k to generate a connected network. In these cases, the generated networks are dense, which in turn can cause: inefficiency in the processing procedure that leads to high computational time; the weak performance of the algorithm due to the homogeneity of the network or even both.

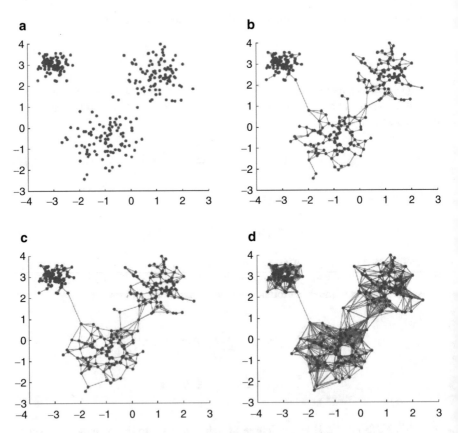

Fig. 4.6 Networks formed by the k-NN in a data set with three distinct groups. (**a**) Original network with 300 vertices. Results for k-NN when (**b**) $k = 5$, (**c**) $k = 20$, and (**d**) $k = 33$. Reproduced from [31] with permission from the author

In [15], the authors proposed a method that is based on data clustering heuristic to perform the network formation process. Specifically, they employ the single-link method to overcome the aforementioned problem that happens with traditional network formation techniques. The clustering heuristics for network formation is capable of constructing connected and sparse networks and, at the same time, tends to keep the cluster structure of the original data set. The method consists of the following steps:

1. Generate an initial totally disconnected network, where each vertex represents a data instance. In this way, we have V vertex groups (each vertex is in an isolated group).
2. Construct a dissimilarity matrix using any distance measure, for example, the Euclidean distance, to represent distances between all pairs of groups. According to the single-linkage criteria, the dissimilarity between two groups is computed as the dissimilarity between the two closest vertices.
3. Identify the two closest groups and denote them by G_1 and G_2, respectively.
4. Calculate the average dissimilarity among vertices (data instances) within each group G_1 and G_2, and denote them by d_1 and d_2, respectively.
5. Select the k-most similar pairs of vertices between G_1 and G_2, and generate an edge between each of the k selected pairs if their dissimilarity is smaller than the threshold: $d_c = \lambda \max(d_1, d_2)$, where $\lambda > 0$.
6. Update the dissimilarity matrix considering the merging result in Step 5.
7. If the number of groups is larger than one, return to Step 3;

The condition in Step 7 guarantees that the final network is connected.

In order to illustrate the effectiveness of the method, Fig. 4.7 shows the resulting networks for an artificial data set composed of three clusters of different sizes and densities. In this simulation, the following values for k are used: 3, 5 and 20. We see that the generated networks are connected and relatively sparse. At the same time, the original cluster features are well preserved. Figure 4.8 shows the network construction results by varying the threshold parameter λ. Again, we see the good performance of the method.

4.3.7 Network Formation Using Overlapping Histogram Segments

This network formation technique is introduced in [42]. The network formation using overlapping histogram segments always produces a connected graph with vertices of the same community densely interconnected and with vertices of different communities sparsely interconnected.

As we have discussed, the k-NN network formation technique is widely employed in the machine learning domain. However, it suffers from several problems, among which we can highlight: (1) the constructed network may not

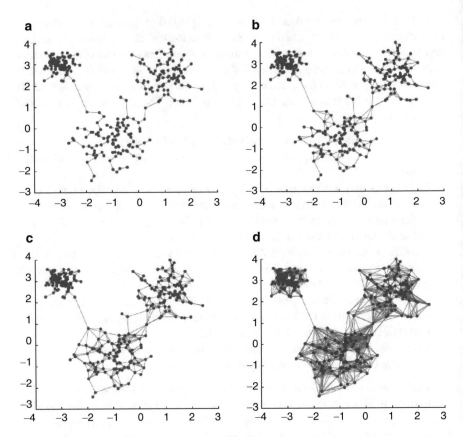

Fig. 4.7 Networks formed by the application of the network formation technique based on clustering heuristics on the data set in Fig. 4.6a with $\lambda = 3$. (**a**) $k = 1$, (**b**) $k = 3$, (**c**) $k = 5$, (**d**) $k = 20$. Reproduced from [15] with permission from Elsevier

necessarily be connected, and even worse (2) the constructed network may not reliably represent the data distribution. For instance, consider a data set with two clusters: a very large cluster and another very small. If we select a large k value, these two clusters would be heavily interconnected, as each of the vertices belonging to the small cluster would be compelled to connect to vertices from the large cluster. Conversely, if k is small, the large cluster may get fragmented into several small communities. Both situations are undesirable in data analysis. The network formation using overlapping histogram segments addresses these problems.

Define a mapping function $h : \mathscr{X} \mapsto [0,1]$, which receives a data item with $P > 0$ attributes and converts it into a scalar value in the unitary interval.[9] The function h should be smooth in the sense that similar items receive approximately

---

[9]We use the unitary interval with no loss of generality.

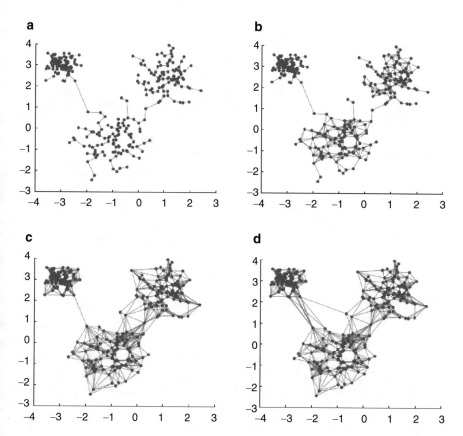

Fig. 4.8 Networks formed by the application of the network formation technique based on clustering heuristics on the data set in Fig. 4.6a with $k = 5$. (**a**) $\lambda = 1$, (**b**) $\lambda = 2$, (**c**) $\lambda = 4$, (**d**) $\lambda = 8$. Reproduced from [31] with permission from the author

similar scalar values. For instance, in a gray-level image in which each vertex is a pixel, $P = 1$ and h is simply the identity function. When $P > 1$, we can use for instance a linear weighted combination of the attributes in the feature vector.

Once transformed the vector-based data set, we construct the histogram of distribution h that resides inside the interval $[0, 1]$ by definition. We also define the set of overlapping intervals $\mathscr{I}$ that are constructed using overlapping histogram segments as follows:

$$\mathscr{I} = \{[0, d]; [d - \kappa, 2d]; \ldots; [(M - 1) \times d - \kappa, 1]\}, \tag{4.39}$$

in which M denotes the number of intervals, $d > 0$ is the non-overlapping window width over adjacent intervals, and $\kappa \geq 0$ is the overlapping factor. The overlapping factor κ is essential for the network formation. It serves the purpose of not letting the resulting network become disconnected.

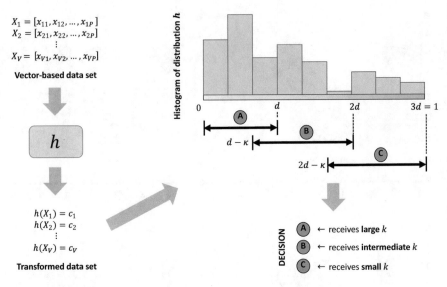

Fig. 4.9 Illustration of the network formation using overlapping histogram segments. In the histogram, we have $d = 3$ and $\kappa = 1$

Let S_i represent the quantity of vertices that is inside interval $i \in \mathscr{I}$. The network is formed by connecting each vertex in i to its k_i most similar neighbors. That is, k_i is adaptive as it varies from interval to interval. Mathematically, k_i is given by:

$$k_i = S_{\max}^2 - (S_{\max} - S_i)^2, \qquad (4.40)$$

in which $S_{\max} = \max(S_1, S_2, \ldots, S_M)$. Note that vertices in regions with several vertices are given a large k.[10] Vertices inside regions that have few other vertices are given a small k. In our previous example, the vertices in the large clusters would have large k and those in the small cluster, small k. This behavior gives two nice properties for the resulting network: (1) the number of intercommunity links is expected to be small and (2) the large cluster is not fragmented.

We illustrate the entire network formation process in Fig. 4.9.

We follow the methodology in [42] and apply the network formation process in pixel clustering. In this case, each pixel in the image corresponds to a vertex. We use grayscale images, so h is the identity function. For the clustering technique, we use the greedy modularity function.[11]

---

[10]Read k in the sense of the k-nearest neighbors in which k is global and static.

[11]In brief, the modularity measures how good a particular network division is in terms of communities. It ranges from 0 to 1. The larger the modularity, the better defined are the communities. See Chap. 6 for more information.

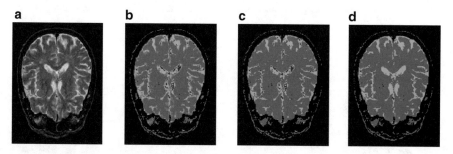

Fig. 4.10 Pixel clustering of a human brain. (**a**) Brain image. (**b**) Results for five clusters. (**c**) Results for four clusters. (**d**) Results for three clusters. The *colors* in (**b**)–(**d**) represent the clusters. We use $d = 0.008$ and $\kappa = 0.5d$

We illustrate the potentialities of the network formation technique based on overlapping histogram segments with pixel clustering tasks. Figure 4.10 shows the results for pixel clustering of a human brain. As the greedy modularity technique is a hierarchical technique, we can see the communities at different levels of granularity. Figures 4.10b–d portray the clustering results for five, four, and three communities, respectively. The network formation process attains a maximum modularity of 0.81 when there are five communities, suggesting that the network communities are well-defined. The histogram of the human brain is supplied in Fig. 4.11. To note the difference of the discussed technique with the k-nearest neighbors technique, we plot the number of vertices as a function of k values in Fig. 4.12. In the original k-NN, the k value is static and global. In contrast, the network formation technique based on overlapping histogram segments has adaptive k.

As another illustrative pixel clustering example, consider the image in Fig. 4.13, in which there are two dogs in the grass. We use a grayscale version of this image in our simulations. The pixel clustering results are given in Fig. 4.14a–d. The maximum modularity is 0.74 and is achieved when four communities are found. The histogram of Fig. 4.13 in grayscale is displayed in Fig. 4.15. To inspect how the underlying network is formed, we also plot the value of k assumed by the vertices in Fig. 4.16. Note that there are two peaks corresponding to the dogs' color and the background composed of grass. Note that the great advantage of this technique is that it can adapt k in terms of community sizes such as to construct well-defined communities (large modularity).

4.3.8 More Advanced Network Formation Techniques

In the previous section, we have presented the k-nearest neighbor, ϵ-radius and other dynamical network formation techniques. The way that they have been introduced assumes that we have no information about the data relationships except the raw data set itself. This hypothesis is consistent with unsupervised

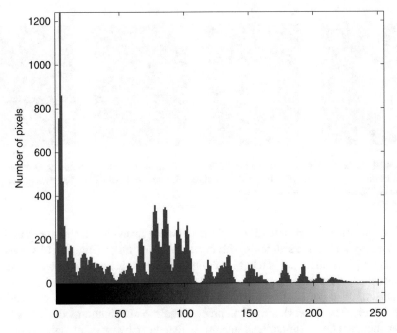

Fig. 4.11 Histogram of the human brain image in Fig. 4.10a. We rescale the h function to the usual pixel interval $[0, 255]$ to facilitate the understanding of the grayscale tones. For instance, the *black color* in a pixel is represented by the value 0 and the *white color* is denoted by 255

learning techniques. Methods that lie in the semi-supervised and supervised learning paradigms, however, are provided with additional information, other than the data set itself. As we have seen, we term that additional (external) information as labels or targets. Each data item has a corresponding label indicating the specific class to which it belongs in the analyzed domain. When these labels are discrete, we have a semi-supervised or supervised classification task. When the labels are continuous, we term the learning process as semi-supervised or supervised regression.

A natural refinement in these network formation techniques is to also consider the labels of the data items when creating links. For instance, we may be interested in creating links between pairs of vertices that only belong to the same class. This constraint would lead to the emergence of more than one network component, each of which with vertices of the same class. In other applications, it would be desirable to have connections between members of different classes. In this case, each network component would possibly have a mixture of members of several different classes.

The way we employ different constraints in our network formation procedure defines the topological properties of the resulting network. In addition, the constraints that we are able to bring are not limited to the labels in case of semi-supervised or supervised learning techniques. For instance, if we have a technique that potentially can create several network components, we can create links if

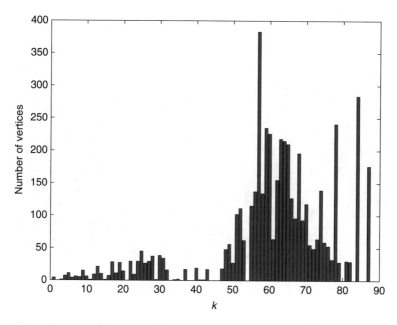

Fig. 4.12 Number of vertices as a function of their assumed k value (network formation parameter) for the human brain image depicted in Fig. 4.10a

the new member that is being tested meets some topological requirements of all of the current members of that component. The topological requirements can be fashioned using local, mixed, and global information. Examples of local information are similar in- and out-degree or in- and out-strength. Network measurements that carry mixed information include similar clustering coefficient, closeness, and betweenness values. Among examples of global information, we can highlight assortativity, network diameter, and the rich club effect.[12] In fact, in one of our case studies in Chap. 8, we discuss more advanced network formation techniques that use both network topological aspects and labels of the data items to construct and to evolve the network in a self-learning process.

4.4 Transforming Time Series Data into Networks

Various time series network construction techniques have been studied in [17]. A time series is a sequence of data points, normally consisting of successive measurements made over a time interval [22]. Examples of time series are ocean tides, counts of sunspots, and the daily closing values of stock markets. Time series

---

[12]C.f. Sect. 2.3.5 for a more comprehensive classification of network measurements.

Fig. 4.13 Picture of two dogs in the grass. Photo by Liang Zhao

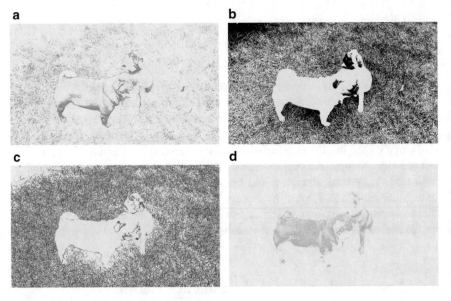

Fig. 4.14 Pixel clustering results of the image portrayed in Fig. 4.13. We use $d = 0.008$ and $\kappa = 0.5d$. (**a**) Community 1. (**b**) Community 2. (**c**) Community 3. (**d**) Community 4

are largely employed in any domain of applied science and engineering that involves temporal measurements, such as in: statistics, signal processing, pattern recognition, econometrics, mathematical finance, weather forecasting, intelligent transport and

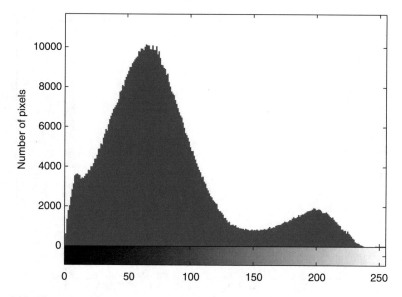

Fig. 4.15 Histogram of the image in Fig. 4.13. We rescale the h function to the usual pixel interval $[0, 255]$ to facilitate the understanding of the grayscale tones. For instance, the *black color* in a pixel is represented by the value 0 and the *white color* is denoted by 255

trajectory forecasting, earthquake prediction, electroencephalography, control engineering, astronomy, and communications engineering [6, 7, 19, 22, 32, 46].

One interesting phenomenon arising in time series analysis is of state recurrence. Formally, in dynamical systems that model time series, there is recurrence of state x_i whenever that system reaches state x_j at another time j that is sufficiently similar to that initial state ($x_i \approx x_j$). Normally, time series data is a sequence in the phase space $\mathcal{T} = \{x_i \in \mathbb{R}^P\}_{i=0}^{\infty}$ that is periodically sampled at intervals Δi, where x_i denotes the time series state in an arbitrary P-dimensional phase space. The set $\mathcal{T}$ is termed as the trajectory of the phase space representing the time series.

Suppose we have that trajectory of the dynamical system in its phase space. The corresponding recurrence plot (RP) is represented by the following recurrence matrix:

$$\mathbf{R}_{ij} = \begin{cases} 1, & \text{if } \|x_i - x_j\| <= \epsilon \\ 0, & \text{otherwise} \end{cases} \qquad (4.41)$$

in which $\epsilon > 0$ is a constant that enables state equality up to a small error. Essentially, the recurrence matrix compares the system states at times i and j. If they are similar enough, then $\mathbf{R}_{ij} = 1$. Conversely, when states i and j are rather different, the corresponding entry in the recurrence matrix is $\mathbf{R}_{ij} = 0$. So the recurrence matrix tells us when similar states of the underlying system occur.

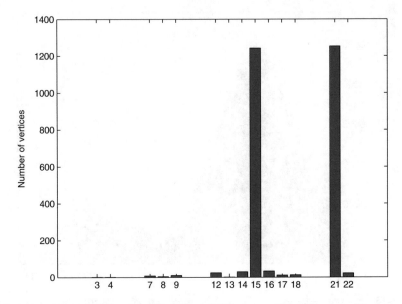

Fig. 4.16 Number of vertices as a function of their assumed k value (network formation parameter) for the image depicted in Fig. 4.13

Several approaches for transforming time series data into complex network have been proposed in the literature. The main approaches are the proximity networks and the transition networks [17]. The former is constructed using mutual proximity of different segments of a time series, while the latter considers transition probabilities between discrete states. The network connectivity of proximity networks is defined in a data-adaptive local fashion, where distinct regions centered at an arbitrary reference vertex in the phase space are considered. The idea can be understood as an adaptive ϵ-radius technique, which we have seen in Sect. 4.3. For transition networks, in contrast, the corresponding regions are rigid, i.e., they are determined by a fixed coarse-graining parameter evaluated in the phase space. In turn, proximity networks are characterized using mutual closeness or similarity of the time series trajectory.

In the following, we briefly discuss some network formation techniques for time series data.

4.4.1 Cycle Networks

The research by Zhang and Small [53] was pioneering in studying topological features of pseudo-periodic time series. In their technique, individual cycles in a time series are identified by the vertices of an undirected network. Links are established between pairs of vertices if cycles behave similarly. Zhang et al. [54]

introduce a generalization of the correlation coefficient applicable to cycles of possibly different lengths to quantify the proximity of cycles in the phase space. The correlation index is defined as the maximum of the cross-correlation between two signals when the shortest of both is slid relative to the longest one. Suppose we compare two cycles $C_1 = \{x_1, x_2, \ldots, x_\alpha\}$ and $C_2 = \{y_1, y_2, \ldots, y_\beta\}$, where $\alpha \leq \beta$.[13] Then, we compute:

$$\rho(C_1, C_2) = \max_{i=0,\ldots,\beta-\alpha} \langle (x_1, x_2, \ldots, x_\alpha), (y_{1+i}, y_{2+i}, \ldots, y_{\alpha+i}) \rangle, \qquad (4.42)$$

in which $\langle ., . \rangle$ represents the standard correlation coefficient of two α-dimensional vectors. In this case, we evaluate the (i, j)-th entry of the adjacency matrix $\mathbf{A}$ of the network as follows:

$$\mathbf{A}_{ij} = \Theta \left(\rho(C_i, C_j) - \rho_{\max} \right) - \mathbb{1}_{[i=j]}, \qquad (4.43)$$

in which $\Theta(.)$ is the Heaviside function that outputs 1 if the argument is positive and 0, otherwise. $\mathbb{1}_{[.]}$ is the Kronecker delta function that yields 1 if the argument is true and 0, otherwise. $\rho_{\max}$ is the maximum attainable correlation between any two given cycles (vertices). The Kronecker delta function is introduced to prevent self-loops in the resulting network. Another measure to quantify the proximity of cycles is based on phase space distance [53], which is expressed as:

$$D(C_1, C_2) = \min_{i=0,\ldots,\beta-\alpha} \frac{1}{\alpha} \sum_{j=1}^{\alpha} \| x_j - y_{j+i} \|. \qquad (4.44)$$

Using the phase space distance, we compute the adjacency matrix as:

$$\mathbf{A}_{ij} = \Theta \left(D_{\max} - D(C_i, C_j) \right) - \mathbb{1}_{[i=j]}. \qquad (4.45)$$

Cycle networks are robust against additive noise and have the advantage that explicit time delay embedding is avoided.

4.4.2 Correlation Networks

Consider each time series represented by a state vector x_i. That is, we have a set of time series for which we want to construct a representative network. The Pearson correlation coefficient can be calculated:

---

[13] We set the length assumption without loss of generality. If it is not the case, we can simply re-define C_1 and C_2 such as that the hypothesis holds true.

$$r_{ij} = \frac{\text{cov}(x_i, x_j)}{\sigma_{x_i} \sigma_{x_j}}, \tag{4.46}$$

in which $\text{cov}(x_i, x_j)$ denotes the covariance between the time series vectors x_i and x_j, and σ_{x_i}, the standard deviation of vector x_i.

The factor $1 - r_{ij}$ therefore is a proximity or similarity measure in the context of time series data. In order to construct the network, we establish a link between states or vertices i and j whenever $1 - r_{ij}$ is larger than a given threshold r [20, 51]:

$$\mathbf{A}_{ij} = \Theta\left(r - r_{ij}\right) - \mathbb{1}_{[i=j]}. \tag{4.47}$$

4.4.3   Recurrence Networks

Recurrence networks are complex networks whose adjacency matrices are given by the recurrence matrices of time series, as computed in (4.41). We define the adjacency matrix of a recurrence network by:

$$\mathbf{A}_{ij} = \mathbf{R}_{ij} - \mathbb{1}_{[i=j]}. \tag{4.48}$$

Because the recurrence matrix can be defined in different ways, there are distinct subtypes of recurrence networks that are characterized by somewhat different structural properties, such as the k-nearest neighbors networks and ϵ-recurrence networks [18, 20, 35]. In k-nearest neighbors networks, we consider every observation vector as a vertex i, which is then connected to its k nearest neighbors in the phase space. In ϵ-recurrence networks, the neighborhood of a vertex (time series) is all times series within a predefined phase space distance ϵ.

4.4.4   Transition Networks

In this method, we build a network from a single time series. The first step in order to build transition networks from time series is to find the amplitude of the analyzed time-varying signal. With that interval at hand, we then discretize it into a suitable set of K classes $\mathscr{S} = \{S_1, \ldots, S_K\}$. The transition probabilities measure the likelihood of the signal jumping from one class (region) to another [37, 41]. Mathematically, $\pi_{\alpha\beta} = P(x_{i+1} = S_\beta | x_i = S_\alpha)$ indicates the probability of the signal to reach region S_β given that it is currently at region S_α. This approach is equivalent to applying a symbolic discretization with static grouping [16] to the phase space of the studied system.

Unlike proximity networks, the resulting transition networks explicitly make use of the temporal order of observations, i.e., their connectivity represents causality relationships contained in the dynamics of the observed system. By introducing a cutoff $p < 1$ to the transition probability $\pi_{\alpha\beta}$ between pairs of discrete states S_α and S_β, we obtain a non-weighted network representation, which is, however, still directed. Note that for a trajectory that does not leave a finite volume in phase space, there is only a finite number of discrete states S_i with a given minimum size in phase space. This implies the presence of absorbing or recurrent states in the resulting transition network.

The transition probability approach is well suited for identifying the states that have a special importance for the causal evolution of the studied system in terms of betweenness centrality and related measures. However, its main disadvantage is a significant loss of information on small amplitude variations [16].

4.5 Classification of Network Formation Techniques

In this section, we classify the network formation techniques with respect to the type of information they use in their construction processes. The types of network formation techniques are classified as using:

- *Quasi-local information*: these techniques are often restricted to geometrical issues of the data, such as shortest distances between pairs of vertices. In this way, they only employ information from a small set of vertices to construct the links.
- *Long-range information*: these techniques use not only local geometrical information, but long-range information such as the trajectory of shortest paths. That is, we not only account for distances between the endpoints of pairwise shortest paths, but we also use information from the trajectory itself from those shortest paths. In contrast, note that network formation techniques classified as using quasi-local information would only use the distance of the shortest path.
- *Global information*: these techniques use information from all of the data items at once to construct the network. For instance, it may rely on the data distribution itself to adjust parameters of the network formation process.

Tables 4.2 and 4.3 report the classification of the vector-based and time series network formation techniques, respectively. Techniques that are classified as using quasi-local information often share properties such as simplicity and generality, as they can be applied to any data set and for any machine learning task. In this way, they ignore global characteristics that are embedded within the data relationships and are not specialized to a domain-specific task. Techniques that are classified as using long-range and global information are more sophisticated in the sense that they are able to capture local and global characteristics of the data relationships. However, they are often employed to specific purpose tasks. For instance, the network formation technique using clustering heuristics tends to construct a network with the goal of data clustering. In other words, these techniques anticipate the desired result in the network construction phase (before the inference or clustering phases). Techniques that use global information are a tendency in this research topic.

Table 4.2 Classification of vector-based network formation techniques

Definition	Description	Classification
Section 4.3.1	k-NN	Quasi-local information
Section 4.3.1	ϵ-radius	Quasi-local information
Section 4.3.2	Combination of k-NN and ϵ-radius	Quasi-local information
Section 4.3.3	b-matching networks	Long-range information
Section 4.3.4	Linear neighborhood networks	Long-range information
Section 4.3.5	Relaxed linear neighborhood networks	Long-range information
Section 4.3.6	Network formation using clustering heuristics	Long-range information
Section 4.3.7	Network formation using overlapping histogram segments	Global information

The classes are designed using the type of information these techniques employ in the construction process

Table 4.3 Classification of time series network formation techniques

Definition	Description	Classification
Section 4.4.1	Cycle networks	Quasi-local information
Section 4.4.2	Correlation networks	Quasi-local information
Section 4.4.3	Recurrence networks	Quasi-local information
Section 4.4.4	Transition networks	Quasi-local information

The classes are designed using the type of information these techniques employ in the construction process

4.6 Challenges in Transforming Unstructured Data to Networked Data

In this chapter, we have discussed several network formation techniques. When applying network-based machine learning methods to given sets of data points, there are several choices to be made: the type of the network to be constructed and the network formation parameters. However, the question how these choices should be made has received little attention in the literature. This is not so severe in the domain of supervised learning, where parameters can be set using cross-validation. However, it poses a serious problem in unsupervised and semi-supervised learning. While different researchers use different heuristics to set these parameters, few systematic empirical studies have been conducted. For instance, it is important to know how sensitive the results are to the parameters that define the network formation technique. The problem becomes even more severe when we try to find theoretical models that justify the use of one parameter value in detriment to others.

In the domain of unsupervised learning, the study in [34] analyzes clustering measures, such as the normalized cut, in nearest neighbors networks (k-nearest neighbors and ϵ-radius). The investigation shows that, depending on the selected criteria to construct the network, the normalized cut criterion converges to different limit results. The fact that all of these nearest neighbors networks lead to different clustering criteria shows that we cannot study these criteria isolated from the network that they are applied to.

The picture is not different in the semi-supervised learning domain. According to [55], there is no reliable approach for model selection if only a few labeled instances are available. Unfortunately, this is often the case as labeling is expensive. We find a plethora of studies that deal with regularization frameworks that essentially solve constrained optimization processes. These frameworks are roughly composed of two abstract terms: the loss and the regularization functions. While the loss function penalizes decisions that flip labels of labeled vertices, the regularization function models the costs of propagating labels to unlabeled instances. These algorithms mainly adjust or propose new ways of modeling these two functions [56, 57]. The label propagation process, which strongly depends on the network topology, has been extensively studied in the literature. Up to now, little has been done for the network topology analysis *per se*, which is a product of a network formation process. Therefore, a relevant question that arises is whether the resulting network really represents the transformed (vector-based) data. If it is not the case, then the label diffusion process will probably be flawed, as the network topology will be incorrect. Even though most of them lack theoretical framework, we find some few empirical investigations in the literature about network formation, as follows [56]:

- *Construction of networks using domain knowledge.* For instance, the research in [3] builds networks for video surveillance using strong domain knowledge, where the network of web-camera images consists of time edges, color edges and face edges. Such networks reflect a deep understanding of the problem structure and how unlabeled data is expected to help. We note that constructing domain knowledge requires an active work of human experts. Recall that the labeling process of human experts is expensive and time-consuming. Moreover, edge construction process is even more slower as it is a problem with a mapping function of the type $\mathcal{V} \times \mathcal{V} \mapsto \mathbb{R}$, i.e., we consider the weight of every potential edge between arbitrary pairwise vertices.[14] In this way, even though domain knowledge may improve the performance of machine learning algorithms, it definitely turns into an unfeasible procedure as the number of the data items grows.

- *Construction of nearest neighbors networks.* Empirically, weighted k-NN network with small k tends to perform better. We can also build near complete networks, using, for instance, correlation or kernel Gaussian functions as similarity functions. In the sparsification process, we can apply several tricks. The investigation in [8] builds robust networks from multiple minimum spanning trees by perturbation and edge removal. When using a Gaussian function as edge weights, the bandwidth of the Gaussian needs to be carefully chosen. In turn, the study in [52] derives a cross-validation approach to tune the bandwidth for

---

[14]In contrast, the vertex labeling is a mapping task $\mathcal{V} \mapsto \mathcal{Y}$, which is much quicker than edge labeling.

each feature dimension, by minimizing the leave-one-out mean squared error of predictions and given labels on labeled points.

- *Construction of networks using local fit procedures.* The investigation in [23, 24] proposes an algorithm to de-noise points sampled from a manifold. That is, data points are assumed to be noisy samples of some unknown underlying manifold. They used the de-noising algorithm as a preprocessing step for network-based semi-supervised learning, so that the network can be constructed from better separated data points. Such preprocessing results in better semi-supervised classification accuracy.

4.7 Chapter Remarks

In this section, we have reviewed the main ingredients involved in constructing networks from vector-based and time series data. In special, we have seen that we need a proper similarity function and a suitable strategy for creating links in the network. Several similarity functions have been discussed. The shortcomings and advantages of the most well-known network formation techniques have been explored. We also visited the inherent challenges that we face when building networks that reliably maintain the data distribution.

References

1. Aggarwal, C.C.: Mining text data. In: Data Mining, pp. 429–455. Springer International Publishing, New York (2015)
2. Baeza-Yates, R., Ribeiro-Neto, B.: Modern Information Retrieval. ACM Press, New York (1999)
3. Balcan, M.F., Blum, A., Choi, P.P., Lafferty, J., Pantano, B., Rwebangira, M.R., Zhu, X.: Person identification in webcam images: an application of semi-supervised learning. In: ICML 2005 Workshop on Learning with Partially Classified Training Data, vol. 2. ACM Press (2005)
4. Baroni-Urbani, C., Buser, M.W.: Similarity of binary data. Syst. Zool. **25**, 251–259 (1976)
5. Belkin, M., Niyogi, P.: Laplacian eigenmaps for dimensionality reduction and data representation. Neural Comput. **15**(6), 1373–1396 (2003)
6. Bloomfield, P.: Fourier Analysis of Time Series: An Introduction. Wiley, New York (1976)
7. Brockwell, P.J., Davis, R.A.: Introduction to Time Series and Forecasting. Springer, New York (1996)
8. Carreira-Perpiñán, M.A., Zemel, R.S.: Proximity graphs for clustering and manifold learning. In: Saul, L.K., Weiss, Y., Bottou, L. (eds.) Advances in Neural Information Processing Systems, MIT Press, Cambridge, MA, vol. 17, pp. 225–232 (2004)
9. Celikyilmaz, A., Hakkani-Tur, D.: A graph-based semi-supervised learning for question semantic labeling. In: Proceedings of the NAACL HLT 2010 Workshop on Semantic Search, pp. 27–35. Association for Computational Linguistics, Los Angeles, CA (2010)
10. Celikyilmaz, A., Thint, M., Huang, Z.: A graph-based semi-supervised learning for question-answering. In: Proceedings of the Joint Conference of the 47th Annual Meeting of the ACL and the 4th International Joint Conference on Natural Language Processing of the AFNLP: Volume 2 - Volume 2, ACL '09, pp. 719–727. Association for Computational Linguistics (2009)

11. Cha, S.H.: Comprehensive survey on distance/similarity measures between probability density functions. Int. J. Math. Models Methods Appl. Sci. **1**, 300–307 (2007)
12. Chao, A., Chazdon, R.L., Colwell, R.K., Shen, T.J.: Abundance-based similarity indices and their estimation when there are unseen species in samples. Biometrics **62**(2), 361–371 (2006)
13. Cock, M.D., Kerre, E.: On (un)suitable relations to model approximate equality. Fuzzy Sets Syst. **133**, 137–153 (2003)
14. Colwell, R., Coddington, J.: Estimating terrestrial biodiversity through extrapolation. Philos. Trans. R. Soc. B Biol. Sci. **345**, 101–118 (1994)
15. Cupertino, T.H., Huertas, J., Zhao, L.: Data clustering using controlled consensus in complex networks. Neurocomputing **118**, 132–140 (2013)
16. Donner, R., Hinrichs, U., Scholz-Reiter, B.: Symbolic recurrence plots: A new quantitative framework for performance analysis of manufacturing networks. Eur. Phys. J. Spec. Top. **164**(1), 85–104 (2008)
17. Donner, R.V., Small, M., Donges, J.F., Marwan, N., Zou, Y., Xiang, R., Kurths, J.: Recurrence-based time series analysis by means of complex network methods. Int. J. Bifurcation Chaos **21**(4), 1019–1046 (2010)
18. Donner, R.V., Zou, Y., Donges, J.F., Marwan, N., Kurths, J.: Recurrence networks – a novel paradigm for nonlinear time series analysis. New J. Phys. **12**, 033025 (2010)
19. Durbin, J., Koopman, S.J.: Time Series Analysis by State Space Methods. Oxford University Press, Oxford (2001)
20. Gao, Z., Jin, N.: Flow-pattern identification and nonlinear dynamics of gas-liquid two-phase flow in complex networks. Phys. Rev. E **79**, 066303 (2009)
21. Giguère, S., Laviolette, F., Marchand, M., Tremblay, D., Moineau, S., Liang, X., Biron, A., Corbeil, J.: Machine learning assisted design of highly active peptides for drug discovery. Public Libr. Sci. Comput. Biol. **11**(4), e1004074 (2015)
22. Hamilton, J.D.: Time Series Analysis. Princeton University Press, Princeton, NJ (1994)
23. Hein, M., Maier, M.: Manifold denoising. In: Advances in Neural Information Processing Systems, vol. 19, pp. 561–568. MIT Press, Cambridge (2007)
24. Hein, M., Maier, M.: Manifold denoising as preprocessing for finding natural representations of data. In: Association for the Advancement of Artificial Intelligence, pp. 1646–1649. AAAI Press, San Jose (2007)
25. Huang, B.C., Jebara, T.: Loopy belief propagation for bipartite maximum weight b-matching. In: International Conference on Artificial Intelligence and Statistics, pp. 195–202 (2007)
26. Jaccard, P.: Etude comparative de la distribution florale dans une portion des Alpes et des Jura. Bull. Soc. Vaud. Sci. Nat. **37**, 547 (1901)
27. Jebara, T., Wang, J., Chang, S.F.: Graph construction and b-matching for semi-supervised learning. In: Proceedings of the 26th Annual International Conference on Machine Learning, ICML '09, pp. 441–448. ACM, New York, NY (2009)
28. Koleff, P., Gaston, K.J., Lennon, J.J.: Measuring beta diversity for presence-absence data. J. Anim. Ecol. **72**(3), 367–382 (2003)
29. Kumar, B.V.K.V., Hassebrook, L.G.: Performance measures for correlation filters. Appl. Opt. **29**, 2997–3006 (1990)
30. Libbrecht, M.W., Noble, W.S.: Machine learning applications in genetics and genomics. Nat. Rev. Genet. **16**, 321–332 (2015)
31. Lopez, J.P.H.: Análise de dados utilizando a medida de tempo de consenso em redes complexas (2011). Master Thesis, Instituto de Ciências Matemáticas e de Computação, Universidade de São Paulo (USP)
32. Luetkepohl, H.: Introduction to Multiple Time Series Analysis. Springer, New York (1991)
33. MacCuish, J.D., MacCuish, N.E.: Clustering in Bioinformatics and Drug Discovery. CRC Press, Boca Raton (2010)
34. Maier, M., von Luxburg, U., Hein, M.: Influence of graph construction on graph-based clustering measures. Neural Inf. Process. Syst. **22**, 1025–1032 (2009)
35. Marwan, N., Donges, J.F., Zou, Y., Donner, R.V., Kurths, J.: Complex network approach for recurrence analysis of time series. Phys. Lett. A **373**, 4246–4254 (2009)

36. Morisita, M.: Measuring of interspecific association and similarity between communities. Mem. Fac. Sci. Kyushu Univ. Ser E (Biology) **3**, 65–80 (1959)
37. Nicolis, G., Cantú, A.G., Nicolis, C.: Dynamical aspects of interaction networks. Int. J. Bifurcation Chaos **15**(11), 3467–3480 (2005)
38. Orozco, J., Belanche, L.: Towards a mathematical framework for similarity and dissimilarity. Technical Report, University of Sevilla (2005)
39. Roweis, S.T., Saul, L.K.: Nonlinear dimensionality reduction by locally linear embedding. Science **290**, 2323–2326 (2000)
40. Santini, S., Jain, R.: Similarity measures. IEEE Trans. Pattern Anal. Mach. Intell. **21**(9), 871–883 (1999)
41. Shirazi, A.H., Reza Jafari, G., Davoudi, J., Peinke, J., Tabar, M.R.R., Sahimi, M.: Mapping stochastic processes onto complex networks. J. Stat. Mech: Theory Exp. **2009**, P07046 (2009)
42. Silva, T.C., Zhao, L.: Pixel clustering by using complex network community detection technique. In: Proceedings of 7th International Conference on Intelligent Systems Design and Applications, pp. 925–932. IEEE Computer Society (2007)
43. Silva, T.C., Zhao, L.: Network-based high level data classification. IEEE Trans. Neural Netw. Learn. Syst. **23**(6), 954–970 (2012)
44. Silva, T.C., Zhao, L.: High-level pattern-based classification via tourist walks in networks. Inf. Sci. **294**(0), 109–126 (2015). Innovative Applications of Artificial Neural Networks in Engineering
45. Sørensen, T.: A method of establishing groups of equal amplitude in plant sociology based on similarity of species and its application to analyses of the vegetation on Danish commons. Biol. Skr. **5**, 1–34 (1948)
46. Tsay, R.S.: Analysis of Financial Time Series. Wiley Series in Probability and Statistics. Wiley-Interscience, Hoboken, NJ (2005)
47. Wang, F., Zhang, C.: Label propagation through linear neighborhoods. IEEE Trans. Knowl. Data Eng. **20**(1), 55–67 (2008)
48. Williams, J., Steele, N.: Difference, distance and similarity as a basis for fuzzy decision support based on prototypical decision classes. Fuzzy Sets Syst. **131**, 35–46 (2002)
49. Wolda, H.: Similarity indices, sample size and diversity. Oecologia **50**(3), 296–302 (1981)
50. Xu, Z., Xia, M.: Distance and similarity measures for hesitant fuzzy sets. Inf. Sci. **181**(11), 2128–2138 (2011)
51. Yang, Y., Yang, H.: Complex network-based time series analysis. Physica A **387**, 1381–1386 (2008)
52. Zhang, X., Lee, W.S.: Hyperparameter learning for graph based semi-supervised learning algorithms. In: The Conference on Neural Information Processing Systems (NIPS) (2006)
53. Zhang, J., Small, M.: Complex network from pseudoperiodic time series: topology versus dynamics. Phys. Rev. Lett. **96**, 238701 (2006)
54. Zhang, J., Luo, X., Small, M.: Detecting chaos in pseudoperiodic time series without embedding. Phys. Rev. E **73**, 016216 (2006)
55. Zhou, D., Bousquet, O., Lal, T.N., Weston, J., Schölkopf, B.: Learning with local and global consistency. In: Advances in Neural Information Processing Systems, vol. 16, pp. 321–328. MIT Press, Cambridge (2004)
56. Zhu, X.: Semi-supervised learning literature survey. Technical Report 1530, Computer Sciences, University of Wisconsin-Madison (2005)
57. Zhu, X., Goldberg, A.B.: Introduction to Semi-Supervised Learning. Synthesis Lectures on Artificial Intelligence and Machine Learning. Morgan and Claypool Publishers, San Francisco (2009)

Chapter 5
Network-Based Supervised Learning

Abstract In this chapter, we focus on supervised learning algorithms that act on networked environments. These methods utilize external information in the form of labels to induce or train their models. Generally, the learning process is composed of two serial steps denominated training and classification phases. While in the first the algorithm attempts to learn from the data according to some external aid, such as of a human expert, in the latter the algorithm is tested against unseen data to verify its generalization power. In network-based methods, both phases take place in a network by navigating through it or updating its structure according to new information originated from the human expert. In the test phase, normally the network structure remains static as new data items are classified. However, some algorithms attempt to update the learned network structure in a process classified as self-learning. In this chapter, we present some of the shortcomings and advantages of using the network-based approach to conduct supervised learning. Representative network-based methods are discussed.

5.1 Introduction

Network-based unsupervised and semi-supervised learning techniques have been extensively studied in the literature [4, 10]. There are still, however, few reported network-based supervised learning techniques [3]. In this regard, there is a big space for the proposal and discovery of new ways of supervised learning in networked environments. Presumably, several network-based semi-supervised inductive methods, such as those presented in [2, 5, 19], can be converted into a supervised learning scheme when a reasonable number of labeled instances is provided. However, these methods aim at considering unlabeled instances during the training phase and a network-based approach is employed to model the data into a manifold in order to first propagate the labels to all of the unlabeled instances. In this case, if the majority of instances in the data set is labeled, there is no space for label propagation in a data network. Thus, a regular supervised approach that uses only labeled instances in the learning process would be preferable [3].

Another type of network-based classification approach refers to relational classification. Such type of supervised classification deals with data that differs from the typical data because they violate the instance-independence assumption, which

© Springer International Publishing Switzerland 2016
T.C. Silva, L. Zhao, *Machine Learning in Complex Networks*,
DOI 10.1007/978-3-319-17290-3_5

means that the class label of an instance might not depend only on its own attributes, but also on the labels of its neighbors [14]. This kind of data is usually presented in a network form (also termed within-network data) with some of the vertices labeled and the rest unlabeled. The task is to infer the labels of the unlabeled vertices. Relational classification techniques can be applied to solve a wide range of problems, such as in the discovery of molecular pathways in gene expressions [17], classification of linked scientific research papers [13], link prediction [12], among others. For example, in link prediction on social networks [1, 7, 9, 12], the task is to predict the likelihood of a future association between two vertices, knowing that there is no association between the vertices in the current state of the network [9]. Such approach has a wide variety of further applications, among which we highlight in: recommendation systems, identification of probable professional or academic associations in e-commerce sites or scientific collaboration networks, and of structures of criminal networks and structural analysis in the field of microbiology or biomedicine. All of these applications demand for much more efficient and versatile approaches for link prediction, thereby making it an important and scientifically attractive research topic. Another application that is also related to relational classification is defined as the detection of small connected subgraphs that best capture the relationship between two vertices in a social network. In this respect, the research in [8] proposed an efficient algorithm based on electrical circuit laws to find the connected subgraph from large social networks. It has also been shown that a connected subgraph can be used to effectively compute several topological feature values for the supervised link prediction problem, especially when the network is very large [9].

State-of-the-art approaches that spread labels inside the network to infer labels of interrelated vertices in a jointly manner are known as collective inference models [16]. This kind of inference can significantly reduce classification error when compared to traditional inference techniques [11]. Collective inference methods may use both data attributes and data relational features to perform classification. Traditionally, vector-based methods have treated data items as independent ones, which makes it possible to infer class membership on an instance per instance basis. With networked data, the class membership of one data item (vertex) may have an influence on the class membership of a related vertex. Furthermore, vertices that are not directly linked may be related by chains of links, fact that suggests that it may be beneficial to infer the class memberships of all of the vertices simultaneously. Collective inference in relational data makes simultaneous statistical judgments regarding the values of an attribute or attributes of multiple entities in a network for which some attribute values are not known [16].

In the literature, some algorithms have been proposed that only employ collective inference on specific phases of the learning process. For example, one may employ a local classifier, such as Naïve Bayes or relational probability trees, to predict labels for each unlabeled vertex and further use a collective inference algorithm, such as

ICA [13] or Gibbs sampling [11], to restate the class labels of vertices that are employed in the next iteration. Such kinds of methods are called local classifiers. Another kind of approach, called global formulation-based methods, does not use a separate local classifier, but it uses the entire algorithm for the training and inference, also using relational and non-relational data. Such an approach conducts training with the objective of optimizing a global objective function. Examples of these algorithms include loopy belief propagation and relaxation labeling [18]. In search of a unification on relational data classification in networks, the research in [15] proposed a general supervised learning network-based framework. The framework builds a model considering three components: a local classifier, which makes use of a training set to estimate the probability distribution of the classes; a relational classifier, which also aims to estimate a probability distribution but now considering the neighboring relations in the network; and a collective inference component, which further refines the class predictions.

While collective inference presents some advantages, in some cases, inferring labels collectively causes uncertainties that may actually lead to lower classification accuracies when compared to non-relational approaches. For example, an incorrectly predicted label may influence the predictions of its neighbors in future iterations, possibly cascading this error through long chains of vertices [16]. On one hand, there is a tendency to represent data by networks; on the other hand, some approaches consider transforming networked data into raw, vector-based data in order to apply classical methods, such as SVM and neural networks. This kind of method requires extracting features from the networked data in order to construct a trainable vector-based set. The task of feature extraction from a given relational data can be divided according to the presence or absence of labels in the vertices, and named label-dependent and label-independent extraction, respectively. The former uses both network structure and label information throughout the neighboring vertices and the latter exclusively considers the network structure [16]. An approach to estimate the similarity between edges in a network or between two networks as a whole is graph kernel. Briefly, these kinds of methods use a kernel to establish a similarity measure on networks. In this approach, the main difficulty is in defining a kernel that is suitable for the network structure and reasonably efficient to be evaluated.

5.2 Representative Network-Based Supervised Learning Techniques

In the following, we present several representative network-based supervised learning techniques.

5.2.1   Classification Using k-Associated Graphs

This technique is introduced in [3]. As usual for network-based methods, the basis of the k-associated graphs technique lies on representing the training set as a network, more specifically a directed network referred to as k-associated graph. Such a network is built from a vector-based data set by abstracting data items to vertices and pairwise similarities to edges. After a k-associated graph is constructed for a given k, the purity measure for every component in the network is computed and is used to determine the optimal network for classification, both on the training and the test phases. The edges in a k-associated graph are established in accordance with a modified version of the k-nearest neighbor technique. In this peculiar network formation heuristic, only vertices that share the same label or class are permitted to interconnect. This simple rule generates class components in the overall network.

The purity is defined for each component (isolated subgraph) in the network as follows: given a parameter k, which is used to construct the networks using the modified version of the k-nearest neighbor technique, a vertex can have at most $2k$ connections. Since the resulting networks are digraphs or directed networks, each vertex will have degrees ranging from k to $2k$.[1] The purity measure explores this feasible range of degree values that each vertex can assume. In essence, it quantifies the proportion of edges that has effectively been created between vertices of the same class over the total number of possible connections per each vertex, $2k$. In mathematical terms, the purity ϕ of component α, $\phi^{(\alpha)}$, is then defined as:

$$\phi^{(\alpha)} = \frac{\bar{k}^{(\alpha)}}{2k}, \tag{5.1}$$

in which $\bar{k}^{(\alpha)}$ denotes the average degree of component $\alpha \in \mathscr{G}$. In general, a purity value close to 1 indicates that a large portion of edges are shared among vertices in the network component, resulting in a high-density component, while lower values reveal high levels of class mixture between components of different classes. It is for this reason $\phi^{(\alpha)}$ is called a purity measure for component α: large values indicate purer components in terms of connections. The purity measure can be conceived as the *a priori* probability of connections within a component. This property is explored by the classifier to decide the classes of each of the test instances.

In the referred technique, one can note that parameter k plays a key role in the learning process during the training phase, as its value has considerable implications on the resulting network topology. By virtue of this, a procedure for estimating the value of k for each of the components has been developed in [3]. It is intuitive that some networks may have better components than others according to the purity measure. Rarely, the network obtained using a uniform and unique value of k for

---

[1]In an undirected network, in contrast, each vertex will always have a degree of $2k$, because whenever vertex $j \in \mathscr{V}$ is one of the k-nearest neighbors of $i \in \mathscr{V}$, then the reciprocal is always true.

all of the network component produces the best configuration of vertices into class components. A single value of k produces components with nearly the same size, therefore structure and purity are restrained to only one possible value of k at a time. Consequently, it would be better to allow for multiples values of k to represent the same data space. In this way, each class component can then decide the best fit of k in accordance with the observed data distribution, producing therefore class-dependent component sizes and purity values. Bearing this in mind, a suggestive idea is to obtain a network with the best organization of the data into components with different and localized k, i.e., each component has its own optimal k. The optimality is obtained by choosing a class-dependent k such that the purity of each network component is the highest. The network component together with its optimal class-dependent k value is termed the k-associated optimal network.

To obtain the optimal k-associated graph, the rationale is to increase k while keeping the best components found so far starting from the 1-associated graph. For each k and network component, the purity measure is calculated and is used to compare between components of different k-associated graphs formed within different values of k. The component with the highest purity value is maintained, while the others are discarded.

Once the optimal k-associated graph has been properly obtained, then the classification phase begins. In this phase, the authors use a Bayes classifier in order to predict. Specifically, the *a priori* probabilities are calculated using a normalized purity value of each of the class components, rather than the traditional size proportions that we encounter in the literature.

A great potentiality of this technique is that no parameters are needed to be adjusted, which eliminates the step of external model selection. However, since the algorithm must create a network from the vector-based data set, then its time complexity is at least of the order of $\mathcal{O}(V^2)$.

5.2.2 Network Learning Toolkit (NetKit)

This work is introduced in [15]. This is a within-network inference technique, rather than an across-network inference method. In within-network inference methods, the training data items are connected directly to the test entities whose labels or classes are to be estimated. In contrast, in across-network inference, we often learn from one network and apply the learned models to a separate, presumably similar network. In essence, the toolkit is composed of three terms, each of which focusing on different perspectives or visions of the data items. The modules are:

1. *Non-relational ("local") module*: This component consists of a (learned) model that only uses local information of the data items, namely the information about their attributes, to estimate their labels or classes. The local models can be used to generate priors that comprise the initial state for the relational learning and collective inference components. They also can be used as one source of evidence

during collective inference. These models are typically produced by traditional machine learning methods.

2. *Relational module*: In contrast to the non-relational component, the relational model makes use of the relations in the network as well as of the attribute values of related entities, possibly through long chains of relations. Relational models also may use local attributes of the data items.

3. *Collective inference module*: The collective inference component determines how the unknown values are estimated together, possibly influencing each other in a collective manner.

Depending on the choices of each of the three aforementioned components, one can get new types of classifiers. Some of these choices result in well-known classifiers in the related community. For example, using a Naïve Bayes classifier as the local model, a Naïve Bayes Markov Random Field classifier for the relational model, and relaxation labeling for the collective inference module form the system used by Chakrabarti *et al.* [15].

It is worth registering that the collective inference component can explore relational autocorrelation, which is a widely observed characteristic of relational data. This phenomenon may reveal that a variable for one instance is highly correlated with the value of the same variable on another instance. By making inferences about multiple data instances simultaneously, collective inference can significantly reduce classification error in some cases.

The importance of NetKit is threefold: (i) it generalizes several existing methods for classification in networked data, thereby making comparison to existing methods possible; (ii) it enables the creation and use of many new algorithms by its modularity and extensibility; and (iii) it enables the analysis/comparison of individual components and configurations. These contributions are welcomed by the literature, because, since then, there has been no systematic study of machine learning methods for within-network classification that compares various algorithms on several data sets.

5.2.3   Classification Using Ease of Access Heuristic

This network-based supervised classification technique is proposed in [6]. The intuition of this method is to perform the classification task using a heuristic called *ease of access* in a networked environment. The measurement of ease of access is built upon the concept of limiting probabilities in the Markov chain theory. First, a set of labeled instances is mapped as vertices of a network. Recalling that a network can be conceived as a discrete Markov chain, each vertex then represents a state in the Markovian process. When classifying an unlabeled data, this network of only labeled vertices is modified by a specific link weight composition, which takes into account the bias information of that unlabeled instance. The bias information alters the network structure in a way that, after the computation of the modified limiting probabilities, the most easily reached labeled instances represent the class label of that unlabeled instance.

The classification problem requires a given labeled data set, $\mathscr{L} = \{x_1, x_2, \ldots, x_L\}$, where each instance is described by P attributes $x_i = (x_{i1}, x_{i2}, \ldots, x_{iP})$. Each instance in this set has a single assigned label $y \in \mathscr{Y}$. It is also given an unlabeled data set, $\mathscr{U} = \{x_{L+1}, x_{L+2}, \ldots, x_{L+U}\}$, containing instances whose labels are to be estimated. There are L labeled instances and U unlabeled instances. The classification technique is divided into the two classical phases: training and classification.

5.2.3.1 Training Phase

In the training phase, a weighted and undirected network $\mathscr{G} = \langle \mathscr{V}, \mathscr{E} \rangle$ is constructed without self-loops. Vertices represent labeled data instances, $\mathscr{V} = \mathscr{L}$, and link weights are established using a similarity function (cf. Sect. 4.2) and a network formation strategy (cf. Sect. 4.3). At the end of this phase, we get a network $\mathscr{G}$ called training network.

5.2.3.2 Classification Phase

To classify an unlabeled instance $x \in \mathscr{U}$, a weight vector $s = [s_1, s_2, \ldots, s_L]$ is first calculated, in which each entry s_i contains the similarity of that unlabeled data to the labeled vertex i. That is, vertex x is inserted into the training network $\mathscr{G}$ by calculating the link weights to all of the other labeled vertices into this network, and is subsequently removed from $\mathscr{G}$. Then, the weighted and asymmetric adjacency matrix with L vertices is perturbed as follows:

$$\hat{A} = A + \epsilon \hat{S}, \tag{5.2}$$

in which A and $\hat{A}$ are the original and the perturbed adjacency matrices, respectively; ϵ is a non-negative parameter; and $\hat{S}$ is the following $L \times L$ matrix:

$$\hat{S} = \begin{bmatrix} S_{(1)} \\ S_{(2)} \\ \vdots \\ S_{(L)} \end{bmatrix}, \tag{5.3}$$

in which $S_{(i)}$ is a row-vector $L \times 1$ whose entries are all s_i. It can be observed in (5.2) that the weight biases of the unlabeled instance x, encoded in matrix $\hat{S}$, are applied over all of the links in the original adjacency matrix A of the training network $\mathscr{G}$, that is, the weight of each link is linearly added up with its corresponding weight bias. The idea behind this operation is that the distance between any pair of vertices is modified due to the new weights of network routes introduced by the insertion of the link biases from the unlabeled instance links. The higher the similarity between the unlabeled instance and a vertex, say vertex i, the more strengthened are the connections from all of the other vertices to vertex i after this operation. The parameter ϵ controls the influence of the weight biases in the training network. The larger the value of parameter ϵ is, the greater is the influence of the bias weights.

The perturbed adjacency matrix $\hat{\mathbf{A}}$ is termed as the classification network. By using this network, it is now possible to apply the random walk limiting probabilities over the states represented by the network vertices. The transition probabilities can be found by means of the matrix $\hat{\mathbf{A}}$. To compute the entries of the transition matrix $\mathbf{P}$, the entries of matrix $\hat{\mathbf{A}}$ are normalized:

$$\mathbf{P}_{ij} = \hat{\mathbf{A}}_{ij} / \sum_{j \in \mathscr{V}} \hat{\mathbf{A}}_{ij}. \tag{5.4}$$

With the above matrix $\mathbf{P}$ at hand, the limiting probabilities can be calculated by using one of two possible ways: finding the eigenvector corresponding to the unit eigenvalue of matrix $\mathbf{P}$ or iterating the system

$$p_{i+1} = p_i \mathbf{P} \tag{5.5}$$

to the stationary state, where p is the state distribution. Under the constraints discussed in Sect. 2.4.1, the limiting or stationary probability is unique and independent on the system's initial state and has the form:

$$\mathbf{p}^\infty = \pi = [\pi_1, \pi_2, \ldots, \pi_L], \tag{5.6}$$

in which each element represents a state, and each entry p_i can be interpreted as the probability of x to belong to the class of state i.

As the final step, the classification of x is accomplished by assigning it the most representative label from the set of states. To accomplish that, a set $\mathscr{T}$ containing the t states with the largest limiting probabilities are selected and the most representative class in $\mathscr{T}$ is associated to x.

5.3 Chapter Remarks

The literature in network-based supervised learning is still very scarce, as very few methods have been developed so far. This is mainly due to the fact that the large percentage of labeled data makes the data network almost fixed. Then, there is not enough space for label propagation. Up to now, the developed techniques are based on two main ideas: (1) collective inference with the data network, as done in k-associated graphs [3] and the network learning toolkit [15]; (2) classification using the pattern formation of the entire network, as performed in the classification that uses the ease of access heuristic [6]. At the conceptual level, the within-network techniques do not differ much from traditional data classification ones. However, we still have a large space to explore the across-network approach. This is because we have many ways to characterize the global patterns of the data network. In Chap. 8, we walk through a pioneer across-network supervised learning technique.

References

1. Barabási, A.L., Jeong, H., Neda, Z., Ravasz, E., Schubert, A., Vicsek, T.: Evolution of the social network of scientific collaborations. Phys. A Stat. Mech. Appl. **311**(3–4), 590–614 (2002)
2. Belkin, M., Niyogi, P., Sindhwani, V.: Manifold regularization: a geometric framework for learning from labeled and unlabeled examples. J. Mach. Learn. Res. **7**, 2399–2434 (2006)
3. Bertini J.R. Jr., Zhao, L., Motta, R., Lopes, A.A.: A nonparametric classification method based on K-Associated graphs. Inf. Sci. **181**, 5435–5456 (2011)
4. Chapelle, O., Schölkopf, B., Zien, A. (eds.): Semi-supervised learning. Adaptive Computation and Machine Learning. MIT, Cambridge (2006)
5. Chen, J., Fang, H.R., Saad, Y.: Fast approximate kNN graph construction for high dimensional data via recursive lanczos bisection. J. Mach. Learn. Res. **10**, 1989–2012 (2009)
6. Cupertino, T.H., Zhao, L., Carneiro, M.G.: Network-based supervised data classification by using an heuristic of ease of access. Neurocomputing **149**(Part A), 86–92 (2015)
7. Dorogovtsev, S.N., Mendes, J.F.F.: Evolution of Networks: From Biological Nets to the Internet and WWW (Physics). Oxford University Press, Oxford (2003)
8. Faloutsos, C., Mccurley, K.S., Tomkins, A.: Fast discovery of connection subgraphs. In: Proceedings of the 2004 ACM SIGKDD international conference on Knowledge discovery and data mining (KDD), pp. 118–127. ACM, New York (2004)
9. Hasan, M.A., Chaoji, V., Salem, S., Zaki, M.: Link prediction using supervised learning. In: Proceedings of SDM 06 workshop on Link Analysis, Counterterrorism and Security (2006)
10. Jain, A.K.: Data clustering: 50 years beyond K-Means. Pattern Recogn. Lett. **31**, 651–666 (2010)
11. Jensen, D., Neville, J., Gallagher, B.: Why collective inference improves relational classification. In: In Proceedings of the 10th ACM SIGKDD International Conference on Knowledge Discovery and Data Mining, pp. 593–598 (2004)
12. Liben-Nowell, D., Kleinberg, J.: The link-prediction problem for social networks. J. Am. Soc. Inf. Sci. Technol. **58**(7), 1019–1031 (2007)
13. Lu, Q., Getoor, L.: Link-based Classification using Labeled and Unlabeled Data. In: Proceedings of the ICML 2003 Workshop on The Continuum from Labeled to Unlabeled Data. Washington (2003)
14. Macskassy, S.A., Provost, F.: A simple relational classifier. In: Proceedings of the Second Workshop on Multi-Relational Data Mining (MRDM-2003) at the Knowledge Discovery and Data Mining Conference (KDD), pp. 64–76 (2003)
15. Macskassy, S.A., Provost, F.: Classification in networked data: a toolkit and a univariate case study. J. Mach. Learn. Res. **8**, 935–983 (2007)
16. McDowell, L., Gupta, K.M., Aha, D.W.: Cautious collective classification. J. Mach. Learn. Res. **10**, 2777–2836 (2009)
17. Segal, E., Wang, H., Koller, D.: Discovering molecular pathways from protein interaction and gene expression data. In: Proceedings of the Eleventh International Conference on Intelligent Systems for Molecular Biology, Brisbane, pp. 264–272 (2003)
18. Sen, P., Namata, G.M., Bilgic, M., Getoor, L., Gallagher, B., Eliassi-Rad, T.: Collective classification in network data. Artif. Intell. Mag. **29**(3), 93–106 (2008)
19. Sindhwani, V., Niyogi, P., Belkin, M.: Beyond the point cloud: from transductive to semi-supervised learning. In: Proceedings of the 22nd international conference on Machine learning (ICML), pp. 824–831. ACM, New York (2005)

Chapter 6
Network-Based Unsupervised Learning

Abstract In this chapter, we present representative state-of-the-art unsupervised learning techniques that rely on networked environments to conduct the learning process. In a typical unsupervised task, no external knowledge is presented to the algorithm. As such, the learning process is guided by the provided data, since no prior knowledge about the existing groups is supplied. For network-based methods, the learning procedure is performed by navigating in networks that are constructed from the input data set according to some similarity criterion. As networks naturally embody topological information of data relationships, network-based methods take advantage over algorithms that make use of raw, vector-based data. Moreover, network-based methods can be conceived as a general solution for unsupervised learning tasks even for data sets that are not represented by networks. In this case, we can apply network formation techniques on that data set to generate a network from the input data. Once the network is constructed, all of the network-based techniques described in this chapter can effectively be employed.

6.1 Introduction

In this chapter, we shift our attention to network-based unsupervised methods. The data representation as networks enables us to systematically investigate the topology and function of data relationships using well-understood graph-theoretical concepts that can be employed to uncover structural and dynamical properties of the underlying constructed network.

One of the main tasks of unsupervised learning is data clustering. In essence, data clustering can be considered as a community detection problem once a network is constructed from the original data set. In this transformation, each vertex corresponds to a data item and connections are established according to a certain similarity measure. The clusters in a community detection task are often denominated communities. A community is defined as a subgraph whose vertices are densely connected within itself, but sparsely connected with the remainder of the network. Figure 6.1 illustrates typical processes in data clustering and in community detection. In the former, unstructured or raw data are received by a data clustering procedure that finds similar groups in accordance with a similarity criterion. In the latter, the community detection procedure uncovers communities in the network.

© Springer International Publishing Switzerland 2016
T.C. Silva, L. Zhao, *Machine Learning in Complex Networks*,
DOI 10.1007/978-3-319-17290-3_6

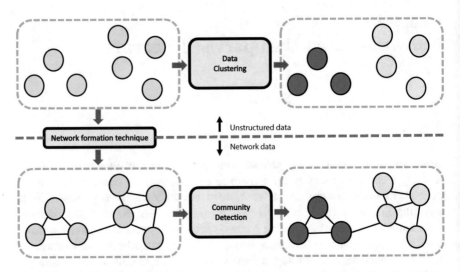

Fig. 6.1 Similarities between data clustering and community detection tasks. The *dotted horizontal line* represents the frontier of unstructured data and networked data. A network formation method interfaces between unstructured and networked data. Note that each of the data items is represented by a vertex in the networked domain

Topological information of the data, such as direct or indirect neighborhoods, can be readily employed by the community detection method. Observe that the network formation method serves as interface between unstructured and networked data.

Network-based methods are specially useful when we deal with clusters of arbitrary shape, proximity, orientation, and varying densities [36]. Since in unsupervised learning methods we usually do not know how the clusters are shaped nor how many of them exist, network-based methods stand as good candidates for tasks related to data clustering. Consider that we use as input the data set depicted in Fig. 6.2a in the schematic shown in Fig. 6.1. For the data clustering method in unstructured data, we choose the well-known K-Means procedure with a number of clusters calibrated to 2. For the community detection task in networked data, we use the Chameleon technique [36], which is a network-based unsupervised learning method that we discuss in this chapter. We employ the k-nearest neighbor technique with $k = 7$ as the network formation technique that interfaces between unstructured and networked data. The clustering result for the K-Means technique is displayed in Fig. 6.2b, while the outcome of Chameleon is portrayed in Fig. 6.2c. While K-Means has difficulty in clustering arbitrary-shaped clusters due to its strong bias on circular-shaped items, network-based methods can provide robust results as they are guided by the network topology in the learning process. This is because network-based methods use the network topology to derive its decisions, in a way that we do not need assumptions about the data distribution nor about the number of clusters or communities. Consequently, we prevent the insertion of wrong biases over the data distributions that can severely hamper the quality of the learning process.

Fig. 6.2 Comparison of vector- and network-based methods in data clustering and community detection tasks, respectively. We use the K-Means algorithm with $K = 2$ in Fig. 6.2b. In Fig. 6.2c, we first construct the network from the unstructured data in Fig. 6.2a using $k = 7$ and then apply the Chameleon. (**a**) Initial state (vector-based data); (**b**) Results for vector-based learning method; (**c**) Results for network-based learning method

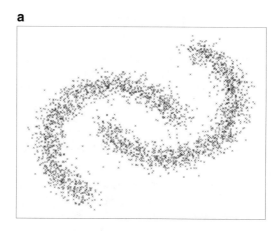

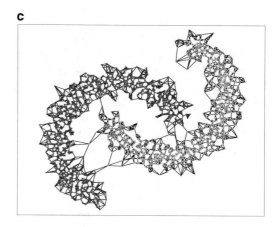

6.2   Community Detection

In this section, we introduce the main concepts of community detection, as well as a brief description of the related state-of-the-art techniques. In addition, we present some broadly accepted community detection benchmarks.

6.2.1   Relevant Concepts and Motivations

Complex networks are found in fields as diverse as the Internet, the World Wide Web, food webs, and biological and social organizations [7]. Even though the main features of complex networks have been properly described at the microscale level, such as strict-local properties of network vertices, and also at the macroscale level, such as global properties of the entire network, some of the characteristics lying at a mesoscale level are still elusive.

Nonetheless, modern science related to networks brought a substantial advance in understanding complex networks. One of the features evident and prominent in complex networks is the presence of mesoscale structures called *communities*. These communities can carry functional, relational, or even social common concepts. Though the formal definition of a community is controversial in the literature, the essence of a community is straightforward: each community is defined as a subgraph whose vertices are densely interconnected, and, at the same time, these vertices have few links with the remainder of the network. Figure 6.3 portrays a network in which four well-defined communities can be observed, because the quantity of edges between members of the same community is perceptively larger than the number of edges connecting different communities. The community detection task in complex networks has become an important topic in graphs and data mining [16, 22, 57]. In graph theory, community detection corresponds to the graph partitioning problem, which is an NP-complete problem [22].

The study of community detection is very important for understanding various phenomena in complex networks [32]. Modular structure introduces important heterogeneities in complex networks. Each module, for example, can have different local statistics [55]; some modules may have many connections, while other modules may be sparse. When there is large variation among communities, global values of statistical measures can be misleading. The presence of modular structure may also alter the way in which dynamical processes (e.g. spreading processes and synchronization [3]) unfold on the network. In biological networks, communities correspond to functional modules in which module members function coherently to perform essential cellular tasks. Both metabolic networks [69] and protein phosphorylation networks [35], for instance, have modular structures.

A promising computational approach to discovering functions of genes and proteins is to identify functional modules in biological networks. Since modules are sets of genes or proteins that perform biological processes together, it is

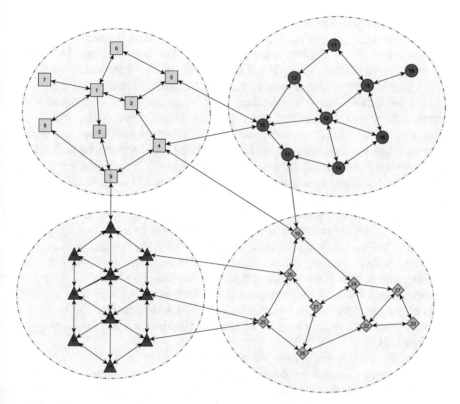

Fig. 6.3 A network that presents four well-defined communities. The *different vertices' colors or formats* denote the communities to which each of them belong

possible to classify proteins with unknown functions by determining to what module they belong [63]. Correct identification of functional modules has also important biotechnological and drug design applications. In many cases, the deletion of a certain function may be necessary and this can be achieved by removing the entire functional module.

Several distinct ways of detecting modules in complex networks have been proposed [22]. One popular approach considers communities as sets of adjacent motifs [63], other methods are inspired by information theory [71], message passing [26], or Bayesian principles [34, 58]. A widely used class of algorithms is based on the optimization of a quantity called modularity [57].

Another important aspect related to community structure is of the hierarchical organization displayed by most networked systems in the real world [22]. Real networks are usually composed of communities including smaller communities, which in turn include even smaller communities, and so on. The human body offers a paradigmatic example of hierarchical organization: it is composed by organs, organs are composed by tissues, tissues by cells, etc. Another example is represented by business firms, which are characterized by a pyramidal organization, going from the workers to the president, with intermediate levels corresponding to work groups, departments and management. Other example is the network formed by all human acquaintances. While at a local scale we expect to find many communities formed by families and friends, on a larger scale, the expected communities turn into cities, regions, followed by countries, and, finally, probably continental areas. The generation and evolution of systems organized in interrelated stable subsystems are much quicker than unstructured systems. One evidence corroborating this fact is that it is much easier to assemble the smallest sub-parts of a structured system first and then use them as building blocks to build up larger structures, until the entire system is assembled. In view of these examples, it is clear that the study of community presence in networks plays an important role in understanding natural concepts encountered in various branches of science.

Another interesting topic is of overlapping communities. We have seen that the identification of modules and their boundaries enables us to classify vertices according to their structural positions in those modules. So, vertices with a central position in their clusters, i.e., which share large numbers of edges with other group partners, may have important functions of control and stability within the group. Notwithstanding, vertices lying at the boundaries between these modules also play an important role of mediation and lead the relationships and exchanges between different communities. These kinds of vertices are termed as overlapping vertices [22].

Formally, overlapping vertices are defined as those vertices that are members of more than one community or class at the same time [63]. For example, in a network of semantic association concepts [38], the term "brilliant" may be a member of several classes, such as the one representing the concepts related to "light," to "astronomy," "color," and so on [63]. In a social network, each person naturally belongs to the company where he/she works and also to the group representing the members of his/her family. Given this scenario, the discovery of overlapping vertices and communities is important for data analysis in general.

6.2.2   Mathematical Formalization and Fundamental Assumptions

Unsupervised learning methods are guided exclusively by the intrinsic structure of the data items throughout the learning process, i.e., without any sort of external knowledge. Consider that $\mathscr{X} = \{x_1, x_2, \ldots, x_N\}$ is a data set, where $N = |\mathscr{X}|$ is the total number of data items involved in the learning process. Techniques that are members of the network-based unsupervised learning paradigm always accept as input a network. In this respect, we can face the following scenarios:

- The items in the data set are already in the network format, i.e., the vertex set $\mathcal{V}$ coincides with the set of data items $\mathcal{X}$ and the set of edges $\mathcal{E}$ is given. In this case, no preprocessing is needed. Well-known examples that already are in the form of networks include: WWW, Internet, transport and financial networks. Data sets of this type are inherent candidates to serve as input to network-based unsupervised learning methods.
- The items in the data set are presented in a raw, vector-based format. Normally, $\mathcal{X} = \mathcal{V}$, but we can also use compacted or expanded sets of $\mathcal{X}$ to build up $\mathcal{V}$. The edge set $\mathcal{E}$ is unknown and must be estimated using a network construction technique. Normally, the set of edges is constructed according to some similarity criteria that are imposed by the network construction process. Figure 6.1 illustrates this process. In Chap. 4, we have presented several manners to deal with this problem. Here, we assume that there exists such a function of network formation technique that simply transforms the vector-based format to a network.

Suppose the network $\mathcal{G} = \langle \mathcal{V}, \mathcal{E} \rangle$ is obtained from the input data items. Then, the unsupervised learning problem is now posed in a network-based form. Recall that data clustering turns into a community detection task when the network structure of the data distribution is well-conditioned.

Though intuitive at first sight, the problem of community detection is actually not well defined. The main elements that make up the community detection task *per se*, that is, the concepts of community and partition, are not rigorously defined. In view of that, one must accept some degree of arbitrariness or common sense [22]. In fact, some ambiguities are hidden and there are often many equally legitimate ways of resolving them. It is not surprising, thus, that there are plenty of recipes in the literature and that people do not even try to ground the problem on shared definitions.

One point that is at least common sense in the literature is of the identification of the structural constraint for the existence of communities. In this regard, the existence of structural and well-defined communities is only possible when graphs are sparse. Sparseness arises when the number of edges E is of the order of the number of vertices V in non-weighted graphs, i.e. $E = \mathcal{O}(V)$. If $E \gg V$, the distribution of edges among the vertices is too homogeneous for communities to make sense. In this case, the problem turns into something rather different, close to data clustering, as the network structure does not convey relevant information to identify the community structures. The main difference between a community detection and data clustering task is that, while communities in graphs are related, explicitly or implicitly, to the concept of edge density (inside versus outside the community), in data clustering communities are sets of points which are "close" to each other, with respect to a measure of distance or similarity, defined for each pair of points [22].

6.2.3   Overview of the State-of-the-Art Techniques

Given that the task of accurately solving a problem of community detection is NP-complete, many efforts have been expended towards the development of approximate and efficient solutions. Some of these solutions include the spectral method [54], the betweenness-based technique [57], modularity greedy optimization [52], detection of communities based on the Potts model [70], synchronization [3], information theory [24], and random walks [92]. A thorough review on this topic is presented in [22].

Regarding the techniques which aim at detecting overlapping vertices and communities, various methods have been proposed in the literature [19, 43, 59, 63, 77, 79, 90]. In the research in [90], the authors combine the idea of the modularity function Q, spectral relaxation, and fuzzy C-Means clustering in order to build a new modularity function based on a generalized Newman and Girvan's Q function, which is an approximate mapping of the network vertices into the Euclidean space. In the study in [63], the community structure is uncovered by means of a k-clique percolation and the overlaps among communities are guaranteed by the fact that one vertex can participate in more than one clique. However, the k-clique percolation method gives rise to an incomplete cover of the network, i.e., some vertices may not belong to any community. In addition, the hierarchical structure may not be revealed for a given k. In contrast, the investigation in [43] introduces an algorithm that concomitantly finds both overlapping communities and the hierarchical structure based on a fitness function and a resolution parameter. In turn, the research in [19] proposes a method to recognize the overlapping community structure by partitioning a graph built from the original network. A perceptive drawback of the majority of these techniques resides in the fact that the detection of the overlapping characteristics of the input network is performed as a separated or dedicated process apart from the standard community detection technique. In this way, additional computational time is required. As a result, the whole process may have high computational complexity.

6.2.4   Community Detection Benchmarks

In this section, we introduce two community detection benchmarks, which are frequently used for comparing different competing techniques.

Benchmark of Girvan and Newman [28] This benchmark uses an agglomerative method that groups V initially isolated vertices into M communities. This is managed by creating links between two vertices with probability p_{in}, if they belong to the same community, or with probability p_{out}, if they belong to distinct communities. The values of p_{in} and p_{out} can be arbitrarily chosen to control the number of intracommunity and intercommunity links, z_{in} and z_{out}, respectively, for an arbitrary average network degree $\bar{k}$. On the basis of these parameters, we are able to define the fraction of intracommunity links $z_{in}/\bar{k}$ and, likewise, the fraction of intercommunity links $z_{out}/\bar{k}$. The quantity $z_{out}/\bar{k}$ defines the mixture of the communities, i.e., as $z_{out}/\bar{k}$ increases, the communities become more mixed and harder to be identified.

The benchmark works by varying the mixture of communities, i.e., $z_{out}/\bar{k}$, for a fixed network comprising V vertices and M communities. For each run, the community detection accuracy is registered. After all of the runs have been properly performed, a curve is plotted in a two-dimensional graph. This curve serves the purpose of comparing the community detection performance of a control algorithm in relation to competing techniques.

Benchmark of Lancichinatti et al. [42] The Girvan-Newman's benchmark in its original form suffers from several drawbacks, among which we can highlight:

- Each community has necessarily a random network topology. Therefore, the vertices have similar degrees and therefore have trivial link relationships; and
- Communities are forced to be of the same size.

Motivated by the fact that real-world networks are characterized by heterogeneous distributions of vertex degree, whose tails often decay as power laws, the benchmark of Lancichinatti et al. generates artificial networks with properties that overcome the size homogeneity of communities and the random network topology of the Girvan-Newman's benchmark.

The constructed networks assume that both degree and community size distributions follow a power law function, with exponents γ and β, respectively. Typical values of real-world networks are: $2 \leq \gamma \leq 3$ and $1 \leq \beta \leq 2$. Moreover, a mixing parameter μ is employed to interconnect communities in the following manner: each vertex shares a fraction $1 - \mu$ of its links with other vertices of the same community and a fraction μ with vertices of other communities.

The benchmark process consists in varying the mixing parameter μ and evaluating the normalized mutual information index, which is a similarity measure of partitions borrowed from the information theory [16] that measures the mutual dependence of different random variables.

6.3 Representative Network-Based Unsupervised Learning Techniques

In the following, we present representative techniques that are members of the network-based unsupervised learning.

6.3.1   Betweenness

A natural strategy to identify communities in a network is to detect and subsequently remove those edges that connect vertices of different communities, so that the communities eventually get disconnected from each other. In this case, the number of network components represents the number of communities. This is the philosophy of divisive algorithms. The crucial point resides in finding useful properties of intercommunity edges that could allow for their identification.

The most popular algorithm is that proposed by Girvan and Newman [28, 57]. In the edge removal process, the algorithm selects edges according to the values of edge centrality, estimating the importance of edges according to some property or process running on the network. The steps of the algorithm are:

1. Computation of the centrality for all of the edges;
2. Removal of the edge with the largest centrality: in case of ties with other edges, one of them is picked at random;
3. Recalculation of centralities on the modified network (network without that removed edge);
4. Iteration of the cycle from Step 2.

Girvan and Newman focused on the concept of betweenness, which is a variable expressing the frequency of the participation of edges to a process. They considered three alternative definitions: geodesic edge betweenness, random-walk edge betweenness and current-flow edge betweenness. In the following we shall refer to them as edge betweenness, random-walk betweenness and current-flow betweenness, respectively.

The betweenness of an edge is the number of shortest paths between all of the vertex pairs that run along that edge. It is an extension to edges of the popular concept of site betweenness, introduced by Freeman in 1977 [25] and expresses the importance of edges in processes like information spreading, where information usually flows through shortest paths. It is intuitive that intercommunity edges have large values of edge betweenness, because many shortest paths connecting vertices of different communities pass through them. As in the calculation of vertex betweenness, if there are two or more geodesic paths with the same endpoints that run through an edge, the contribution of each of them to the betweenness of the edge must be divided by the multiplicity of the paths, as one assumes that the signal/information propagates equally along each geodesic path.

In random-walk betweenness, one could imagine that signals flow across random rather than geodesic paths. In this case, the betweenness of an edge is given by the frequency of passages of a random walker across that edge. A random walker moving from a vertex follows each adjacent edge with equal probability. The algorithm works by first choosing a pair of vertices at random, say $s \in \mathscr{V}$ and $t \in \mathscr{V}$. The walker starts at s and keeps moving until it finally reaches t, where it stops. We then compute the probability that each edge in the network is crossed by that random walker. We perform this process for every given pair of network vertices s and t and take the average values. In this process, it is meaningful to compute the net crossing probability, which is proportional to the number of times the walk crossed an edge in one direction. In this way one neglects back and forth passages that are accidents of the random walk and tell us nothing about the centrality of that edge.

In current-flow betweenness, the network is considered as a resistor network, with edges having unit resistance. If a voltage difference is applied between any two vertices, each edge carries some amount of current, that can be calculated by solving Kirchoff's equations. The procedure is repeated for all of the possible vertex pairs: the current-flow betweenness of an edge is the average value of the current carried by the edge. It is possible to show that this measure is equivalent to random-walk betweenness, as the voltage differences and the random walks net flows across the edges satisfy the same equations [53].

In practical applications, the Girvan-Newman algorithm with edge betweenness gives better results than adopting the other centrality measures and is also much faster to compute than current-flow or random walk betweenness [51]. Nevertheless, the algorithm is still quite slow and is not applicable to large-scale graphs. In the original version of the Girvan-Newman algorithm [28], the authors had to deal with the entire hierarchy of partitions, as they had no procedure to say which partition is the best. In a successive refinement [57], they incorporate the process of selecting the best partition into the algorithm by employing the largest value of modularity.

Chen and Yuan [10] pointed out that considering all of the possible shortest paths in the evaluation of the edge betweenness may lead to unbalanced partitions, with communities of very different sizes. In order to overcome that problem, they proposed to count only non-redundant paths, i.e. those paths whose endpoints are all different from each other: the resulting betweenness yields better results than standard edge betweenness for mixed clusters on the benchmark graphs of Girvan and Newman.

6.3.2   Modularity Maximization

The scientific community considers the modularity algorithm as a seminal work in community detection. This class of algorithms relies on the fact that maximizing modularity is a good strategy for obtaining well-established communities. Before we discuss some representative methods that maximize modularity, we first recap the concept of network modularity, which has already been introduced in Definition 2.50.

The modularity measure quantifies how good a particular division of a network is [13, 55] and is designed to measure the strength of division of a network into modules (also called groups, clusters or communities). Generally, it ranges from 0 to 1. When the modularity is near 0, it means that the network does not present community structure, suggesting that the links are disposed at random in the network. As the modularity grows, the community structure gets more and more defined, that is, the mixture between communities gets smaller and therefore the fraction of links inside communities is larger than that between different communities. Mathematically, the network modularity is given by:

$$Q = \frac{1}{2E} \sum_{i,j \in \mathcal{V}} \left(A_{ij} - \frac{k_i k_j}{2E} \right) \mathbb{1}_{[c_i = c_j]}, \tag{6.1}$$

in which E represents the total number of edges in the network; A_{ij} indicates the edge weight linking i to j; k_i stands for the degree of the vertex i; c_i is the community of vertex i; and $\mathbb{1}_{[c_i = c_j]}$ indicates the Kronecker's Delta or the indicator function, which produces 1 if $c_i = c_j$ and 0, otherwise. Essentially, the modularity captures how well the network structure fits to a given set of communities. In the computation, random chances are canceled out by subtracting the edge quantity that is expected within a community from an equivalent random network.

Modularity has been used to compare the quality of the partitions obtained by different methods, but has also been used as an objective function to be optimized [52]. Unfortunately, exact modularity optimization is a problem that is computationally hard [6] and so approximation algorithms are necessary when dealing with large networks.

The first proposed method to perform modularity optimization was done by Clauset et al. [13]. Since then, several other versions have been proposed [6, 11, 31, 66, 83]. The greedy algorithm proposed by Clauset et al. may produce modularity values that are significantly lower than what can be found by using, for instance, simulated annealing [31]. Moreover, the method proposed in [13] has a tendency to produce super-communities that contain a large fraction of the vertices, even on synthetic networks that have no significant community structure. This artefact also has the disadvantage to slow down the algorithm considerably and makes it inapplicable to networks of more than a million vertices. The Louvain method [6] is the fastest modularity optimization algorithm proposed so far. In addition, the mechanism underlying the Louvain algorithm circumvents the undesired effect of unbalanced communities encountered in Clauset et al. by introducing tricks in order to balance the size of the communities being merged, thereby speeding up the running time and making it possible to deal with networks that have a few million vertices.

In the following, we first discuss the traditional modularity optimization method proposed by Clauset et al. [13] and then the Louvain method [6].

6.3.2.1 Clauset et al. Algorithm

At each time step of the modularity maximization, the algorithm of Clauset et al. [13] chooses to merge two communities that lead to the largest increase in the modularity Q, i.e., it finds the largest modularity increment ΔQ. In the initial step, the increment in the network modularity if communities i and j are joined is:

$$\Delta Q_{ij} = \begin{cases} \frac{1}{2E} - \frac{k_i k_j}{(2E)^2}, & \text{if } i \text{ and } j \text{ are connected.} \\ 0, & \text{otherwise.} \end{cases} \quad (6.2)$$

Two communities, say i and j, are merged, in such a way that their merge causes the largest increment (or the least decrement) of the modularity at a particular step. The algorithm is agglomerative and each vertex represents a community in the initial configuration. If one wants to stop the merges when the network configuration reaches its maximum modularity, one can use the stop criterion as follows: once a negative increment is encountered in this greedy process, the maximum global value associated to the modularity has been reached and subsequent merges will only monotonically decrease the modularity of the network. Therefore, by looking at the signal of ΔQ_{ij} at each iteration, it is sufficient to know when to stop merging. In addition, no restrictions on the communities to be merged are specified by the original model.

A major advantage of the modularity greedy algorithm is that no model selection is required, as no parameters need to be adjusted. Moreover, we have a nice stopping criterion for the algorithm due to the behavior of the modularity curve.

A drawback of the original modularity algorithm is in its resolution limit. Several studies have shown that it is unable to detect very small communities [23, 39, 41]. Roughly speaking, the modularity compares the number of edges inside a community with the expected number of edges that one would find in the community if the network were a random network with the same number of vertices, each of which with the same degree, but with edges randomly reattached. This random null model implicitly assumes that each vertex can get attached to any other vertex of the network. Such assumption is however unreasonable if the network is very large, as the horizon of a vertex includes a small part of the network, ignoring most of it. Moreover, this null model implies that the expected number of edges between two groups of vertices decreases if the size of the network increases. So, if a network is large enough, the expected number of edges between two groups of vertices in the modularity's null model may be smaller than one. If this happens, a single edge between the two communities would be interpreted by modularity as a sign of a strong correlation between these two communities, and the modularity optimization procedure would lead to the merge of them, independently of the communities' features. So, even weakly interconnected complete graphs, which have the highest possible density of internal edges, and represent the best identifiable communities, would be merged by the modularity optimization process if the network is sufficiently large. For this reason, optimizing modularity in large networks would fail to identify small communities, even when they are well defined. This bias is inevitable for methods like modularity optimization, which rely on a global null model.

6.3.2.2 Louvain Algorithm

The Louvain algorithm [6] is divided into two phases that are repeated iteratively. Assume that we start with a weighted network of V vertices. First, we assign a different community to each vertex. So, in this initial partition, there are as many communities as there are vertices. Then, for each vertex i, we consider the neighbors j of i and we evaluate the gain of modularity that would take place by removing i from its community and by placing it in the community of j. Vertex i is then placed in the community for which this gain is maximum, but only if this gain is positive. If no positive gain is possible, vertex i stays in its original community. This process is applied repeatedly and sequentially for all of the vertices until no further improvement can be achieved. When the equilibrium is reached, the first phase of the Louvain algorithm is then complete. Note that a vertex may be, and often is, considered several times in this community flipping process. This first phase stops when a local maximum of the modularity is attained, i.e., when no individual move can improve the modularity. One should also observe that the output of the algorithm depends on the order in which the vertices are processed. Preliminary results on several test cases seem to indicate that the ordering of the vertices does not have a significant influence on the achieved maximum modularity. However, the ordering can influence the computation time. The problem of choosing an order is thus worth studying since it could give good heuristics to enhancing the computation time.

Part of the efficiency of the algorithm results from the fact that the gain in modularity ΔQ obtained by moving an isolated vertex i into a community m can easily be computed by:

$$\Delta Q = \left[\frac{\Sigma_{in} + s_{i,in}}{2E} - \left(\frac{\Sigma_{tot} + s_i}{2E} \right)^2 \right] - \left[\frac{\Sigma_{in}}{2E} - \left(\frac{\Sigma_{tot}}{2E} \right)^2 - \left(\frac{s_i}{2E} \right)^2 \right], \qquad (6.3)$$

in which Σ_{in} is the sum of link weights inside community m, Σ_{tot} is the sum of link weights incident to vertices in community m, s_i is the sum of the link weights incident to vertex i (in-strength), $s_{i,in}$ is the sum of link weights from i to vertices in community m, and E is the sum of link weights in the network. A similar expression is used to evaluate the change of modularity when i is removed from its community. In practice, one therefore evaluates the modularity change by removing i from its community and then by moving it into a neighboring community.

The second phase of the algorithm consists in building a new network whose vertices are now the communities found during the first phase. To do so, the weights of the links between the new vertices are given by the sum of the link weights vertices in the corresponding two communities [4]. Links between vertices of the same community lead to self-loops for this community in the new network. Once this second phase is completed, it is then possible to reapply the first phase of the algorithm to the resulting weighted network and to iterate.

This simple algorithm has several advantages. First, the procedure is intuitive and easy to implement, and the outcome is unsupervised. Moreover, the algorithm is extremely fast, i.e., computer simulations on large ad-hoc modular networks suggest that its complexity is linear on typical and sparse data. This is due to the fact that the possible gains in modularity are easy to compute with the above formula and that the number of communities decreases drastically just after a few passes so that most of the running time is concentrated on the first iterations. The so-called resolution limit problem of modularity is also circumvented due to the intrinsic multi-level nature of the algorithm.

6.3.3 Spectral Bisection Method

Spectral graph theory is concerned with graph properties such as its characteristic polynomial, eigenvalues, and eigenvectors of matrices associated to the adjacency matrix or the Laplacian matrix of the graph. We define the spectrum of a finite graph $\mathcal{G}$ as the spectrum of the adjacency matrix $\mathbf{A}$, that is, its set of eigenvalues and their multiplicities together with the set of orthonormal eigenvectors. The Laplace spectrum of a finite undirected graph without loops is the spectrum of the Laplace matrix $\mathbf{L}$.

An undirected network with real-valued edges, for example, has a symmetric adjacency matrix and therefore has real eigenvalues. The set of all of these eigenvalues and the corresponding complete set of orthonormal eigenvectors make up the graph spectrum. While the adjacency matrix depends on the vertex labeling or ordering, its spectrum is graph invariant. The spectral bisection method is one type of algorithm that falls into this category.

Spectral methods for graph partitioning have been known to be robust but computationally expensive.

The use of spectral methods to compute cuts in graphs was first considered by Donath and Hoffman [18] who first suggested using the eigenvectors of adjacency matrices of graphs to find partitions. Fiedler [12] associated the second smallest eigenvalue of the Laplacian matrix with its connectivity and suggested partitioning the graph by splitting vertices according to their values in the corresponding eigenvector. Thus, the eigenvector corresponding to the second smallest eigenvalue (i.e., the algebraic connectivity) of the Laplacian matrix of a graph $\mathcal{G}$ is termed as the Fiedler vector, while the corresponding eigenvalue, the Fiedler value. Since then, spectral methods for computing and analyzing graph properties have received increasing attention by the community [2, 37, 56, 91].

In one of these spectral methods [54], the spectral bisection method defines the cut size R of a graph partition into two groups as:

$$R = \frac{1}{2} \sum_{i,j \in \mathcal{V}} \mathbf{A}_{ij} \mathbb{1}_{[c_i \neq c_j]}, \tag{6.4}$$

in which the indicator function makes sure that only those edges crossing different communities are considered in the computation of the cut size R.

Consider the index vector s, whose component s_i is $+1$ if vertex i is in one group and -1 if it is in the other group:

$$s_i = \begin{cases} +1, & \text{if vertex } i \text{ belongs to group 1.} \\ -1, & \text{if vertex } i \text{ belongs to group 2.} \end{cases} \tag{6.5}$$

Then, R can be rewritten as:

$$R = \frac{1}{4} \sum_{i,j \in \mathscr{V}} (1 - s_i s_j) \mathbf{A}_{ij}. \tag{6.6}$$

As the degree of vertex i is $k_i = \sum_{j \in \mathscr{V}} \mathbf{A}_{ij}$, then we have $\sum_{i,j \in \mathscr{V}} \mathbf{A}_{ij} = \sum_{i \in \mathscr{V}} k_i = \sum_{i \in \mathscr{V}} s_i^2 k_i = \sum_{i,j \in \mathscr{V}} s_i s_j k_i \mathbb{1}_{[i=j]}$.

Then, R can be rewritten as:

$$R = \frac{1}{4} \sum_{i,j \in \mathscr{V}} s_i s_j (k_i \mathbb{1}_{[i=j]} - \mathbf{A}_{ij}). \tag{6.7}$$

In matrix form, we have:

$$R = \frac{1}{4} s^T \mathbf{L} s, \tag{6.8}$$

in which s^T is the transpose of s and $\mathbf{L} = k_i \mathbb{1}_{[i=j]} - \mathbf{A}_{ij}$ is the Laplacian matrix.

Let us write s as a linear combination of the orthonormal eigenvectors v_i of the Laplacian:

$$s = \sum_{i \in \mathscr{V}} a_i v_i, \tag{6.9}$$

in which $a_i = v_i^T s$. The normalization implies $s^T s = V$ and $\sum_{i \in \mathscr{V}} a_i^2 = V$, where V is the number of network vertices. Then, we have:

$$R = \sum_{i \in \mathscr{V}} a_i v_i^T \mathbf{L} \sum_{j \in \mathscr{V}} a_i v_i = \sum_{i,j \in \mathscr{V}} a_i a_j \lambda_j \mathbb{1}_{[i=j]} = \sum_{i \in \mathscr{V}} a_i^2 \lambda_i, \tag{6.10}$$

in which λ_i is the eigenvalue of $\mathbf{L}$ corresponding to the eigenvector v_i and we have made use of $v_i^T v_j = \mathbb{1}_{[i=j]}$.

Assume that the eigenvalues are labeled in increasing order $\lambda_1 \le \lambda_2 \le \dots \le \lambda_V$. The task of minimizing R can then be equivalently equated as the task of choosing nonnegative quantities a_i^2, in a way to place larger weights to the components that correspond to the smallest eigenvalues in the sum of R.

The sum of every row (and column) of the Laplacian matrix is zero: $\sum_{j \in \mathcal{V}} \mathbf{L}_{ij} = \sum_{j \in \mathcal{V}} (k_i \mathbb{1}_{[i=j]} - \mathbf{A}_{ij}) = \sum_{j \in \mathcal{V}} k_i - k_i = 0$. Thus, the vector $(1, 1, \ldots, 1)$ is always an eigenvector of the Laplacian with eigenvalue zero. The Laplacian is symmetric and hence its eigenvalues are all squares of real vectors, i.e., all eigenvalues of the Laplacian are nonnegative, i.e., $0 = \lambda_1 \leq \lambda_2 \leq \ldots \leq \lambda_V$.

Since the eigenvectors are orthogonal, a good approximate solution can be obtained by choosing s to be as close to parallel with v_2 as possible, i.e., minimizing:

$$|v_2^T s| = \left| \sum_{i \in \mathcal{V}} v_i^{(2)} s_i \right| \leq \sum_{i \in \mathcal{V}} |v_i^{(2)}|. \tag{6.11}$$

A simple choice for defining the clusters ($+1$ or -1) is:

$$s_i = \begin{cases} +1, & \text{if } v_i^{(2)} \geq 0. \\ -1, & \text{if } v_i^{(2)} < 0. \end{cases} \tag{6.12}$$

6.3.4 Community Detection Using Particle Competition

This technique is proposed in [68]. The evolution of this model is very similar to various natural and social processes, such as resource competition, territory exploration by animals, election campaigns, etc. In this model, particles explore a network by combining roles of random and deterministic moving. The investigation of the behavior of this technique reveals that the introduction of a certain level of randomness can yield a big gain in the learning process. This phenomenon is analogous to stochastic resonance in which the performance of a nonlinear deterministic system can be largely enhanced by a certain level of noise. The study shows that learning techniques consisting of only deterministic rules are insufficient. This is because the number of rules required to completely describe even a very specific environment can be prohibitively high. In a dynamical environment, the situation gets worse because the system should keep acquiring new knowledge over time. In this way, a certain level of randomness or chaos is essential for the learning process. The random term models the "I don't know" state and serves as a novelty finder. It can also help learning agents, like particles in this model, in escaping from traps in the physical or learning spaces.

The technique relies on the competition of several particles in a networked environment to identify communities. These particles navigate in the network with the purpose of dominating new vertices, while also trying to defend their previous dominated territory. In the long run, the subsets of vertices that each particle dominates represent the communities.

Two dynamical variables for the j-th particle, denoted by ρ_j, are maintained:

- $\rho_j^v(t)$: it represents the vertex that particle ρ_j is visiting at time t.

- $\rho_j^\omega(t) \in [\omega_{min}, \omega_{max}]$: it indicates the exploration potential of particle ρ_j at time t, where ω_{min} and ω_{max} are scalars that define the minimum and maximum potential that each particle can reach in the learning process, respectively.

The update rules that govern the movement and the exploration potentials of the particles are given by:

$$\rho_j^\omega(t+1) = \begin{cases} \rho_j^\omega(t) & \text{if } v_i^\rho(t) = 0 \\ \rho_j^\omega(t) + (\omega_{max} - \rho_j^\omega(t))\Delta_\rho & \text{if } v_i^\rho(t) = \rho_j \neq 0 \ , \\ \rho_j^\omega(t) - (\rho_j^\omega(t) - \omega_{min})\Delta_\rho & \text{if } v_i^\rho(t) \neq \rho_j \neq 0 \end{cases} \qquad (6.13)$$

in which Δ_ρ controls the exploration level variation that each particle gains or loses, depending on the nature of the vertex that it visits. Specifically, if it visits an already dominated vertex, then the particle's exploration level is strengthened; otherwise, it is decremented. The location of particle ρ_j at $t + 1$, $\rho_j^v(t + 1)$, is determined by sampling from a mixture of deterministic and random walk distributions.

Each vertex v_i in the network is represented by three scalar variables:

- $v_i^\rho(t)$: it defines the proprietary particle of the vertex v_i at time t.
- $v_i^\omega(t)$: it indicates the level of domination imposed by proprietary particle $v_i^\rho(t)$ on vertex v_i at time t.
- $v_i^\gamma(t)$: it symbolizes whether or not vertex v_i is being visited by any of the particles at time t.

With the help of these variables, the dynamical behaviors of the quantities related to the vertices in the network are governed by the following set of equations:

$$v_i^\rho(t+1) = \begin{cases} \rho_j & \text{if } v_i^\gamma(t) = 1 \text{ and } v_i^\omega(t) = \omega_{min} \\ v_i^\rho(t) & \text{otherwise} \end{cases} , \qquad (6.14)$$

$$v_i^\omega(t+1) = \begin{cases} v_i^\omega(t) & \text{if } v_i^\gamma(t) = 0 \\ \max\{\omega_{min}, v_i^\omega(t) - \Delta_v\} & \text{if } v_i^\gamma(t) = 1 \text{ and } v_i^\rho(t) \neq \rho_j \ , \\ \rho_j^\omega(t+1) & \text{if } v_i^\gamma(t) = 1 \text{ and } v_i^\rho(t) = \rho_j \end{cases} \qquad (6.15)$$

in which Δ_v denotes the exploration level fraction lost by a vertex, if a rival particle visits it.

The detection algorithm begins by putting K particles into random vertices. At the beginning of the dynamical process, each particle ρ_j and each vertex v_i have their potentials set to $\rho_j^\omega(0) = \omega_{min}$ and $v_i^\omega(t) = \omega_{min}$, respectively. At each iteration, each particle travels to a neighboring vertex, in accordance with a movement policy that consists in a combination of deterministic and random walks. In the former, the particle randomly visits the neighbors of the currently visited vertex. In the latter, the particle prefers to visit vertices that are already being dominated by the same particle. In the following, we illustrate cases that can be faced when a particle j, ρ_j, is on the process of choosing the next vertex to visit:

- If the visited vertex v_i still does not belong to any particles, then initially $v_i^p(t) = 0$. In this case, such vertex starts to be dominated by the visiting particle, i.e., $v_i^p(t) = \rho_j$. The particle's potential $\rho_j^\omega(t)$ is not altered and the vertex's potential receives the particle's potential: $v_i^\omega(t) = \rho_j^\omega(t)$;
- If the visited vertex is dominated by the same particle, the visiting particle's potential is incremented and v_i receives the new potential of that particle: $v_i^\omega(t) = \rho_j^\omega(t)$;
- If the visited vertex belongs to a rival particle, then the particle's and the vertex's potentials are weakened. If the particle's potential $\rho_j^\omega(t)$ reaches a value lower than ω_{min}, then this particle is reset to a new randomly chosen vertex. If the potential of the vertex $v_j^\omega(t)$ reaches a value lower than ω_{min}, then the vertex becomes no longer owned by the previous particle, i.e., it regresses to the free, non-dominated state: $v_j^\omega(t) = 0$.

Thus, the vertex's level of domination increases if it is visited by the same particle that dominates it at the present moment. In contrast, during the visit of a rival particle, the domination level imposed by the current dominating particle on that vertex is weakened. If this domination is not strong enough, the proprietary particle loses its domination over that vertex. In the long run, it is expected that each particle will dominate a community in the network.

The model proposed in [68] has two noticeable features: (1) high community detection rates and (2) low computational complexity. However, in its original form, only a procedure of particle competition is introduced, without any formal definitions. This precludes any further analyses or predictions on the model's behavior. In Chap. 9, we show a rigorous model for particle competition that is governed by a stochastic competitive dynamical system. That same model is also adapted to a semi-supervised learning environment in Chap. 10, where we also investigate the relevant problem of imperfect learning.

6.3.5 Chameleon

This is a well-known method in the network-based community for data clustering [36]. In general, existing clustering algorithms use static models of the clusters and do not use information about the nature of individual clusters as they are merged or divided. Furthermore, while some schemes ignore the information about the aggregate interconnectivity of data items in two clusters, other schemes ignore information about the closeness of two clusters as defined by the similarity of the closest items across two clusters. By only considering either interconnectivity or closeness, these algorithms can easily select and merge the wrong pair of clusters.

Chameleon is an agglomerative hierarchical clustering algorithm that employs both interconnectivity and closeness features in identifying the most similar pairs of clusters. It is designed to overcome the major limitation of learning methods that assume a static, user-supplied interconnectivity model. Such models are inflexible and can easily lead to incorrect merge decisions when the model under- or overestimates the interconnectivity of the data set or when different clusters exhibit different interconnectivity characteristics. For that, Chameleon uses a combined approach to model the degree of interconnectivity and closeness between each pair of clusters. This approach considers the internal and adaptive characteristics of the clusters themselves. Thus, it does not depend on a static, user-supplied model and can automatically adapt to the internal characteristics of the merged clusters.

Given a vector-based data set, Chameleon first constructs a network using the k-nearest neighbors method, that is, each data sample is represented by a vertex and it is connected to the other k most similar data samples using a similarity metric. Then, Chameleon finds the initial partition of the network using an algorithm that partitions the network into several communities in a way to minimize the edge cut. Since each edge in the k-nearest neighbor graph represents the similarity among data points, a partitioning that minimizes the edge cut effectively minimizes the relationship (affinity) among data points across the partitions. After finding subclusters, Chameleon switches to an algorithm that repeatedly combines these small subclusters, using the cluster similarity measures, which determine the similarity between pairs of clusters by looking at their Relative Interconnectivity (RI) and Relative Closeness (RC). The definitions of these two internal indices are given in the following.

- **Relative interconnectivity.** Relative interconnectivity between clusters C_i and C_j, denoted as $RI(C_i, C_j)$, is defined as the absolute interconnectivity between C_i and C_j normalized with respect to the internal interconnectivity of the two clusters C_i and C_j. The absolute interconnectivity between a pair of clusters C_i and C_j, symbolized as $EC(C_i, C_j)$, is defined as the sum of the weight of the edges that connect vertices in C_i to vertices in C_j. This is essentially the edge-cut of the cluster containing both C_i and C_j such that the cluster is broken into C_i and C_j. The internal interconnectivity of a cluster C_i can easily be captured by the size of its min-cut bisector $EC(C_i)$, which is the weighted sum of edges that partition the graph into two roughly equal parts. Thus, the relative interconnectivity between C_i and C_j is:

$$RI(C_i, C_j) = \frac{|EC(C_i, C_j)|}{\frac{|EC(C_i)| + |EC(C_i)|}{2}}.$$ (6.16)

- **Relative closeness.** The closeness of clusters of C_i and C_j, $RC(C_i, C_j)$, is the average weight of the edges that connect vertices in C_i to those in C_j. It provides a good measure of the affinity between the data items along the interface layer of the two clusters. At the same time, this measure is tolerant to outliers and noise. To get a cluster's internal closeness, we take the average of the edge weights across a min-cut bisection that splits the cluster into two roughly equal parts. The relative closeness between a pair of clusters is the absolute closeness normalized with respect to the internal closeness of the two clusters:

$$RC(C_i, C_j) = \frac{\bar{S}_{EC}(C_i, C_j)}{\frac{|C_i|}{|C_i|+|C_j|}\bar{S}_{EC}(C_i) + \frac{|C_j|}{|C_i|+|C_j|}\bar{S}_{EC}(C_j)}, \qquad (6.17)$$

where $\bar{S}_{EC}(C_i)$ and $\bar{S}_{EC}(C_j)$ are the average weights of the edges that belong in the min-cut bisector of clusters C_i and C_j, and $\bar{S}_{EC}(C_i, C_j)$ is the average weight of the edges that connect vertices in C_i and C_j. Terms $|C_i|$ and $|C_j|$ are the number of data points in each cluster. This equation also normalizes the absolute closeness of the two clusters by the weighted average of the internal closeness of C_i and C_j. This feature discourages merges of small sparse clusters into large dense clusters.

Chameleon selects pairs to merge for which both RI and RC are high. That is, it selects clusters that are well interconnected as well as close together. The merge scheme implemented in Chameleon uses a function to combine the relative inter-connectivity and relative closeness. For this purpose, Chameleon selects the pair of clusters that maximizes

$$RI(C_i, C_j) \times RC(C_i, C_j)^{\alpha}, \qquad (6.18)$$

where α is a user-specified parameter. If $\alpha > 1$, then Chameleon gives a higher importance to the relative closeness, and when $\alpha < 1$, it gives a higher importance to the relative interconnectivity.

The algorithm is well suited for applications in which the volume of the available data is large. For large V, the worst-case time complexity of the algorithm is $\mathcal{O}(V(\log_2 V + M))$, where M is the number of clusters formed after completion of the first phase of the algorithm.

The good performance of the Chameleon is recognized when applied to low-dimensional spaces. However, the performance of Chameleon in high-dimensional spaces is still not thoroughly clarified [87]. The time complexity of the Chameleon algorithm in high-dimensional spaces is $\mathcal{O}(V^2)$.

6.3.6   Community Detection by Space Transformation and Swarm Dynamics

We describe the technique introduced in [17, 61], which is based on collective dynamics. Much interest has been spent in the study of collective motion of biological entities, like schools of fish, flocks of birds, herds of hoof animals or swarms of insects. Swarm behavior is a collective behavior exhibited by animals of similar size that aggregate together, perhaps milling about the same spot or perhaps moving *en masse* or migrating in some direction. The swarm approach seeks methods consisting of a large number of simple and locally interacting agents that collectively present macroscopically complex organizations [29, 81]. Swarm behavior techniques have been successfully applied to solve various optimization problems [14].

The community detection algorithm using space transformation and swarm dynamics uses collective dynamics in a networked environment and consists of two serial steps. In the first step, the method determines how data items are represented as a network. In the second step, it detects clusters or communities by partitioning that constructed network using rules built on neighborhood agreements. This is a divisive hierarchical algorithm, in which we initially consider the entire network as a large cluster and we split it into smaller clusters, until each vertex corresponds to a cluster. Due to its hierarchical nature, we can illustrate the algorithm's result using a dendrogram, a special kind of tree where each vertex represents a cluster. A horizontal cut on the dendrogram represents a partition of the data set.

We summarize these two steps in the following:

1. *Network formation*: In this step, a weighted complete network is constructed using the input data set, in which each vertex represents a data sample. Then, a non-weighted network is generated using the k-NN method, i.e., each vertex is connected to its k most similar vertices. The similarity is determined by calculating the Euclidean distance between pairs of data samples.[1]
2. *Angle's updating rule*: After the network is constructed, the algorithm organizes the vertices on a circle. The displacement of vertices is conducted in a random manner. Thus, each vertex v_i has an initial angle $\theta_i(t = 0)$ that is randomly chosen over the range $[0, 2\pi)$. While the angle's updating rule approximates vertices that belong to the same cluster, it also separates vertices that belong to different clusters. At each time step t, the method updates the angle of each vertex according to the angles of its neighbors. We define the angle's updating rule by the following equation:

$$\theta_i(t + 1) = \theta_i(t) + \eta_i(t) \left[\frac{\sum_{j \in \mathcal{N}(v_i)} \mathbf{A}_{ij} \theta_j(t)}{\sum_{j \in \mathcal{N}(v_i)} \mathbf{A}_{ij}} - \theta_i(t) \right], \tag{6.19}$$

in which $\mathcal{N}(v_i)$ is the set of neighbors of vertex v_i, $\eta_i(t)$ is the moving rate of v_i at time step t, and $\mathbf{A}_{ij}$ is the weight that represents the influence of neighbor v_j on v_i.

The edge weight $\mathbf{A}_{ij}$ aims at approximating vertices that belong to the same cluster. It is composed of two parts: $CN(v_i, v_j)$ and $SN(v_i, v_j)$. Mathematically, $\mathbf{A}_{ij}$ is expressed as:

$$\mathbf{A}_{ij} = CN(v_i, v_j) \times SN(v_i, v_j). \tag{6.20}$$

The idea of the term $CN(v_i, v_j)$ is to model physical proximity between v_i and v_j. As such, it gives more importance to vertex v_j the closer v_i and v_j are. This

---

[1] See Chap. 4 for a thorough review on network formation methods and similarity functions.

kind of behavior can effectively be captured by modeling $CN(v_i, v_j)$ according to the following rule:

$$CN(v_i, v_j) = e^{-\alpha d(v_i, v_j)}, \tag{6.21}$$

in which parameter α controls for the penalization decay rate of the Euclidean distance $d(v_i, v_j)$ from v_i to v_j. The algorithm can change the relative importance of a neighbor by adjusting α. The angle's updating rule can also be applied to non-weighted networks. In this case, $CN(v_i, v_j) = 1$ for all of the pairs of neighbors v_i and v_j.

In contrast to that, the term $SN(v_i, v_j)$ models the topology similarity between v_i and v_j. The hypothesis is: whenever two vertices belong to the same cluster, they are likely to share a large number of common neighbors. With that in mind, we can write $SN(v_i, v_j)$ as follows:

$$SN(v_i, v_j) = \frac{c(v_i, v_j)}{|\mathcal{N}(i)|}, \tag{6.22}$$

in which $c(v_i, v_j)$ is the number of common neighbors shared by v_i and v_j and $|\mathcal{N}(i)|$ is the number of neighbors of i. In this way, $SN(v_i, v_j)$ yields large values for vertices that share a large portion of common neighbors, regardless of the physical distance. Conversely, if they share only a small fraction of common neighbors, $SN(v_i, v_j)$ outputs small values.

Intuitively, the term $CN(v_i, v_j)$ forces angles of neighbor vertices to approximate to that of v_i and $SN(v_i, v_j)$ stops such an approximation between pairs of vertices that possibly belong to different clusters. However, these two mechanisms still cannot eliminate interference between different groups, which may cause the angles of all of the network vertices to approach each other. To mitigate this problem, one solution is to reduce the moving rate $\eta_i(t)$ in (6.19) as a function of how quickly the angles change as follows:

$$\eta_i(t) = \exp-\left(\frac{\beta}{\sigma(v_i)}\right), \tag{6.23}$$

in which $\sigma(v_i)$ is the standard deviation of the angle distribution and β is a user-defined parameter to scale the updating process of $\eta_i(t)$ as a function of $\sigma(v_i)$.

The moving rate parameter $\eta_i(t)$ decreases as the standard deviation $\sigma(v_i)$ among angles decreases. At the beginning, each angle takes a random value. In this way, the standard deviation of the angles distribution $\sigma(v_i)$ is expected to be high in such a way that $\eta_i(t)$ assumes large values, say $\eta_i(t) \approx 1$. In this situation, angles of neighboring vertices approximate freely to form angle bands. As time progresses, $\sigma(v_i)$ and consequently $\eta_i(t)$ assume smaller values. When $\eta_i(t)$ reaches a very small value, say $\eta_i(t) \approx 0$, all of the angles remain steady and a stable state is reached.

To illustrate the algorithm, we use a random clustered network with three unbalanced communities. We inspect the evolution of the angle's update process in Fig. 6.4. In this case, the algorithm identifies three communities in the network as there are three perceptive angle bands in the time series.

Now, we see the performance of the method in a real-world data set, which is a social network describing the associations (interactions) among dolphins [46]. This network has 62 vertices and 159 edges without weights. It presents two well-known communities, formed by 21 and 41 elements, respectively. Inspecting how the vertices' angles are updated in Fig. 6.5, it is possible to identify two distinct groups or communities of angles. In Fig. 6.6, we see the same simulation results but through a dendrogram perspective, where the color of each vertex indicates the community to which it originally belongs.

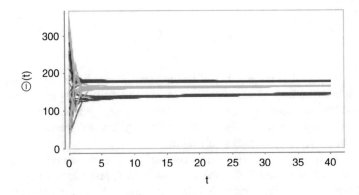

Fig. 6.4 Evolution of the angle's updating process. In the first iterations, the vertices' angles are disordered due to the random arrangements. After some iterations, they converge to stable subgroups. Reproduced from [62] with permission from the author

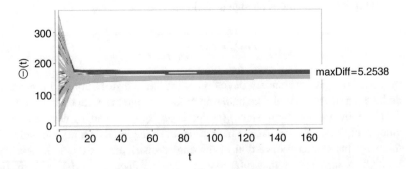

Fig. 6.5 Evolution of the angle's updating process for the social network presented in [46]. Two communities can be clearly identified by inspecting the time series. Reproduced from [62] with permission from the author

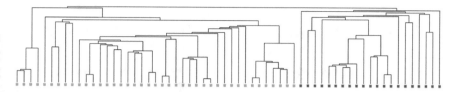

Fig. 6.6 Dendrogram showing the community detection results for the social network presented in [46]. The *dendrogram* reveals the division of data into two original communities, indicated by 41 *green* elements and 21 *red* elements. Reproduced from [62] with permission from the author

6.3.7   Synchronization Methods

Physicists have given increasing attention to the dynamics of a diversity of complex systems. In special, several studies have investigated the paradigmatic analysis of large populations of coupled oscillators [40, 64, 78, 85]. The emergence of synchronization patterns in these systems is closely related to the underlying topology of interactions. In this section, we discuss methods that rely on dynamical processes towards synchronization. In this respect, these models show different patterns over time that are intrinsically connected to the hierarchical organization of communities in complex networks. The ubiquity of synchronization phenomena in the real world makes this approach interesting from a physical and biological perspectives [3].

One of the most successful attempts to understanding synchronization phenomena comes from Kuramoto [40], who analyzed a model of phase oscillators coupled by the sine of their phase differences. The model is rich enough to display a large variety of synchronization patterns and sufficiently flexible to be adapted to many different contexts [1].

The Kuramoto model consists of V coupled phase oscillators, in which the phase of the i-th unit, denoted by $\theta_i(t)$, evolves in time according to the following dynamic:

$$\frac{d\theta_i}{dt} = \omega_i + \sum_{j \in \mathcal{V}} \mathbf{A}_{ij} \sin(\theta_j - \theta_i), \qquad (6.24)$$

for $i \in \mathcal{V}$. The term ω_i stands for the natural frequency of the i-th oscillator and $\mathbf{A}_{ij}$ describes the coupling between units. The coupling weights are extracted from a network, in which each vertex is an oscillator and edge weights denote the coupling strength between different oscillators.

In particular, some works have shown that highly interconnected sets of oscillators synchronize more easily that those with sparse connections [48, 60]. This scenario suggests that, for a complex network with nontrivial connectivity patterns, starting from random initial conditions, those highly interconnected units forming local clusters will synchronize first. Then, in a sequential process, larger and larger spatial structures will do the same until we reach a final state in which the entire

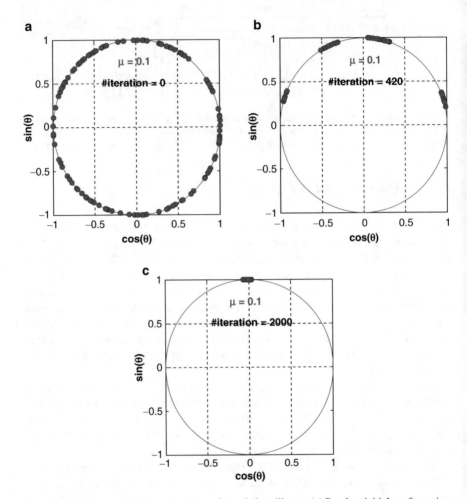

Fig. 6.7 States assumed by the population of coupled oscillators. (**a**) Random initial configuration; (**b**) Intermediate state (four communities); (**c**) Final global synchronization state

population has the same phase. This process is expected to occur at different time scales whenever clear community structures exist. Thus, the dynamical route towards the global attractor reveals different topological structures, presumably those which represent communities.

For an artificial random clustered network with four communities, Figs. 6.7a–c show, respectively, the initial configuration of the oscillators, the formation of four synchronized communities of oscillators, and the global synchronization state.

Li et al. [44] has shown that communities are delineated by interface or overlapping vertices [63], in which the oscillating frequency is intermediate among different modules, in such a way that synchronization techniques cannot clearly group these interface vertices into a single community. From this reported shortcoming,

Wu et al. [86] has developed an alternative method that is capable of detecting these overlapping vertices. Contrasting to the result of Arenas et al. [3], in which the stable state is necessarily reached only by a global synchronization, in the research of Wu et al. [86], the synchronization may occur within modules. Thus, after the method synchronizes the oscillators between different communities, we can understand the phases that are in the valley between different modules to be the overlapping vertices. In order to do so, besides the global coupling supplied by the traditional Kuramoto model, another type of coupling, which is negative, between oscillators that are not connected is applied. The network dynamic can be mathematically expressed as:

$$\frac{d\theta_i}{dt} = \omega_i + \frac{K_p}{V} \sum_{j \in \mathscr{V}} \mathbf{A}_{ij} \sin(\theta_j - \theta_i) - \frac{K_n}{V} \sum_{j \in \mathscr{V}} (1 - \mathbf{A}_{ij}) \sin(\theta_j - \theta_i). \quad (6.25)$$

In this adapted format, the phases of the interconnected oscillators i and j are modeled by a positive coupling (with coupling strength K_p) in accordance with the original expression in (6.24). Thus, their phases evolve together. Non-connected vertices in the network, in contrast, tend to have opposite phases, on account of the negative coupling forced by K_n. In summary, after reaching the dynamic equilibrium, oscillators that make up the same community in the network will indicate similar phase values. Opposed to that, oscillators that represent overlapping vertices will have their phases in-between different modules [44].

6.3.8   Finding Overlapping Communities

The community structure is a fundamental property of most real-world networks, i.e., it is commonly observed that groups of vertices are densely interconnected. It would be oversimplifying, however, if we assumed that communities are well-defined partitions over the entire network. This is a strong assumption that may not be fulfilled in many cases. First, it is very natural for a vertex to participate in more than one community at a time; i.e., communities often overlap. Second, some vertices might not participate in any community; i.e., we might have outliers [33]. An outlier is not necessarily solitary, and it might have some negligible connection with some communities. Finally, some vertices of a community might be special in the sense that they are linked with almost all of the others. In the literature, these vertices are known as hubs, leaders, or centers. Since many real-world networks are huge, the analysis usually starts from the identification of the underlying communities possibly with overlapping characteristics. Needless to say, the community structure will greatly benefit from the simultaneous detection of hubs and outliers [9].

We can easily find overlapping communities in real-world networks. A person, for instance, can be member of a social network, of his/her family community, and also of his/her institutional community. In community detection or network-based data clustering, the detection of overlapping communities is specially interesting for fuzzy clustering.

In this section, we present some popular overlapping community detection techniques.

6.3.8.1   Clique Percolation

The most popular community detection with overlapping vertices is the Clique Percolation Method (CPM) [63]. CPM relies on the assumptions that communities consist of overlapping sets of fully connected subgraphs and that it is unlikely the existence of cliques in intercommunity edges. The general idea of the method is to detect communities by searching for adjacent cliques. It begins by identifying network cliques of size k, termed as k-cliques. Once these have been identified, a new collapsed graph is constructed in such a way that each vertex represents each of these k-cliques. Two vertices in the collapsed graph are connected if the k-cliques that represent them share $k - 1$ members. In this case, we say that these two k-cliques are adjacent. The union of adjacent k-cliques is called k-clique chain. Finally, a k-clique community is the largest connected subgraph obtained by the union of a k-clique and of all k-cliques that are connected to it.

Since a vertex can be in multiple k-cliques simultaneously, the identification of overlapping communities is possible. CPM is suitable for networks with densely connected parts. Empirically, small values of k (typically between 3 and 6) often give good results [30, 42, 63].

CPM has been extended to weighted, directed, and bipartite graphs. For weighted graphs, in principle, one can follow the standard procedure of thresholding the edge weights, and of applying the method on the resulting graph, treating them as non-weighted. Farkas et al. [20] has proposed to threshold the weight of cliques, defined as the geometric mean of the weights of all edges of the clique. The value of the threshold is chosen slightly above the critical value at which a giant k-clique community emerges, in order to get the richest possible variety of clusters.

CPM has a notable drawback in that it assumes that the network has a large number of cliques [22]. As such, CPM may fail to give meaningful covers for graphs with few cliques, like technological networks and some social networks. In contrast, if the network presents many cliques, the method may deliver trivial community structure, like a cover consisting of the entire network as a single giant cluster. A more fundamental issue is the fact that the method does not look for actual communities, consistent with the shared notion of dense subgraphs, but for subgraphs "containing" many cliques, which may be quite different objects than communities. (For instance, they could be "chains" of cliques with low internal edge density.) Another problem is that there are considerable fractions of vertices in real networks that are left out of the communities, like leaves or singletons. One could think of some postprocessing procedure to include them in the communities, but for that it is necessary to introduce a new criterion, outside the framework that inspired the method. Furthermore, besides empirical work, it is not clear *a priori* which values of k one has to choose to identify meaningful structures. Finally, the criterion to choose the threshold for weighted networks and the definition of directed k-cliques are also rather arbitrary.

6.3.8.2 Bayesian Nonnegative Matrix Factorization Algorithm

This method has been described in various works [9, 21, 67, 75]. It relies on a centrality matrix of vertices and a degree matrix of communities. The importance of a vertex to a community is represented by its centrality. The centrality matrix, hence, carries the vertices' importance in each community. An element of the degree matrix of communities, which is diagonal, indicates the degree of the community, and is equivalent to the summation of the expected degree of all vertices of that community. The algorithm then learns these two quantities by the multiplicative updating rule using a nonnegative matrix factorization style. These matrices enable us to rank each vertex's centrality in each community, and use the community degree as a cutting off criterion. Since the communities are retrieved independently, when we are working on a new community, we do not need to care whether or not its vertices belong to previously identified communities. The overlapping communities are thus handled naturally. The importance of a hub in a community ensures that it gets ranked at the top of the community. After all of the communities have been decided, those vertices that have not been included in any of them are declared as outliers. In summary, this algorithm is capable of identifying overlapping communities as well as detecting hubs and outliers simultaneously.

Mathematically, Bayesian nonnegative matrix factorization is an adaptation of the nonnegative matrix factorization technique used in machine learning for dimensionality reduction and feature extraction [89]. This technique factorizes the matrix $V \in \mathbb{R}_+^{V \times V}$ into two matrices $W \in \mathbb{R}_+^{V \times M}$ and $H \in \mathbb{R}_+^{M \times V}$, whose elements are nonnegative, such that $A \approx WH$. Within the context of community detection, A is the adjacency matrix of the network, V is the number of vertices and M is the pre-defined number of communities. Each element of the i-th line or the j-th column of matrix W is the statistical dependence between a vertex i to community j. Due to matrix multiplication, the traditional nonnegative matrix factorization procedure is inefficient with respect to time and memory restrictions. In [9], a hybrid optimization algorithm that relies on a Bayesian optimization process is proposed. In essence, this algorithm optimizes an objective function expressed in terms of the above-mentioned matrices and user-supplied parameters $\beta \in \mathbb{R}^M = [\beta_1, \ldots \beta_M]$, which represent the importance of communities on the interactions of the adjacency matrix. The algorithm involves consecutive updates of W, H, and β until these parameters achieve convergence or until a maximum number of iterations is processed. Matrices W, H and the user-supplied parameters β are calculated as follows:

$$H = \left(\frac{H}{W^T 1 + BH} \right) \cdot \left[W^T \left(\frac{V}{WH} \right) \right], \tag{6.26}$$

$$\mathbf{W} = \left(\frac{\mathbf{W}}{\mathbf{1}\mathbf{H}^{\mathrm{T}} + \mathbf{W}\mathbf{B}} \right) \cdot \left[\left(\frac{\mathbf{V}}{\mathbf{W}\mathbf{H}} \right) \mathbf{H}^{\mathrm{T}} \right], \tag{6.27}$$

$$\beta_i = \frac{V + a - 1}{\frac{1}{2} \left(\sum_i \mathbf{W}_{ik}^2 + \sum_j \mathbf{W}_{ij}^2 \right) + b}, \tag{6.28}$$

in which a and b are fixed parameters of a Gamma distribution and matrices $\mathbf{W}$, $\mathbf{H}$ are initialized with random values. Finally, the columns of $\mathbf{W}$ (or lines $\mathbf{H}$) containing elements with only zero values are removed and the number of communities is given by the number of columns of $\mathbf{W}$ (or the number of lines of $\mathbf{H}$) obtained after the removal.

6.3.8.3 Fuzzy Partition Algorithm

The fuzzy partition algorithm is introduced in [49]. The procedure runs as a constrained optimization problem. In that study, the expression "overlapping vertices" is conceived as "bridges" in the context of social networks, in which it is very common to find individuals that are members of multiple communities at the same time. In a social network context, "bridges" then can be defined as those vertices that cross structural holes between discrete groups of people [8]. It is therefore important to define a quantity that measures the commitment of a vertex to several communities in order to obtain a more realistic view of these networks.

The intuitive meaning of a bridge vertex may differ in different types of networks that exist beyond sociometrics. In protein interaction networks, proteins with multiple roles can be seen as bridge vertices. In cortical networks containing brain areas responsible for different modalities, the cortical areas that assume integrative roles and that provide higher level processing of sensory signals are the bridges vertices. In word-association networks, words with multiple meanings are likely to be bridges.

The overlapping condition is modeled via a fuzzy partition algorithm. A convenient representation of a given partition is the partition matrix $\mathbf{U} = [u_{ik}]$, where i indexes for the fuzzy membership across clusters k, for the data items. In this way, matrix $\mathbf{U}$ has V columns and M rows, where M is the number of subsets or clusters. We observe that $u_{ik} = 1$ if and only if vertex k belongs to the i-th subset in the partition; otherwise, it is zero. For a complete partitioning algorithm, $\sum_{i=1}^{M} u_{ik} = 1$, $\forall k \in \{1, \ldots, V\}$ must hold. The size of community i can then be calculated as $\sum_{k=1}^{V} u_{ik}$, and for any meaningful partition, we can assume that $0 < \sum_{k=1}^{V} u_{ik} < V$. These partitions are traditionally called hard or crisp partitions, because a vertex can belong to one and only one of the detected communities.

The generalization of the hard partition follows by allowing u_{ik} to attain any real value from the interval $[0, 1]$. The constraints imposed on the partition matrix remain the same.

It should be observed that a meaningful partition should group vertices that are somehow similar to each other in the same community. It is reasonable to assume that an edge between vertex v_1 and v_2 implies the similarity of v_1 and v_2, and likewise, the absence of an edge implies dissimilarity. Define $s(\mathbf{U}, i, j)$ as a similarity function that respects the following restrictions:

- $s(\mathbf{U}, i, j) \in [0, 1]$;
- $s(\mathbf{U}, i, j)$ is continuous and differentiable $\forall u_{ij}, i \in \{1, \ldots, c\}, j \in \{1, \ldots, N\}$;
- $s(\mathbf{U}, i, j)$ increases as i and j are more similar. Therefore, $s(\mathbf{U}, i, j)$ assumes its maximum value, $s(\mathbf{U}, i, j) = 1$, when i and j are as similar as possible. Conversely, $s(\mathbf{U}, i, j) = 0$ when i and j are totally dissimilar.

As a shorthand, consider that $s_{ij} = s(\mathbf{U}, i, j)$. Suppose we have a prior assumption about the actual similarity of vertices i and j, denoted by $\tilde{s}_{ij}$. Define the fitness of a given partition $\mathbf{U}$ of the graph by quantifying how precisely it approximates the prescribed similarity values to s_{ij}:

$$D_G(\mathbf{U}) = \sum_{i=1}^{V} \sum_{j=1}^{V} w_{ij} (\tilde{s}_{ij} - s_{ij})^2, \tag{6.29}$$

in which w_{ij} are optional weights. Say that $\mathbf{W} = [w_{ij}]$, $\mathbf{S}(\mathbf{U}) = [s_{ij}]$, and $\tilde{\mathbf{S}}(\mathbf{U}) = [\tilde{s}_{ij}]$. From now on, we assume that $\tilde{\mathbf{S}} = \mathbf{A}$, the adjacency matrix of the graph, which is in accordance with our assumption that the similarity of connected vertex pairs should be close to 1 and the similarity of disconnected vertex pairs should be close to zero. Consider that the similarity function s_{ij} is given as follows:

$$s_{ij} = \sum_{i=1}^{V} \sum_{k=1}^{M} u_{ki} u_{kj} = \mathbf{U}^T \mathbf{U}. \tag{6.30}$$

The community detection problem in this framework boils down to the optimization of $D_G(\mathbf{U})$ defined in accordance with (6.29). We note that the goal is to find a matrix $\mathbf{U}$ such that it minimizes $D_G(\mathbf{U})$. The number of clusters c, the weight matrix $\mathbf{W}$ and the desired similarity matrix $\mathbf{S}$, which is often the adjacency matrix of the network, are supplied by the user. This is a nonlinear constrained optimization problem. Although there exists a set of necessary conditions that restrict the set of possible matrices $\mathbf{U}$ worth evaluating [73], the computationally most feasible approach to optimize $D_G(\mathbf{U})$ is to use a gradient-based iterative optimization method (e.g., simulated annealing).

Consider the following objective function:

$$D_G(\mathbf{U}) = \sum_{i=1}^{V} \sum_{j=1}^{V} w_{ij} (\tilde{s}_{ij} - s_{ij})^2 + \sum_{i=1}^{V} \lambda_i \left(\sum_{k=1}^{M} u_{ki} - 1 \right), \qquad (6.31)$$

in which $\lambda = [\lambda_1, \ldots, \lambda_N]$ are Lagrangian multipliers that simply force the total membership degree for each vertex to be 1 (complete partitioning).

Now we need to find $\mathbf{S}$ to minimize $D_G(\mathbf{U})$ satisfying the above constraints. The partial derivative of $D_G(U)$, with respect to u_{kl} is therefore:

$$\frac{\partial D_G(\mathbf{U})}{\partial u_{kl}} = 2 \sum_{i=1}^{V} (e_{il} + e_{li}) \left(\frac{1}{M} - u_{ki} \right), \qquad (6.32)$$

in which $e_{ij} = w_{ij}(\tilde{s}_{ij} - s_{ij})$.

The simplest gradient-based algorithm for finding a local minimum of D_G is then the following:

1. Start from an arbitrary random partition $\mathbf{U}(0)$ and let $t = 0$.
2. Calculate the gradient vector of D_G according to (6.32) and the current $\mathbf{U}(t)$.
3. If $\max_{k,l} \left| \frac{\partial D_G(U)}{\partial u_{kl}} \right| < \epsilon$, stop the iteration and declare $\mathbf{U}(t)$ a solution.
4. Otherwise, calculate the next partition in the iteration with the following equation:

$$u_{ij}^{(t+1)} = u_{ij}^{(t)} + \alpha^{(t)} \frac{\partial D_G(U)}{\partial u_{ij}}, \qquad (6.33)$$

in which $\alpha^{(t)}$ is a small step size constant chosen appropriately.
5. Increase t and continue from step 2.

6.3.9 Network Embedding and Dimension Reduction

Dimension reduction is an important pre-processing in data analysis and machine learning. It can be considered as a procedure to produce a compact low-dimensional encoding of a given high-dimensional data set [47, 74, 84]. Dimension reduction is specially interesting when we deal with data sets that have many more variables than data samples. For example, microarray data sets usually are composed by thousands of variables (genes) in dozens of samples. The most famous dimension reduction technique is the Principal Component Analysis (PCA) that dates back to Karl Pearson in 1901 [65]. The basic idea is to find a new coordinate system via a linear or a nonlinear transformation in which the input data can be expressed with many less variables without a significant loss. Isomap [80] was originally proposed as a generalization of multidimensional scaling [15]. An alternative method known as Locally Linear Embedding (LLE) [72] was developed that solved a consecutive pair of linear least square optimizations. The kernel method, including graph kernel method, has also been proposed for nonlinear dimension reduction by performing linear operations on kernel mapping functions. Graph kernel methods for data analysis and machine learning are an active research topic and are not covered in this book. The interested readers may refer to [5, 27, 45, 50, 76, 82]. In this book, we just present one technique on this topic. In [88], a graph embedding method has been proposed and is briefly reviewed in the following paragraph.

Consider that it is given a data set $\mathscr{X} = \{x_1, x_2, \ldots, x_V\}$. Each data sample is described by P attributes, that is, a feature vector $x_i = (x_{i1}, x_{i2}, \ldots, x_{iP})^T$. Consider $\mathbf{X}$ as the matrix whose columns denote each data item in $\mathscr{X}$. The goal of the technique is to perform dimensionality reduction in the data items' feature vectors to a smaller number P' of projected attributes. For example, the feature dimension P of images is usually very high, and transforming the data from the original high-dimensional space to a low-dimensional space can alleviate the curse of dimensionality problem. To accomplish that, a technique should find a mapping function F that transforms each feature vector $x \in \mathbb{R}^P$ into the desired low-dimensional representation y, so that $y = F(x)$, $y \in \mathbb{R}^{P'}$. By using an underlying network to find such function F, the dimensionality reduction process can be viewed as a graph-preserving criterion of the following form:

$$Y^* = \arg \min_Y \sum_{\substack{i,j \in \mathscr{V} \\ i \neq j}} \mathbf{A}_{ij} \|y_i - y_j\|^2$$

$$= \arg \min_Y Y^T \mathbf{L} Y, \tag{6.34}$$

constrained to $Y^T \mathbf{B} Y = d$. In this formulation, d is a constant vector, $\mathbf{A}$ is the adjacency matrix of the network, $\mathbf{B}$ is the constraint matrix, and $\mathbf{L}$ is the Laplacian matrix. Recall that the Laplacian matrix can be found via the following operation:

$$\mathbf{L} = \mathbf{D} - \mathbf{A}, \tag{6.35}$$

in which:

$$\mathbf{D}_{ii} = \sum_{\substack{j \in \mathscr{V} \\ j \neq i}} \mathbf{A}_{ij}, \tag{6.36}$$

$\forall i \in \mathscr{V}$.

The constraint matrix $\mathbf{B}$ can be viewed as the adjacency matrix of a penalty network $\mathbf{A}^P$, so that $\mathbf{B} = \mathbf{L}^P = \mathbf{D}^P - \mathbf{A}^P$. The penalty network conveys information about which vertices should not be linked together, that is, which instances should be far apart after the dimensionality reduction process. The similarity preservation property from the graph-preserving criterion has a twofold explanation. For larger similarity between samples x_i and x_j, the distance between y_i and y_j should be smaller to minimize the objective function. Likewise, smaller similarity between x_i and x_j should lead to larger distances between y_i and y_j for minimization. Assume that the low-dimensional attribute space can be found by using a linear projection such as $Y = \mathbf{X}^T w$, where w is the projection vector. The objective function in (6.34) becomes:

$$w^* = \arg\min_w \sum_{\substack{i,j \in \mathcal{V} \\ i \neq j}} \mathbf{A}_{ij} \| w^T x_i - w^T x_j \|^2$$

$$= \arg\min_w w^T \mathbf{X}^T \mathbf{L} \mathbf{X} w, \qquad (6.37)$$

constrained to $w^T \mathbf{X}^T \mathbf{L} \mathbf{X} w = d$. By using the Marginal Fisher Criterion and the penalty network constraint, Eq. (6.35) becomes:

$$w^* = \arg\min_w \frac{w^T \mathbf{X}^T \mathbf{L} \mathbf{X} w}{w^T \mathbf{X} \mathbf{L}^P \mathbf{X}^T w}, \qquad (6.38)$$

which can be solved by the generalized eigenvalue problem by using the equation $\mathbf{X} \mathbf{L} \mathbf{X}^T w = \lambda \mathbf{X} \mathbf{L}^P \mathbf{X}^T w$.

6.4 Chapter Remarks

Clustering is the unsupervised grouping of patterns, such as observations, data items, or feature vectors. The clustering task has been addressed in many contexts and by researchers in many disciplines; this diversity reflects its broad appeal and usefulness as one of the steps in exploratory data analysis. Intuitively, patterns within the same cluster are more similar to each other than they are to a pattern belonging to a different cluster. Clustering is useful in several exploratory tasks, such as in data mining, document retrieval, image segmentation, and pattern classification. Often, there is little prior information (e.g., statistical models) available about the data, and the learning algorithm must make as few assumptions about the data as possible. It is under these restrictions that clustering methodology is particularly appropriate for the exploration of interrelationships among the data points.

In this chapter, we have focused on data clustering in a networked environment, which is often termed as community detection. The study of community detection is very important for understanding various phenomena in complex networks. Modular structure introduces important heterogeneities in complex networks. Each module, for example, can have different local statistics; some modules may have many connections, while other modules may be sparse. When there is large variation among communities, global values of statistical measures can be misleading. The presence of modular structure may also alter the way in which dynamical processes unfold on the network. In biological networks, communities correspond to functional modules in which module members function coherently to perform essential cellular tasks. Hence, the development of efficient community detection methods stands as an important topic in the agenda for the complex network and machine learning communities. Due to that importance, this chapter has dedicated a great part of it to the study of several representative community detection algorithms. For each of them, we have explained the main idea behind the community detection mechanism and also the potentialities and shortcomings of the methods. Community detection benchmarks have also been explored.

The topic of detection of overlapping communities has also been discussed. We can easily find overlapping communities in real-world networks. A person, for instance, can belong to a social network, in his/her family community, and also in his/her institutional community. In network-based unsupervised learning, the detection of overlapping communities is especially interesting for fuzzy clustering. Some representative methods have also been explored.

References

1. Acebrón, J.A., Bonilla, L.L., Vicente, P.C.J., Ritort, F., Spigler, R.: The kuramoto model: A simple paradigm for synchronization phenomena. Rev. Mod. Phys. **77**, 137–185 (2005)
2. Alpert, C.J., Kahng, A.B., Yao, S.Z.: Spectral partitioning with multiple eigenvectors. Discret. Appl. Math. **90**(1-3), 3–26 (1999)
3. Arenas, A., Guilera, A.D., Pérez Vicente, C.J.: Synchronization reveals topological scales in complex networks. Phys. Rev. Lett. **96**(11), 114102 (2006)
4. Arenas, A., Duch, J., Fernández, A., Gómez, S.: Size reduction of complex networks preserving modularity. New J. Phys. **9**(6), 176 (2007)
5. Borgwardt, K.M.: Graph kernels. Ph.D. thesis, Ludwig-Maximilians-Universität München, Germany (2007)
6. Brandes, U., Delling, D., Gaertler, M., Görke, R., Hoefer, M., Nikoloski, Z., Wagner, D.: On modularity clustering. IEEE Trans. Knowl. Data Eng. **20**(2), 172–188 (2008)
7. Buchanan, M.: Nexus: Small Worlds and the Groundbreaking Theory of Networks. W.W. Norton, New York (2003)
8. Burt, R.S.: Structural holes: the social structure of competition. Harvard University Press, Cambridge, MA (1992)
9. Cao, X., Wang, X., Jin, D., Cao, Y., He, D.: Identifying overlapping communities as well as hubs and outliers via nonnegative matrix factorization. Sci. Rep. **3**, 2993 (2013)
10. Chen, J., Yuan, B.: Detecting functional modules in the yeast protein–protein interaction network. Bioinformatics **22**(18), 2283–2290 (2006)
11. Chen, M., Kuzmin, K., Szymanski, B.: Community detection via maximization of modularity and its variants. IEEE Trans. Comput. Soc. Syst. **1**(1), 46–65 (2014)
12. Chung, F.R.K.: Spectral Graph Theory. CBMS Regional Conference Series in Mathematics, vol. 92. American Mathematical Society, Providence, RI (1997)
13. Clauset, A., Newman, M.E.J., Moore, C.: Finding community structure in very large networks. Phys. Rev. E **70**(6), 066111+ (2004)
14. Clerc, M., Kennedy, J.: The particle swarm - explosion, stability, and convergence in a multidimensional complex space. IEEE Trans. Evol. Comput. **6**(1), 58–73 (2002)
15. Cox, T.F., Cox, M.: Multidimensional Scaling. Chapman & Hall/CRC, London/Boca Raton (2000)

16. Danon, L., Díaz-Guilera, A., Duch, J., Arenas, A.: Comparing community structure identification. J. Stat. Mech. Theory Exp. **2005**(09), P09008 (2005)
17. de Oliveira, T., Zhao, L.: Complex network community detection based on swarm aggregation. In: International Conference on Natural Computation, vol. 7, pp. 604–608. IEEE, New York (2008)
18. Donath, W.E., Hoffman, A.J.: Lower bounds for the partitioning of graphs. IBM J. Res. Dev. **17**(5), 420–425 (1973)
19. Evans, T.S., Lambiotte, R.: Line graphs, link partitions, and overlapping communities. Phys. Rev. E **80**(1), 016105 (2009)
20. Farkas, I., Ábel, D., Palla, G., Vicsek, T.: Weighted network modules. New J. Phys. **9**(6), 180 (2007)
21. Févotte, C., Bertin, N., Durrieu, J.L.: Nonnegative matrix factorization with the itakura-saito divergence: with application to music analysis. Neural Comput. **21**(3), 793–830 (2009)
22. Fortunato, S.: Community detection in graphs. Phys. Rep. **486**, 75–174 (2010)
23. Fortunato, S., Barthélemy, M.: Resolution limit in community detection. Proc. Natl. Acad. Sci. **104**(1), 36–41 (2007)
24. Fortunato, S., Latora, V., Marchiori, M.: Method to find community structures based on information centrality. Phys. Rev. E **70**(5), 056104 (2004)
25. Freeman, L.C.: A set of measures of centrality based upon betweenness. Sociometry **40**, 35–41 (1977)
26. Frey, B.J., Dueck, D.: Clustering by passing messages between data points. Science **315**, 972–976 (2007)
27. Gärtner, T.: A survey of kernels for structured data. SIGKDD Explor. **5**(1), 49–58 (2003)
28. Girvan, M., Newman, M.E.J.: Community structure in social and biological networks. Proc. Natl. Acad. Sci. USA **99**(12), 7821–7826 (2002)
29. Golub, T.R., Slonim, D.K., Tamayo, P., Huard, C., Gaasenbeek, M., Mesirov, J.P., Coller, H., Loh, M.L., Downing, J.R., Caligiuri, M.A., Bloomfield, C.D.: Molecular classification of cancer: class discovery and class prediction by gene expression monitoring. Science **286**, 531–537 (1999)
30. Gregory, S.: Finding overlapping communities in networks by label propagation. New J. Phys. **12**(10), 103018 (2010)
31. Guimera, R., Sales-Pardo, M., Amaral, L.: Modularity from fluctuations in random graphs and complex networks. Phys. Rev. E **70**, 025101 (2004)
32. Gulbahce, N., Lehmann, S.: The art of community detection. BioEssays **30**(10), 934–938 (2008)
33. Gupta, M., Gao, J., Aggarwal, C., Han, J.: Outlier detection for temporal data: a survey. IEEE Trans. Knowl. Data Eng. **26**(9), 2250–2267 (2014)
34. Hofman, J.M., Wiggins, C.H.: Bayesian approach to network modularity. Phys. Rev. Lett. **100**(25), 258701+ (2008)
35. Jin, J., Pawson, T.: Modular evolution of phosphorylation-based signalling systems. Philos. Trans. R. Soc. Lond. Ser. B Biol. Sci. **367**(1602), 2540–55 (2012)
36. Karypis, G., Han, E.H., Kumar, V.: Chameleon: hierarchical clustering using dynamic modeling. Computer **32**(8), 68–75 (1999)
37. Kawamoto, T., Kabashima, Y.: Limitations in the spectral method for graph partitioning: detectability threshold and localization of eigenvectors. Phys. Rev. E **91**, 062803 (2015)
38. Kiss, G.R., Armstrong, C., Milroy, R., Piper, J.R.I.: An associative thesaurus of English and its computer analysis. In: The Computer and Literary Studies. University Press, Edinburgh (1973)
39. Kumpula, J.M., Saramäki, J., Kaski, K., Kertész, J.: Limited resolution in complex network community detection with Potts model approach. Eur. Phys. J. B **56** (2007)
40. Kuramoto, Y.: Chemical Oscillations, Waves, and Turbulence. Springer, New York (1984)
41. Lancichinetti, A., Fortunato, S.: Limits of modularity maximization in community detection. Phys. Rev. E **84**, 066122 (2011)
42. Lancichinetti, A., Fortunato, S., Radicchi, F.: Benchmark graphs for testing community detection algorithms. Phys. Rev. E **78**(4), 046110(1–5) (2008)

43. Lancichinetti, A., Fortunato, S., Kertész, J.: Detecting the overlapping and hierarchical community structure in complex networks. New J. Phys. **11**(3), 033015 (2009)
44. Li, D., Leyva, I., Almendral, J.A., Sendina-Nadal, I., Buldu, J.M., Havlin, S., Boccaletti, S.: Synchronization interfaces and overlapping communities in complex networks. Phys. Rev. Lett. **101**(16), 168701 (2008)
45. Liu, W., Principe, J.C., Haykin, S.: Kernel Adaptive Filtering: A Comprehensive Introduction. Wiley, New York (2010)
46. Lusseau, D.: The emergent properties of a dolphin social network. Proc. R. Soc. B Biol. Sci. **270**(Suppl 2), S186–S188 (2003)
47. Ma, Y., Zhu, L.: A review on dimension reduction. Int. Stat. Rev. **81**(1), 134–150 (2013)
48. Moreno, Y., Vazquez-Prada, M., Pacheco, A.F.: Fitness for synchronization of network motifs. Physica A **343**, 279–287 (2004)
49. Nepusz, T., Petróczi, A., Négyessy, L., Bazsó, F.: Fuzzy communities and the concept of bridgeness in complex networks. Phys. Rev. E **77**, 016107 (2008)
50. Neuhaus, M., Bunke, H.: Bridging the Gap Between Graph Edit Distance and Kernel Machines. World Scientific, River Edge, NJ (2007)
51. Newman, M.E.J.: Analysis of weighted networks. Phys. Rev. E **70**, 056131 (2004)
52. Newman, M.E.J.: Fast algorithm for detecting community structure in networks. Phys. Rev. E **69**(6), 066133 (2004)
53. Newman, M.E.J.: A measure of betweenness centrality based on random walks. Soc. Networks **27**, 39–54 (2005)
54. Newman, M.E.J.: Finding community structure in networks using the eigenvectors of matrices. Phys. Rev. E **74**(3), 036104 (2006)
55. Newman, M.E.J.: Modularity and community structure in networks. Proc. Natl. Acad. Sci. **103**(23), 8577–8582 (2006)
56. Newman, M.E.J.: Spectral methods for community detection and graph partitioning. Phys. Rev. E **88**, 042822 (2013)
57. Newman, M.E.J., Girvan, M.: Finding and evaluating community structure in networks. Phys. Rev. Lett. **69**, 026113 (2004)
58. Newman, M.E.J., Leicht, E.A.: Mixture models and exploratory analysis in networks. Proc. Natl. Acad. Sci. USA **104**(23), 9564–9569 (2007)
59. Nicosia, V., Mangioni, G., Carchiolo, V., Malgeri, M.: Extending the definition of modularity to directed graphs with overlapping communities. J. Stat. Mech. Theory Exp. **2009**(03), 03024 (2009)
60. Oh, E., Rho, K., Hong, H., Kahng, B.: Modular synchronization in complex networks. Phys. Rev. E **72**, 047101 (2005)
61. de Oliveira, T., Zhao, L., Faceli, K., de Carvalho, A.: Data clustering based on complex network community detection. In: IEEE Congress on Evolutionary Computation, pp. 2121–2126. IEEE, New York (2008)
62. Oliveira, T.B.S.: Clusterização de dados utilizando técnicas de redes complexas e computação bioinspirada (2008). Master Thesis. Instituto de Ciências Matemáticas e de Computação, Universidade de São Paulo (USP)
63. Palla, G., Derenyi, I., Farkas, I., Vicsek, T.: Uncovering the overlapping community structure of complex networks in nature and society. Nature **435**(7043), 814–818 (2005)
64. Panaggio, M.J., Abrams, D.M.: Chimera states: coexistence of coherence and incoherence in networks of coupled oscillators. Nonlinearity **28**(3), R67 (2015)
65. Pearson, K.: On lines and planes of closest fit to systems of points in space. Philos. Mag. **2**(6), 559–572 (1901)
66. Pons, P., Latapy, M.: Computing communities in large networks using random walks. J. Graph Algorithms Appl. **10**, 284–293 (2004)
67. Psorakis, I., Roberts, S., Ebden, M., Sheldon, B.: Overlapping community detection using bayesian non-negative matrix factorization. Phys. Rev. E **83**, 066114 (2011)
68. Quiles, M.G., Zhao, L., Alonso, R.L., Romero, R.A.F.: Particle competition for complex network community detection. Chaos **18**(3), 033107 (2008)

69. Ravasz, E., Somera, A.L., Mongru, D.A., Oltvai, Z.N., Barabási, A.L.: Hierarchical organization of modularity in metabolic networks. Science **297**(5586), 1551–1555 (2002)
70. Reichardt, J., Bornholdt, S.: Detecting fuzzy community structures in complex networks with a potts model. Phys. Rev. Lett. **93**(21), 218701(1–4) (2004)
71. Rosvall, M., Bergstrom, C.T.: An information-theoretic framework for resolving community structure in complex networks. Proc. Natl. Acad. Sci. **104**(18), 7327–7331 (2007)
72. Roweis, S.T., Saul, L.K.: Nonlinear dimensionality reduction by locally linear embedding. Science **290**, 2323–2326 (2000)
73. Ruszczyński, A.P.: Nonlinear optimization. Princeton University Press, Princeton, NJ (2006)
74. Sarveniazi, A.: An actual survey of dimensionality reduction. Am. J. Comput. Math. **4**, 55–72 (2014)
75. Schmidt, M.N., Winther, O., Hansen, L.K.: Bayesian non-negative matrix factorization. In: Adali, T., Jutten, C., Romano, J.M.T., Barros, A.K. (eds.) Independent Component Analysis and Signal Separation. Lecture Notes in Computer Science, vol. 5441, pp. 540–547. Springer, Berlin, Heidelberg (2009)
76. Shawe-Taylor, J., Cristianini, N.: Kernel Methods for Pattern Analysis. Cambridge University Press, New York (2004)
77. Shen, H., Cheng, X., Cai, K., Hu, M.B.: Detect overlapping and hierarchical community structure in networks. Physica A **388**(8), 1706–1712 (2009)
78. Strogatz, S.H.: Sync: The Emerging Science of Spontaneous Order. Hyperion, New York (2003)
79. Sun, P.G., Gao, L., Shan Han, S.: Identification of overlapping and non-overlapping community structure by fuzzy clustering in complex networks. Inf. Sci. **181**, 1060–1071 (2011)
80. Tenenbaum, J.B., de Silva, V., Langford, J.C.: A global geometric framework for nonlinear dimensionality reduction. Science **290**(5500), 2319–2323 (2000)
81. Topaz, C.M., Andrea, Bertozzi, L.: Swarming patterns in a two-dimensional kinematic model for biological groups. SIAM J. Appl. Math. **65**, 152–174 (2004)
82. Vishwanathan, S.V.N., Schraudolph, N.N., Kondor, R., Borgwardt, K.M.: Graph kernels. J. Mach. Learn. Res. **11**, 1201–1242 (2010)
83. Wakita, K., Tsurumi, T.: Finding community structure in mega-scale social networks: [extended abstract]. In: Proceedings of the 16th International Conference on World Wide Web, WWW '07, pp. 1275–1276 (2007)
84. Wang, F., Sun, J.: Survey on distance metric learning and dimensionality reduction in data mining. Data Min. Knowl. Disc. **29**(2), 534–564 (2015)
85. Winfree, A.T.: The Geometry of Biological Time. Springer, Berlin (2001)
86. Wu, Z., Duan, J., Fu, X.: Complex projective synchronization in coupled chaotic complex dynamical systems. Nonlinear Dyn. **69**(3), 771–779 (2012)
87. Xu, R., II, D.W.: Survey of clustering algorithms. IEEE Trans. Neural Netw. **16**(3), 645–678 (2005)
88. Yan, S., Xu, D., Zhang, B., Zhang, H.J., Yang, Q., Lin, S.: Graph embedding and extensions: a general framework for dimensionality reduction. IEEE Trans. Pattern Anal. Mach. Intell. **29**(1), 40–51 (2007)
89. Zarei, M., Izadi, D., Samani, K.: Detecting overlapping community structure of networks based on vertex-vertex correlations. J. Stat. Mech. Theory Exp. **11**, P11013 (2009)
90. Zhang, S., Wang, R.S., Zhang, X.S.: Identification of overlapping community structure in complex networks using fuzzy C-Means clustering. Physica A **374**(1), 483–490 (2007)
91. Zhang, X., Nadakuditi, R.R., Newman, M.E.J.: Spectra of random graphs with community structure and arbitrary degrees. Phys. Rev. E **89**, 042816 (2014)
92. Zhou, H.: Distance, dissimilarity index, and network community structure. Phys. Rev. E **67**(6), 061901 (2003)

Chapter 7
Network-Based Semi-Supervised Learning

Abstract In this chapter, we present network-based algorithms that run in the semi-supervised learning scheme. The semi-supervised learning paradigm lies somewhere in-between the unsupervised learning paradigm, which does not employ any external information to infer knowledge, and the supervised learning paradigm, which in contrast makes use of a fully labeled set to train models. Semi-supervised learning aims, among other features, to reduce the work of human experts in the labeling process. This feature is quite interesting especially when the labeling process is expensive and time consuming as in video indexing, classification of audio signals, text categorization, medical diagnostics, genome data, among many other applications. In network-based methods, the graph structure is the main driver in propagating labels from labeled vertices to unlabeled vertices. We show that different techniques apply different criteria in their label diffusion processes, generating, as a result, distinct outcomes. In addition, we discuss some of the shortcomings and benefits of the within-graph semi-supervised learning process, also called transductive learning.

7.1 Introduction

Semi-supervised learning is a learning paradigm concerned with the study of how computers and natural systems, such as humans, learn in the presence of both labeled and unlabeled data [39]. Traditionally, learning has been studied either in the unsupervised paradigm (e.g., clustering, outlier detection), in which all of the data are unlabeled, or in the supervised paradigm (e.g., classification, regression), in which all of the data are labeled. The goal of semi-supervised learning is to understand how combining labeled and unlabeled data may change the learning behavior and to design algorithms that take advantage of such a combination. Semi-supervised learning is of great interest in machine learning and data mining because it can use readily available unlabeled data to improve supervised learning tasks when the labeled data is scarce or expensive. Semi-supervised learning also displays potential when conceived as a quantitative tool to understanding categorical human learning, in which most of the input or received information is self-evidently unlabeled [39].

© Springer International Publishing Switzerland 2016

T.C. Silva, L. Zhao, *Machine Learning in Complex Networks*,

DOI 10.1007/978-3-319-17290-3_7

During the last years, the most active area of research in the field of semi-supervised learning has been related to methods based on graphs or networks. The common point of these techniques is in the representation of data items as vertices of the network, while the existence of links between data items depends both on the network formation strategy and on the labels of the labeled vertices [8]. A noticeable advantage of using networks for data analysis is the ability of revealing the topological structure of the data relationships. Thus, network-based methods allow for the detection of classes and groups with arbitrary shapes that are in turn difficult to identify for techniques that do not use structured data representations to perform the learning process [14].

Network-based semi-supervised learning begins by constructing a network with the input vector-based data using network formation strategies.[1] Once the network is built, the learning process consists in assigning a label for every unlabeled vertex in the test set. The inference is done by diffusing labels through the edges that interconnect vertices of the network [8]. The learning process employs both the labeled and unlabeled sets in the label diffusion process. In contrast to the traditional techniques that make use of attribute-value tables to conduct their analyzes on the data, network-based techniques directly use the direct and/or indirect neighborhood structures of the graph constructed from the input data to analyze and predict labels for unlabeled instances. As explained in several researches in the literature [20, 21, 23–25, 36], this feature may generate classifiers that are more robust and efficient.

In a semi-supervised learning process, algorithms can be either inductive or transductive. While inductive techniques construct general rules from the given training set, transductive learning limits the prediction to specific test instances.[2] Most network-based methods are transductive techniques, meaning that they aim at inferring a class for each unlabeled vertex in the test set only; thus they are not required to design a global generalizing function to other new vertices that are not in the test set.

Among the main advantages of network-based semi-supervised learning algorithms, we may highlight [8, 37]:

- The network structure can effectively detect clusters of various forms.
- The learning process does not make decisions based explicitly on distance functions.
- The representation of data sets with multiple classes is facilitated.
- Some problems are naturally represented by networks, for example: protein-protein interaction networks, blood mainstream, Internet, among others. In this case, reversing from a network-based data representation to a vector-based

---

[1] See Chap. 4 for a discussion on different network formation methods.

[2] Intuitively, if the learning problem is an exam, then the labeled data correspond to the few example problems that the teacher solved in class. The teacher also provides a set of unsolved problems. In the transductive setting, these unsolved problems are a take-home exam and you want to do well on them in particular. In the inductive setting, these are practice problems of the sort you will encounter on the in-class exam.

representation would be a lossy transformation. To see this, it would be difficult to model cycles in graphs using a vector-based data. Cycles permit recursiveness in the data relationships and are naturally modeled by networks.

Many semi-supervised learning techniques, such as Transductive SVM, can identify data classes of well-defined forms, but usually fail to identify classes of irregular forms. Thus, assumptions on the class distributions have to be made [8]. Unfortunately, such information is usually unknown *a priori*. In order to overcome this problem, several graph-based methods have been developed in the last years. Among them, we may highlight: mincut [40], local and global consistency [35], local learning regularization [34], local and global regularization [33], manifold regularization [4], semi-supervised modularity [22], D-Walks [7], random walk techniques [12, 29], and label propagation techniques [32, 38]. However, most of the graph-based methods share the same regularization framework, differing basically in the particular choice of the loss function and the regularization function [2, 3, 5, 13, 35, 40], and most of them have cubic order of computational complexity ($\mathcal{O}(V^3)$). This factor makes their applicability limited to small- or middle-sized data sets [36]. As data sets get larger and larger, the development of efficient semi-supervised learning methods is still necessary.

7.2 Network-Based Semi-Supervised Learning Assumptions

Recall that the main difference between supervised and semi-supervised learning is that the latter uses unlabeled data to improve the generalization performance of the classifier. In order to effectively use the unlabeled data in the learning process, we must assume some underlying data structure. Bad matching of the problem structure with the model assumption can lead to degradation of the classifier's performance. For example, quite a few semi-supervised learning methods assume that the decision boundary should avoid regions with high densities of data items. Nonetheless, if data are generated from two heavily overlapping Gaussian distributions, the decision boundary would go right through the densest region, and the majority of the existing methods that rely on such assumption would perform badly. Detecting bad matching in advance, however, is hard and remains an open question [39].

Semi-supervised learning algorithms make use of at least one of the following assumptions [8]:

- *Smoothness assumption*: data points in the attribute space that are close to each other are more likely to share the same label. This is also generally assumed in supervised learning and yields a preference for geometrically simple decision boundaries. In the case of semi-supervised learning, the smoothness assumption additionally yields a preference for decision boundaries in low-density regions, so that it is expected that few points of different classes will reside near the regions where these decision boundaries cross. The smoothness assumption can also be related to the belief that "similar examples ought to have similar labels."

- *Cluster assumption*: data points that can be connected via (many) paths through high-density regions are likely to have the same label. This is a special case of the smoothness assumption and gives rise to feature learning with clustering algorithms.
- *Manifold assumption*: each class lies on a separate manifold of much lower dimension than the input space. In this case, we can attempt to learn the manifold using both the labeled and unlabeled data to avoid the curse of dimensionality. Then learning can proceed using distances and densities defined on the manifold. The manifold assumption is practical when high-dimensional data are being generated by some process that may be hard to model directly, but has only a few degrees of freedom. For instance, human voice is controlled by a few vocal folds [28], and images of various facial expressions are controlled by a few muscles. We would like in these cases to use distances and smoothness in the natural space of the generating problem, rather than in the space of all possible acoustic waves or images respectively.

The typical scenario in a semi-supervised learning task is the following. Denote $\mathscr{T} = \{(x_1, y_1), \ldots, (x_L, y_L)\}$ as the set containing tuples of the form: vertex $x_i \in \mathscr{L}$ and its corresponding label $y_i \in \mathscr{Y}$, where $\mathscr{L}$ symbolizes the set of labeled vertices and $\mathscr{Y}$ stands for the set of class labels. $\mathscr{U} = \{x_{L+1}, \ldots, x_{L+U}\}$ is the unlabeled set, such that $\mathscr{V} = \mathscr{L} \bigcup \mathscr{U}$ is the vertex set. There are $L = |\mathscr{T}| = |\mathscr{L}|$ labeled instances and $U = |\mathscr{U}|$ unlabeled instances. Thus, there is a total of $V = |\mathscr{V}|$ data items in the semi-supervised learning process and $Y = |\mathscr{Y}|$ classes. When $Y = 2$, we have a binary classification problem. When $Y > 2$, we have a multi-class problem. y_i denotes the true label of the i-th data item x_i, while $\hat{y}_i$ represents the approximated label output by the semi-supervised learning algorithm. Frequently, we have few labeled instances and several unlabeled instances, such that the condition $U \gg L$ holds. The goal is to label the unlabeled instances in accordance with some convenient label propagation process using both labeled and unlabeled data in the learning process. Network-based techniques use a graph to approximate the low-dimensional manifold.

The premise that unlabeled data also helps in the learning process is discussed in [26]. Therein, it is shown that, using a finite sample analysis, if the complexity of the distributions under consideration is too high to be learnt using L labeled data points, but is small enough to be learnt using $U \gg L$ unlabeled data points, then semi-supervised learning can improve the performance of a supervised learning task.

One final note is of the existence of multiple manifolds. For instance, in handwritten digit recognition, each digit forms its own manifold in the feature space; in computer vision motion segmentation, moving objects trace different trajectories that are low-dimensional manifolds [31]. These manifolds may intersect or partially overlap, while having different dimensionality, orientation, and density. In graph-based algorithms, if we create a graph that connects points on different manifolds near a manifold intersection, then labels will propagate to other manifolds in an incorrect manner. In this case, we must be aware to construct isolated graph components that do not interconnect, thus avoiding wrong label propagation.

7.3   Representative Network-Based Semi-Supervised Learning Techniques

Semi-supervised methods that rely on networks define a graph in which labeled and unlabeled examples in the data set are represented by vertices, and edges reflect the similarity of those examples. These methods usually assume label smoothness over the graph. In general, graph methods are nonparametric, discriminative, and transductive.

They are generally nonparametric because they make no assumptions about the probability distributions of the data items that are being analyzed (distribution-free). Recall that the difference between parametric and nonparametric models is that the former has a fixed number of parameters, while the number of parameters in the latter grows with the amount of training data [10, 19]. Moreover, we stress that nonparametric models are not the same as none-parametric models: parameters are determined by the training data, not the model. The nonparametric nature of network-based semi-supervised learning algorithms is a positive characteristic, as it prevents the insertion of wrong or misleading biases during the learning process.

Network-based semi-supervised methods are generally discriminative models, because they model the dependence of an unobserved variable y, the class or label in our context, on one or more observed variables x (data items and their similarities). In discriminative models, unlike generative models, there is no room to allow for generating samples from the joint distribution of x and y. For tasks of classification in which that joint distribution is not needed, discriminative models can yield superior performance [15, 17, 27]. In contrast, generative models are generally more flexible than discriminative models in expressing dependencies in complex learning tasks.

Network-based semi-supervised methods usually employ transductive inference because they estimate labels or classes from observed, specific data items (labeled and unlabeled set) to only specific items (unlabeled set). In contrast, inductive inference is reasoning from observed training cases to general rules, which are then applied not only to the unlabeled set but also to other new test cases.

An extensive review on network-based semi-supervised learning techniques can be found in [8, 36, 39].

Many graph-based methods can be expressed in terms of a regularization framework, in which the goal is to minimize a cost or energy function C that is composed of two complementary terms:

$$C = f_{\text{loss}} + f_{\text{reg}}. \qquad (7.1)$$

Each term in (7.1) serves different purposes, which are:

1. *Loss function* (f_{loss}): it leads the algorithm to penalize decisions that flip labels of pre-labeled vertices. Practically, to minimize this term, it is enough to prevent the change of pre-labeled vertices.

2. *Regularization function* (f_{reg}): it is responsible for modeling the cost of propagating labels to unlabeled vertices. Given that many algorithms rely on the smoothness assumption, this function must be smooth in dense regions of the network.

One implicit assumption here is that labeled data items are totally reliable. In imperfect learning, where there are noisy or wrongly labeled data items, the loss function would force algorithms that rely on the regularization framework to propagate wrong labels to the unlabeled data. Depending on the rate of imperfect data, the diffusion of bad labels could easily overwhelm the one of correct labels.

In the next sections, we explore several representative network-based semi-supervised learning techniques.

7.3.1   Maximum Flow and Minimum Cut

This method is presented in [5]. The original method classifies in a binary way, i.e., there are only two classes and the labels are confined in the set $y_i \in \{0, 1\}$, $\forall i \in \mathcal{V}$. The semi-supervised learning classification is posed as a graph mincut problem. In the binary case, positive labels act as sources, and negative labels act as sinks. The objective is to find a minimum set of edges whose removal (cut) would block all flow from the sources to the sinks. After the cut, the vertices connected to the sources are then labeled positive, and those to the sinks are labeled negative. We can view the term f_{loss} as a quadratic loss function with infinity weight:

$$f_{loss} = \lim_{w \to \infty} w \sum_{i \in \mathscr{L}} (\hat{y}_i - y_i)^2, \tag{7.2}$$

so that the influence of the regularization term in labeled data is effectively disabled. Consequently, the values of labeled data are in fact fixed at their true labels.[3] We are left to discuss the label diffusion process applied to the unlabeled data, which is governed by the following regularization function:

$$f_{reg} = \frac{1}{2} \sum_{i,j \in \mathcal{V}} \mathbf{A}_{ij} |\hat{y}_i - \hat{y}_j| = \frac{1}{2} \sum_{i,j \in \mathcal{V}} \mathbf{A}_{ij} \left(\hat{y}_i - \hat{y}_j \right)^2, \tag{7.3}$$

in which $\mathbf{A}_{ij}$ is the edge weight linking i to j and $\hat{y}_i$ is the estimated label of vertex $i \in \mathcal{V}$. Note that the second equality only holds due to the binary nature of $\hat{y}_i$, $\forall i \in \mathcal{V}$. Substituting (7.2) and (7.3) into (7.1), we get our objective function:

---

[3]Otherwise, the objective function that is composed of the loss and regularization terms is infinity. To note that, observe that if $\hat{y}_i \neq y_i$, $i \in \mathscr{L}$, then $f_{loss} \to \infty$.

$$C = \lim_{w \to \infty} w \sum_{i \in \mathscr{L}} (\hat{y}_i - y_i)^2 + \frac{1}{2} \sum_{i,j \in \mathscr{V}} A_{ij} (\hat{y}_i - \hat{y}_j)^2, \tag{7.4}$$

subject to the constraint $\hat{y}_i \in \{0, 1\}$, $\forall i \in \mathscr{V}$ (crisp labeling). Effectively, the estimation of labeled instances $i \in \mathscr{L}$ must coincide with their initial label beliefs, in a way that we are only allowed for deciding the labels of unlabeled instances in $\mathscr{U}$.

The classical graph mincut approach has a number of attractive properties [6]. First, it can be computed in polynomial time using network flow tools. Second, the learning process can be viewed as providing the most probable configuration of labels in the associated Markov random field. Lastly, it can also be motivated from sample-complexity considerations.

The mincut algorithm, however, also suffers from several drawbacks. One noticeable drawback of mincut is that it only outputs hard or crisp classification without confidence intervals. In statistical terms, it only computes the mode, rather than marginal probabilities. For instance, in the research in [6], the graph is perturbed by adding random noise to the edge weights. Mincut is applied to multiple perturbed graphs, and labels are determined by a majority vote criterion. The procedure is similar to bagging, and effectively creates a "soft" mincut. The research in [13] gives a method based on spectral partitioning that produces an approximate minimum ratio cut in the graph. Another shortcoming comes from a practical perspective. A graph may have many minimum cuts, and the mincut algorithm produces just one, typically the "leftmost" one using standard network flow algorithms. For instance, a line of V vertices between two labeled points i and j has $V - 1$ cuts of size 1, and the leftmost cut will be especially unbalanced.

7.3.2   Gaussian Field and Harmonic Function

One of the main limitations of the mincut is that the algorithm only classifies using binary crisp labels. Data that are located in bordering or overlapping regions, however, may be labeled with less confidence than those data items that are in the core of their respective classes. Gaussian random fields and harmonic function methods [38, 40] try to address these problems. These techniques can be viewed as a form of nearest neighbors approach, in which the nearest labeled examples are computed in terms of a random walk on the graph. These learning methods have intimate connections with random walks, electric networks, and spectral graph theory, in particular with heat kernels and normalized cuts.

In the context of networks, harmonic functions estimate the label of an unlabeled vertex according to a weighted average of the labels of vertices in the neighborhood. In this way, the classification becomes smooth. Gaussian random fields and harmonic function methods [38, 40] are a continuous relaxation to the complex discrete Markov random fields. Likewise mincut, they employ a quadratic loss function with infinity weight, so that labeled data are clamped to their pre-defined labels and the regularization function is based on a quadratic form of the graph Laplacian $\mathbf{L}$. The cost function, therefore, is expressed as:

$$
\begin{aligned}
C &= \lim_{w \to \infty} w \sum_{i \in \mathscr{L}} (\hat{y}_i - y_i)^2 + \frac{1}{2} \sum_{i,j \in \mathscr{V}} \mathbf{A}_{ij} (\hat{y}_i - \hat{y}_j)^2 \\
&= \lim_{w \to \infty} w \sum_{i \in \mathscr{L}} (\hat{y}_i - y_i)^2 + \frac{1}{2} \hat{Y}^T \mathbf{L} \hat{Y},
\end{aligned}
\tag{7.5}
$$

in which $\hat{y}_i \in [0, 1]$, $\hat{Y} = [y_1, y_2, \ldots, y_V]^T$ is a vector that stores the estimated labels of all of the vertices, and $\mathbf{L}$ is the graph Laplacian. The fuzziness of $\hat{Y}$ is a key relaxation towards the mincut technique that only allows for crisp classification, i.e., $\hat{y}_i \in \{0, 1\}$, $i \in \mathscr{V}$. Recall that the (i,j)-th entry of the graph Laplacian $\mathbf{L}$ with adjacency matrix $\mathbf{A}$ is:

$$
\mathbf{L}_{ij} = \mathbf{D}_{ij} - \mathbf{A}_{ij} = \begin{cases} k_i, & \text{if } i = j \\ -\mathbf{A}_{ij}, & \text{otherwise,} \end{cases}
\tag{7.6}
$$

in which k_i is the degree of vertex i computed using the adjacency matrix $\mathbf{A}$. $\mathbf{D}$ is the degree matrix that is computed as:

$$
\mathbf{D}_{ij} = \begin{cases} k_i, & \text{if } i = j. \\ 0, & \text{otherwise.} \end{cases}
\tag{7.7}
$$

While it is clear that the solution form of the loss function in (7.5) is the one that does not flip labels of pre-labeled instances, it may not be clear the solution of the regularization term $\hat{Y}^T \mathbf{L} \hat{Y}$ that assures smoothness in the labeling process. We elaborate on that in the following. From (7.5), we see that:

$$
\hat{Y}^T \mathbf{L} \hat{Y} = \sum_{i,j \in \mathscr{V}} \mathbf{A}_{ij} (\hat{y}_i - \hat{y}_j)^2.
\tag{7.8}
$$

The regularization term in (7.8) is a measure of non-smoothness of the estimated labels according to the network topology. For estimated labels that are not similar in the neighborhood, the term $\hat{Y}^T \mathbf{L} \hat{Y}$ yields large values. For estimated labels that are similar in the neighborhood, the term $\hat{Y}^T \mathbf{L} \hat{Y}$ produces small values.

Consider the eigenequation:

$$
\mathbf{L} v = \lambda v,
\tag{7.9}
$$

in which v is one of the eigenvectors of $\mathbf{L}$ and λ is the associated eigenvalue of that eigenvector. If we right-multiply (7.10) by the transpose of the eigenvector v, we get:

$$v^T \mathbf{L} v = v^T \lambda v$$

$$v^T \mathbf{L} v = \lambda v^T v$$

$$v^T \mathbf{L} v = \lambda, \tag{7.10}$$

in which the second equality comes from the factor that λ is a scalar and therefore we can rearrange vector v^T, while the third equality holds from the orthonormality of v, i.e., $v^T v = 1$.

If we consider that the eigenvector v is one of the solutions of regularization function as in (7.8), i.e., $v = y$, we get:

$$y^T \mathbf{L} y = \lambda, \tag{7.11}$$

that is, the eigenvalue λ associated to the solution y of the Laplacian $\mathbf{L}$ gives us an idea of the non-smoothness of the estimated labels (y). As we select eigenvector solutions that have larger and larger associated λ values, the less smooth are the estimated labels. Considering that the loss function in (7.5) cannot be changed by the learning process as our goal is to minimize C, we are effectively minimizing the regularization term constrained to the given labeled instances. Therefore, the estimated labels must be near or equal eigenvectors that have associated eigenvalues with small magnitude.

7.3.3 Tikhonov Regularization Framework

The Tikhonov regularization algorithm in [2] uses a general form of loss function:

$$f_{\text{loss}} = \frac{1}{L} \sum_{i \in \mathscr{L}} V(\hat{y}_i, y_i), \tag{7.12}$$

in which $V(\hat{y}_i, y_i)$ is some loss function. For instance, $V(\hat{y}_i, y_i) = (\hat{y}_i - y_i)^2$, then we have a regularized least squares technique, while $V(\hat{y}_i, y_i) = \max(0, 1 - \hat{y}_i y_i)$ leads to the SVM algorithm. Note that now the loss function allows for changes in the prior labeled set.

The regularization function is:

$$f_{\text{reg}} = \hat{Y}^T \mathbf{S} \hat{Y}, \tag{7.13}$$

in which $\mathbf{S}$ is a smoothness matrix, such as the Laplacian matrix $\mathbf{L}$.

In this way, the cost function of the regularization framework becomes:

$$F = \frac{1}{L} \sum_{i \in \mathscr{L}} (\hat{y}_i - y_i)^2 + \mu \hat{Y}^T \mathbf{S} \hat{Y}, \tag{7.14}$$

in which μ is a regularization parameter that modulates the influences played by the loss and regularization functions. For stability purposes, the prior belief of labeled instances is subtracted from its mean.

The Tikhonov regularization framework has some interesting advantages:

- It eliminates the need of computing multiple eigenvectors or complicated graph invariants (mincut, max flow etc.). There is also a simple closed form solution for the optimal regressor. The problem is reduced to a single, usually sparse, linear system of equations whose solution can be computed efficiently. One of the algorithms proposed (interpolated regularization) is extremely simple with no free parameters.
- The generalization error can be bounded and related to properties of the underlying graph using arguments from algorithmic stability.
- If the graph arises from the local connectivity of data obtained from sampling an underlying manifold, then the approach has natural connections to regularization on that manifold.

7.3.4 Local and Global Consistency

This method has been proposed by Zhou et al. [35] and is one of the first studies in network-based semi-supervised learning techniques. This method considers the general problem of learning from labeled and unlabeled data by means of constructing a classification function that is sufficiently smooth with respect to the intrinsic labeled and unlabeled data structures.

The technique considers the evolution of a set of matrices $\mathcal{M}$ with dimensions $V \times Y$, all of which with nonnegative entries. The matrix $\hat{\mathbf{Y}} = [\hat{Y}_1^T, \ldots, \hat{Y}_V^T]^T \in \mathcal{M}$ corresponds to the fuzzy classification of the data items $\mathcal{V}$, such that, for each labeled or unlabeled vertex $x_i \in \mathcal{V}$, we designate a label in accordance with the expression $\hat{y}_i = \arg\max_{y \in \mathcal{Y}} \mathbf{Y}_{iy}$. One can think of $\hat{\mathbf{Y}}$ as a vectorial function that attributes, for each unlabeled data x_i, the maximum value of $\hat{\mathbf{Y}}_{iy}, y \in \mathcal{Y}$. Define also the matrix $\mathbf{Y}$ with dimensions $V \times Y$, such that $\mathbf{Y}_{iy} = 1$ if x_i is labeled as $y \in \mathcal{Y}$, and $\mathbf{Y}_{iy} = 0$, otherwise. The algorithm evolves as follows:

1. Generate the adjacency matrix $\mathbf{A}$ according to the Gaussian kernel, which is given by $\mathbf{A}_{ij} = \exp\left(\frac{\|x_i - x_j\|^2}{2\sigma^2}\right)$ if $i \neq j$, and $\mathbf{A}_{ii} = 0$, otherwise;
2. Construct the matrix $\mathbf{S} = \mathbf{D}^{-\frac{1}{2}} \mathbf{A} \mathbf{D}^{-\frac{1}{2}}$, in which $\mathbf{D}$ is a diagonal matrix with each entry (i, i) equivalent to the sum of the i-th row of $\mathbf{A}$;
3. Iterate $\hat{\mathbf{Y}}(t + 1) = \alpha \mathbf{S} \hat{\mathbf{Y}}(t) + (1 - \alpha)\mathbf{Y}$ until it converges, where $\alpha \in (0, 1)$;
4. Consider that $\hat{\mathbf{Y}}^*$ denotes the limit of the sequence $\{\hat{\mathbf{Y}}(t) : t \in \mathbb{N}\}$. Then, label each unlabeled vertex x_i following the formula: $\hat{y}_i = \arg\max_{j \in \mathcal{Y}} \hat{\mathbf{Y}}_{ij}^*$.

Moreover, it can be shown that the sequence $\mathfrak{I} = \{\hat{\mathbf{Y}}(t) : t \in \mathbb{N}\}$ converges and assumes the following closed formula:

$$\hat{\mathbf{Y}}^* = \lim_{t \to \infty} \hat{\mathbf{Y}}(t) = (\mathbf{I} - \alpha\mathbf{S})^{-1}\mathbf{Y}. \tag{7.15}$$

Still in [35], a regularization framework is molded with the aforementioned dynamics. In this framework, one aims at minimizing a cost or energy expression. The encountered expression, here written as $C(\hat{\mathbf{Y}})$, is given as:

$$C(\hat{\mathbf{Y}}) = \frac{1}{2} \left(\sum_{i,j \in \mathcal{V}} \mathbf{A}_{ij} \left\| \frac{1}{\sqrt{\mathbf{D}_{ii}}}\hat{\mathbf{Y}}_i - \frac{1}{\sqrt{\mathbf{D}_{jj}}}\hat{\mathbf{Y}}_j \right\|^2 + \mu \sum_{i \in \mathcal{V}} \|\hat{\mathbf{Y}}_i - \mathbf{Y}_i\|^2 \right), \tag{7.16}$$

in which $\mu > 0$ is a regularization parameter. In this case, the optimal values for the classification function become:

$$\hat{\mathbf{Y}}^* = \arg\min_{F \in \mathcal{M}} C(\hat{\mathbf{Y}}). \tag{7.17}$$

The first term in (7.16) enforces smoothness decisions by the classifier, meaning that a good classification function must not have large derivatives in high-density areas. This is exactly the definition of a regularization function. The second term symbolizes the adjustment restriction, revealing that a good classification function also must not exchange the labels from already labeled data. In this case, this definition perfectly fits into the description of a loss function. The counterweight between these two conflicting quantities is given by the positive constant μ.

The advantage of this technique is its simplicity. As one can see, the propagation is done by utilizing a linear update rule and convergence issues have been fully described, enabling one to understand the dynamics of such model in the long run. However, the algorithm suffers from some drawbacks: since the propagation is done utilizing a linear function, nonlinear characteristics of the data may pass unseen by the algorithm. Moreover, since a matrix inversion is involved to find the optimal solution, the algorithm requires $\mathcal{O}(V^3)$ to run, which is unfeasible for large-scale networks.

7.3.5   Adsorption

The adsorption technique was first introduced in [1]. It was then further extended and given a theoretical analysis in [30]. Adsorption has many desirable properties, among which we can highlight:

- Possibility to perform multiclass classification ($Y > 2$).
- Definition can be stated in terms of a parallelized implementation, thus enabling its application on large-scale data sets.
- Mechanism to deal with imperfect training data.[4]

Likewise other label propagation algorithms that work in a networked environment, adsorption propagates label information from the labeled examples to the entire set of vertices via the edges (network topology). The labeling is represented using a non-negative score for each label, in which high scores are attributed to those labels that indicate the highest associations or similarities to unlabeled vertices. If these scores are additively normalized, they can be thought of as a conditional distribution over the labels given the unlabeled data.

Let $\hat{\mathbf{Y}} = [\hat{Y}_1^T, \ldots, \hat{Y}_V^T]^T \in V \times Y$ be a matrix in which the v-th row $\hat{\mathbf{Y}}_v$ corresponds to the fuzzy classification of $v \in \mathcal{V}$ towards the Y possible classes. That is, $\hat{\mathbf{Y}}_{vy}$ is the fuzzy classification of the data item v with respect to class y. Similarly $\mathbf{Y}$ encodes the initial belief of all of the vertices in the network towards the existent classes. At the initial phase of the algorithm, we must supply the belief for all of the vertices $v \in \mathcal{V}$. If v is an unlabeled vertex, we can simply set the belief of v as the zero-valued vector. Adsorption outputs an estimated belief or fuzzy classification label for each class in the vector $\hat{\mathbf{Y}}_v$, $v \in \mathcal{V}$.

We can view the learning mechanism performed by the adsorption algorithm as controlled random walks that are conducted in accordance with the network topology. The control over the random walk is realized via three possible actions: *inject, continue, abandon*, each of which with occurrence probabilities on vertex $v \in \mathcal{V}$ of $p_v^{(\text{inject})}$, $p_v^{(\text{continue})}$, and $p_v^{(\text{abandon})}$, respectively. For a valid transition probability, one must have:

$$p_v^{(\text{inject})} + p_v^{(\text{continue})} + p_v^{(\text{abandon})} = 1. \tag{7.18}$$

To label each unlabeled or even already labeled vertex $v \in \mathcal{V}$, we first initiate a random walk starting at v. At each time step, the random walk is allowed to choose over three actions:

1. With probability $p_v^{(\text{inject})}$, the random walker stops and returns the pre-defined initial belief Y_v, i.e., $\hat{\mathbf{Y}}_v = \mathbf{Y}_v$. A further constraint is also imposed to force label diffusion to unlabeled instances in the graph. Whenever v is unlabeled, we fix $p_v^{(\text{inject})} = 0$, so that the random walk cannot output the initial belief of an unlabeled vertex as its classification decision.
2. With probability $p_v^{(\text{abandon})}$, the random walker abandons the labeling diffusion process and returns the zero-valued vector as the classification decision, that is, $\hat{\mathbf{Y}}_v$.

---

[4]Imperfect training data arises when the labeled instances are not totally reliable. We explore in detail another semi-supervised learning algorithm that deal with the detection and prevention of labels that are diffused by possibly wrong labeled data in Chap. 10.

3. With probability $p_v^{(continue)}$, the random walker continues to navigate in the graph, specifically to one of the neighbors of v with a probability proportional to the edge weight. The transition probability follows the transition matrix of a random walk process as explored in Sect. 2.4.1. For convenience, we rewrite the transition matrix as follows:

$$\mathbf{P}[u \mid v] = \mathbf{P}_{vu} = \frac{\mathbf{A}_{vu}}{\sum_{i \in \mathcal{V}} \mathbf{A}_{vi}}. \tag{7.19}$$

Considering this three-way dynamics of the walker, the expected score $\hat{\mathbf{Y}}_v$, $v \in \mathcal{V}$, is given by:

$$\hat{\mathbf{Y}}_v = p_v^{(inject)} Y_v + p_v^{(continue)} \sum_{u \in \mathcal{N}(v)} \mathbf{P}[u \mid v] \hat{\mathbf{Y}}_u + p_v^{(abandon)} 0_Y, \tag{7.20}$$

in which 0_Y is the zero-valued vector with Y entries and $\mathcal{N}(v)$ returns the set of neighbors of v.

Alternatively, in order to guarantee the positiveness of $\hat{\mathbf{Y}}_v$, a slight modification can be introduced whenever the random walker abandons the walk. Instead of returning a zero-valued vector, we can create a dummy label $y_d \notin \mathcal{Y}$ and designate that dummy label as the estimated label of v. We can conceive this additional dummy class as encoding ignorance or uncertainty about the correct label of v. With this modification, at least one of the three terms in (7.20) always assumes a positive value. Thus, $\hat{\mathbf{Y}}_v$ is positive.

The smoothness assumption of the adsorption algorithm is modeled by the second term in the RHS of (7.20). Note that the estimated label of v, $\hat{\mathbf{Y}}_v$, is composed of a weighted linear combination of the estimated labels in the neighborhood of v. This averaging view then defines a set of fixed-point equations to update the predicted labels. Since the past trajectories or states of the walkers do not need to be maintained, the adsorption algorithm is memoryless. As such, it scales to large-scale data sets with possibly dense configurations and also can be easily parallelized [1].

Some heuristics have been proposed to estimate $p_v^{(inject)}$, $p_v^{(continue)}$, and $p_v^{(abandon)}$ [1, 30]. Effectively, these heuristics suggest that:

$$p_v^{(continue)} \propto c_v,$$
$$p_v^{(inject)} \propto d_v. \tag{7.21}$$

The first quantity $c_v \in [0, 1]$ is a value that monotonically decreases with the number of neighbors of vertex v. That is, the more neighbors v has in the network topology, the smaller is c_v. Intuitively, if v connects with several other vertices, it is probably a difficult vertex to be classified. Hence, the idea is to prevent further label propagation that comes from it. This mechanism assures preferential trajectories that pass through vertices with small degrees.

The other quantity $d_v \geq 0$ is a value that monotonically increases with the entropy (for labeled vertices), and in this case we prefer to use the prior belief rather than the computed quantities from the neighbors. The entropy of vertex v is evaluated using the transition matrix, as follows:

$$H(v) = - \sum_{u \in \mathcal{N}(v)} \mathbf{P}[u \mid v] \log \mathbf{P}[u \mid v]. \tag{7.22}$$

Once computed, we pass the entropy through the following monotonically decreasing function:

$$f(x) = \frac{\log(\beta)}{\log(\beta + e^x)}, \tag{7.23}$$

and the term c_v is defined as:

$$c_v = f(H(v)), \tag{7.24}$$

and the term d_v:

$$d_v = \begin{cases} (1 - c_v) \sqrt{H(v)}, & \text{if } v \text{ is labeled.} \\ 0, & \text{otherwise.} \end{cases} \tag{7.25}$$

Finally, to ensure that (7.18) holds, we set:

$$p_v^{(\text{continue})} = \frac{c_v}{z_v}, \tag{7.26}$$

$$p_v^{(\text{inject})} = \frac{d_v}{z_v}, \tag{7.27}$$

$$p_v^{(\text{abandon})} = 1 - \frac{c_v}{z_v} - \frac{d_v}{z_v}. \tag{7.28}$$

in which z_v is a normalization constant given by:

$$z_v = \max(c_v + d_v, 1). \tag{7.29}$$

7.3.6 Semi-Supervised Modularity Method

This algorithm has been proposed by Silva and Zhao [22] and is inspired by the modularity greedy algorithm, which we have introduced in Sect. 6.3.2.1. In the original modularity greedy algorithm, at each time step, two communities, say i

and j, are merged, in such a way that the largest increment (or least decrement) of the modularity occurs at a particular step. No restrictions on the communities to be merged are specified by the original model.

In order to adapt the modularity greedy algorithm for the context of semi-supervised learning, we make the following modifications:

1. Initially, we have L labeled vertices in the network. The task consists in propagating their labels to the unlabeled vertices. Once an unlabeled vertex receives a label, it cannot be changed.
2. At each step, we merge the communities (at the beginning, each community encompasses only one vertex) in such a way that the modularity increment is maximal. However, such merge is subjected to some constraints: in light of mimicking the propagation of labels in the network, a merge only occurs if at least one of the candidate communities has been labeled before. Suppose that communities c_i and c_j have been selected to be merged, each of which carrying the labels y_i and y_j. Let $\emptyset$ denote the unlabeled class. Then, one of the following four cases occurs:

Case 1. The merge does not occur if $y_i \neq y_j$, provided that $y_i \neq \emptyset$ and $y_j \neq \emptyset$. This case represents a clash between two different classes that have been previously labeled.

Case 2. The merge occurs if $y_i \neq \emptyset$ and $y_j = \emptyset$, or $y_i = \emptyset$ and $y_j \neq \emptyset$. This case represents the traditional label propagation from a labeled community to an unlabeled one. c_j receives the label from c_i in the first case and c_i receives the label from c_j in the second case.

Case 3. The merge occurs if $y_i = y_j$, provided that $y_i \neq \emptyset$ and $y_j \neq \emptyset$. In this case, the merge process just puts two communities of the same class together, maximizing the modularity.

Case 4. The merge does not occur if $y_i = y_j = \emptyset$, since no label is being propagated.

If the merge does not occur, then we select other two communities that have the second largest entry in the modularity increment matrix $\Delta\mathbf{Q}$ to be potentially merged, i.e., the Step 2 is repeated, and so on, until a valid merge takes place.

Keeping in mind that the modularity algorithm tries to maximize the number of edges among vertices of the same community, while also attempting to minimize the same quantity among distinct communities, the dynamic of the procedure will propagate labels so as to maintain the cluster and smoothness assumptions. In this way, the modified modularity greedy algorithm performs the work of propagating the labels in an optimized manner, provided that the network is strongly connected among vertices of the same class and weakly connected among vertices of distinct classes.

For the stop criterion of this algorithm, it simply needs to be run until no unlabeled vertex remains, regardless of the value of the modularity, since we are not looking for a good network division, but for an *ordered* way of labeling vertices. The mechanism of maximizing the modularity does this job for us. The convergence is guaranteed to happen and a proof has been provided in [22].

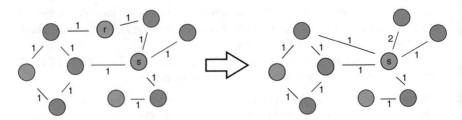

Fig. 7.1 Process of coalescence of vertices s and r. After the merge, s consumes r and becomes a super-vertex. All neighbors of r are connected to s during this procedure. Reproduced from [22] with permission from Elsevier

Additionally, in an attempt to make feasible the application of this semi-supervised algorithm on large-scale networks, a network reduction technique is explored. Let $\varphi(y) \in [0, 1]$ denote the proportion of reduction to be performed over the class pertaining to label $y \in \mathscr{Y}$. Denote also $\Psi_y(t)$ as set of data items that belongs to class $y \in \mathscr{Y}$. Then, the reduction is done by the following procedure, step-by-step:

1. Randomly choose pairs of pre-labeled vertices $r \in \Psi_y(t)$ and $s \in \Psi_y(t)$ to coalesce. In this process, r is removed from the network and s is entitled as a super-vertex, by virtue of the fact that it now represents more than one vertex in the network. In this process, all of the links that are connected to r are redirected to s. Suppose a connection between w and s already exists and w is also a neighbor of r, then we strengthen the connection between w and s by adding the weight from the former edge (w, r) to (w, s). Figure 7.1 illustrates this idea. This is done until $|\Psi_y(t + \Delta_t)| = (1 - \varphi(y))|\Psi_y(t)|$, where $\Delta_t > 0$ is a parameter bounded by the upper limit provided by the convergence proof in [22]. Essentially, $|\Psi_y(t+\Delta_t)|$ denotes the size of Ψ_y in a future time that can be reached in a finite number of steps. If $\varphi(y) = 1$, then we deliberately continue to merge process until $|\Psi_y(t + \Delta_t)| = 1$, i.e., until only one element remains of that class.
2. All of the self-loops, brought into the network by the reduction process, are removed. This prevents the modified modularity greedy algorithm from trying to merge a certain community with itself.
3. This process of reduction is performed for every class $y \in \mathscr{Y}$ that exists in the network.

In general, at the end of the reduction process, it is expected that the network will shrink, because $(V-L)+\sum_{y\in\mathscr{Y}} |\Psi_y(t + \Delta_t)| = (V-L)+\sum_{y\in\mathscr{Y}} [1 - \varphi(y)]|\Psi_y(t)| \leq V$, since $0 \leq \varphi(y) \leq 1$ and thus $\sum_{y\in\mathscr{Y}} [1 - \varphi(y)]|\Psi_y(t)| \leq L$. If the proportion of labeled vertices is large, then this process greatly reduces the network size, provided that $\varphi(y), \forall y \in \mathscr{Y}$, is large. Usually, the quantity of labeled vertices is small, because the task of labeling is generally expensive and cumbersome.

The main advantage of the aforementioned technique is that it does not require any kinds of parameters in order to work. Considering that the task of parameter tuning often takes a considerable time to complete and that lots of traps are involved in this investigation, such as the presence of local maxima, an expert may be needed such as to set feasible initial kickoff values for the parameters tuning procedure.

Therefore, the feature of having no free parameters makes the applicability of the semi-supervised modularity technique in real-world applications easy. A major drawback of the aforementioned technique is that it suffers from the inherent resolution problem that the original modularity greedy algorithm presents.[5] This often leads to bad results when there are classes with very distinct sizes.

7.3.7   Interaction Forces

The interaction forces technique was presented in [9]. This method is nature-inspired and relies on attraction forces. It models data instances as points in a P-dimensional space and performs their motion accordingly to the resultant force applied upon them. The labeled instances act as attraction points, while unlabeled instances receive forces and move towards these attraction points. Under some circumstances, unlabeled instances receive labels from labeled points and thereafter become new attraction points.

The use of attraction forces between labeled and unlabeled instances can provide a model for semi-supervised learning that fits well the smoothness and cluster assumptions. Labeled instances are fixed attraction points that apply attraction forces on unlabeled instances. As a result, unlabeled instances move towards the direction of the resultant force. Eventually, they converge to an attraction point. Once an unlabeled instance gets close enough to a labeled instance, say inside a circle with radius δ, the label from that attraction point propagates to the unlabeled instance located at its surroundings. As a consequence of the labeling process, it then becomes a new fixed attraction point. At the end of the process, it is expected that all of the instances will converge to some attraction point. By means of the attraction forces, instances are kept together in their dense groups (clusters), while different labeled points are responsible for dividing the space under the smoothness assumption.

We need two considerations in order to accomplish the above mentioned behavior and classify the unlabeled instances correctly. One of them is to guarantee that the process is stable, and the other is to certify that the labels propagate adequately through the unlabeled instances, in the sense that the algorithm converges and achieves good classification accuracy. The stability issue can be treated using similar approaches from swarm aggregation methods [11, 16], while the label propagation dynamic can be analyzed in terms of the parameters that underpin the attraction forces.

The motion or the stepwise differential movement of an unlabeled point v_i at step t, denoted here as $\dot{v}_i$, is governed by the following system:

$$\dot{v}_i(t) = \sum_{j \in \mathscr{L}} f[v_j(t) - v_i(t)], \qquad (7.30)$$

---

[5] See Sect. 2.3 for details.

$\forall i \in \mathcal{U}$, where function f is the attraction force among instances. As described by (7.30), each unlabeled instance v_i receives attractive forces from all of the labeled instances and the resultant force is the sum of all individual forces. Thus, the direction and magnitude of $v_i(t)$'s motion are determined by the forces applied by the labeled instances.

The attraction function between an unlabeled item $i \in \mathcal{U}$ and a labeled item $j \in \mathcal{L}$ (attractor) is defined as a Gaussian field with parameters α and β:

$$f[v_j(t) - v_i(t)] = [v_j(t) - v_i(t)] \frac{\alpha}{e^{\beta \|(v_j(t) - v_i(t))\|^2}}. \tag{7.31}$$

The attraction function guarantees that the closer an unlabeled point is to an attractor labeled point, the stronger is the force. Moreover, the parameters of the Gaussian field provide an easy way to adjust the function amplitude and range.

Algorithm 1 summarizes the interaction forces method. The method is performed iteratively in four steps (from 2 to 5), until all of the instances are properly labeled.

7.3.8 Discriminative Walks (D-Walks)

This technique is introduced in [7]. D-walks rely on random walks performed on the input graph seen as a Markov chain. For a review on Markov chain theory, see Sect. 2.4.1. More precisely, D-Walks are essentially computed as a betweenness measure that is based on passage times during constrained random walks of bounded lengths. Unlabeled vertices are assigned to the category for which the betweenness is the highest. The D-walks approach has the following properties:

Algorithm 1 : Interaction forces technique

Input:
 $\mathcal{L}$: labeled data set
 $\mathcal{U}$: unlabeled data set
Output:
 l_i : estimated class for each $\mathbf{x}_i \in \mathcal{U}$
Initialization:
  1.(α, β, δ) = Initialize parameters
Classification:
DO
  2. Calculate distances among points
  3. Calculate attraction forces
  4. Update points' positions
  5. Update labels
WHILE (there are unlabeled instances)

- It has a linear time complexity with respect to the number of edges, the maximum walk length and the number of classes; such a low complexity allows the technique to deal with very large sparse graphs.
- It can handle directed or undirected graphs.
- It can deal with multi-class problems.
- It has a unique hyper-parameter, the walk length, that can be tuned efficiently.

We first transform the (weighted) adjacency matrix into a transition matrix, which is a row-stochastic matrix, using the following operation:

$$P[X_t = q' \mid X_{t-1} = q] = \mathbf{P}_{qq'} \triangleq \frac{\mathbf{A}_{qq'}}{\sum_{k \in \mathscr{V}} \mathbf{A}_{qk}}, \tag{7.32}$$

in which $\mathbf{A}_{qq'}$ stands for the edge weight linking q to q'. Each vertex in the network corresponds to a state in the Markov chain system. The graph can be directed or undirected, weighted or non-weighted.

We now introduce the discriminative random walks (D-Walks). Essentially, a D-walk is a random walk starting in a labeled vertex and ending when any vertex having the same label (possibly the starting vertex itself) is reached for the first time.

Definition 7.1. D-Walk Given a Markov chain defined on the state set $\mathscr{V}$ and a class label $y \in \mathscr{Y}$, a D-Walk is a sequence of states $q_0, \ldots, q_\lambda$, $\lambda > 0$, such that $y_{q_0} = y_{q_\lambda} = y$ and $y_{q_t} \neq y, 0 < t < \lambda$.

The notation $\mathscr{D}^y$ refers to the set of all D-Walks starting and ending in vertices of class y.

The betweenness function $B(q, y)$ measures the extent an unlabeled vertex $q \in \mathscr{U}$ is located "in-between" vertices of class $y \in \mathscr{Y}$. The betweenness $B(q, y)$ is formally defined as the expected number of times vertex q is reached during D-Walks on $\mathscr{D}^y$.

Definition 7.2. D-Walks Betweenness Given an unlabeled vertex $q \in \mathscr{U}$ and a class $y \in \mathscr{Y}$, we define the D-Walk betweenness function $\mathscr{U} \times \mathscr{Y} \to \mathbb{R}^+$ as:

$$\mathbf{B}(q, y) \triangleq \mathbb{E}\left[\mathrm{pt}(q) \mid \mathscr{D}^y\right], \tag{7.33}$$

in which $\mathrm{pt}(q)$ is the passage time of vertex $q \in \mathscr{V}$ whose formal definition has been reviewed in Sect. 2.4.1.

Vertices belonging to class y are first duplicated such that the original vertices are used as absorbing states and the duplicated ones as starting states. The transition matrix $\mathbf{P}$ is augmented as follows:

1. We duplicate the rows of $\mathbf{P}$ corresponding to labeled vertices of class $y \in \mathscr{Y}$ at the bottom of the matrix.
2. We add columns full of zeroes at the right of $\mathbf{P}$. The number of added columns is equal to the number of added rows in the previous step.
3. We define $p_{qq'} = 1 \iff q = q'$ and 0 otherwise, for all of the vertices belonging to y. The augmented matrix is denoted here by $\mathbf{P}^y$. The initial distribution vector is adapted accordingly, resulting in the vector $p(0)^y$.

The betweenness is finally computed as follows:

$$\mathbf{B}(q, y) = \left[(p(0)_T^y)'(\mathbf{I} - \mathbf{P}_T^y)^{-1} \right]_q,\tag{7.34}$$

in which $\mathbf{P}_T^y$ and $p(0)_T^y$ denote, respectively, the transition matrix and the initial distribution vector restricted to transient states. Matrix inversion, however, is computed in $\mathcal{O}(V^3)$, limiting the use of the technique for large-scale graphs.

The authors in [7] instead propose to use bounded walks. Bounding the walk length systematically provides a better classification rate with the additional benefit that the betweenness can be computed very efficiently using forward and backward recurrences. Let $\mathscr{D}_\lambda^y$ denote to the set of all D-Walks of length exactly equal to λ. Moreover, consider that $\mathscr{D}_{\leq \lambda}^y$ refer to the set of all bounded D-Walks up to a given length λ. We define the bounded betweenness measure $\mathbf{B}_\lambda(q, y)$ as follows.

Definition 7.3. Bounded D-Walks Betweenness Given an unlabeled vertex $q \in \mathscr{U}$ and a class $y \in \mathscr{Y}$, we define the bounded D-Walk betweenness function $\mathscr{U} \times \mathscr{Y} \rightarrow \mathbb{R}^+$ as:

$$\mathbf{B}_\lambda(q, y) \triangleq \mathbb{E}\left[\mathrm{pt}(q) \mid \mathscr{D}_{\leq \lambda}^y \right].\tag{7.35}$$

Following the authors in [7], limiting the random walk length brings to major advantages in a classification process:

- The algorithm presents better accuracy rates in relation to unbounded D-Walks.
- We can compute the bounded betweenness measure very efficiently.

An efficient way to evaluate the bounded betweenness measure is to use forward-backward variables, similar to those employed in the Baum-Welch algorithm for hidden Markov models [18]. Given a state $q \in \mathscr{V}$ and a time $t \in \mathbb{N}$, the forward variable $\alpha^y(q, t)$ computes the probability of reaching state q after t steps without visiting vertices of class $y \in \mathscr{Y}$, while starting from any state in class y. We can evaluate the forward variables using the following recurrence:

$$\begin{aligned}
&(\text{case } t = 1)\ \alpha^y(q, 1) = \frac{1}{V_y} \sum_{q' \in \mathscr{L}_y} p_{q'q} \\
&(\text{case } t \geq 2)\ \alpha^y(q, t) = \sum_{q' \in \mathscr{U}} \alpha^y(q', t-1) p_{q'q}
\end{aligned}\tag{7.36}$$

in which $\mathscr{L}_y$ is the set of labeled vertices of class y and $V_y = |\mathscr{L}_y|$. The initial recurrence (case $t = 1$) assumes that the walker can start at any vertex that is member of class y with uniform probability $\frac{1}{V_y}$. Thus, the equation provides the probability that q is visited in the next iteration. We also observe that, while in the first recurrence we loop through members of class y, we forbid visits in members of class y when $t \geq 2$. In fact, a D-Walk finishes whenever the walk visits a vertex with label coincident with the label of the starting vertex.

In an opposite perspective, the backward variable $\beta^y(q, t)$ computes the probability that state q is attained by the process t steps before reaching any vertex labeled y for the first time. We evaluate the backward variables using the following recurrence:

$$
\begin{aligned}
&(\text{case } t = 1)\ \beta^y(q, 1) = \sum_{q' \in \mathcal{L}_y} p_{qq'} \\
&(\text{case } t \geq 2)\ \beta^y(q, t) = \sum_{q' \in \mathcal{U}} \beta^y(q', t-1) p_{qq'}
\end{aligned}
\tag{7.37}
$$

To compute $B_\lambda(q, y)$, we first calculate the mean passage time in a vertex $q \in \mathcal{U}$ during $\mathscr{D}_\lambda^y$. The length-conditioned passage time function $\mathrm{pt}(q)$, $\mathbb{E}\left[\mathrm{pt}(q) \mid \mathscr{D}_\lambda^y\right]$, can be decomposed as a sum of indicator variables: $\mathrm{pt}(q) = \sum_{t=1}^{\lambda-1} \mathbb{1}_{[X_t=q]}$. Consequently,

$$
\begin{aligned}
\mathbb{E}\left[\mathrm{pt}(q) \Big| \mathscr{D}_\lambda^y\right] &= \mathbb{E}\left[\sum_{t=1}^{\lambda-1} \mathbb{1}_{[X_t=q]} \Big| \mathscr{D}_\lambda^y\right] \\
&= \sum_{t=1}^{\lambda-1} \mathbb{E}\left[\mathbb{1}_{[X_t=q]} \mid \mathscr{D}_\lambda^y\right] \\
&= \sum_{t=1}^{\lambda-1} P\left(X_t = q \mid \mathscr{D}_\lambda^y\right) \\
&= \sum_{t=1}^{\lambda-1} \frac{P\left(X_t = q \wedge \mathscr{D}_\lambda^y\right)}{P\left(\mathscr{D}_\lambda^y\right)},
\end{aligned}
\tag{7.38}
$$

in which the second equality comes from the linearity of the expectation operator, the third equality holds because $\mathbb{E}[\mathbb{1}_{[A]}] = P(A)$, and the fourth equality is true due to Bayes theorem.

We can compute the joint probability in the numerator of (7.38) as:

$$
P\left(X_t = q \wedge \mathscr{D}_\lambda^y\right) = \alpha^y(q, t)\beta^y(q, \lambda - t),
\tag{7.39}
$$

which is the probability to start in any vertex of class y, to reach the unlabeled vertex q at time t and then to complete the walk $\lambda - t$ steps later.

The denominator of (7.38) accounts for the probability to perform D-Walks with length λ and can be computed as:

$$
P\left(\mathscr{D}_\lambda^y\right) = \sum_{q' \in \mathcal{L}_y} \alpha^y(q', \lambda).
\tag{7.40}
$$

Plugging (7.39) and (7.40) into (7.38), we get:

$$\mathbb{E}\left[\mathrm{pt}(q)\Big|\mathscr{D}_\lambda^y\right] = \frac{\sum_{t=1}^{\lambda-1} \alpha^y(q,t)\beta^y(q,\lambda-t)}{\sum_{q'\in\mathscr{L}_y} \alpha^y(q',\lambda)}. \tag{7.41}$$

The bounded betweenness measure based on walks up to length λ is obtained as an expectation of the betweennesses for all length $1 \le l \le \lambda$:

$$\mathbf{B}_\lambda(q,y) = \sum_{l=1}^{\lambda} \frac{P\left(\mathscr{D}_l^y\right)}{Z} \mathbb{E}\left[\mathrm{pt}(q)\Big|\mathscr{D}_l^y\right]$$

$$= \frac{\sum_{l=1}^{\lambda}\sum_{t=1}^{l-1} \alpha^y(q,t)\beta^y(q,l-t)}{\sum_{l=1}^{\lambda}\sum_{q'\in\mathscr{L}_y} \alpha^y(q',l)}. \tag{7.42}$$

Finally, the decision process consists in classifying unlabeled vertices using a maximum a posteriori (MAP) decision rule from the betweenness computed for each class $y \in \mathscr{Y}$.

7.4 Chapter Remarks

Semi-supervised learning concerns with how computers and natural systems such as humans learn in the presence of both labeled and unlabeled data. Semi-supervised learning is of great interest in machine learning and data mining because it can use readily available unlabeled data to improve supervised learning tasks when the labeled data is scarce or expensive. Semi-supervised learning also displays potential when conceived as a quantitative tool to understanding categorical human learning, in which most of the input or received information is self-evidently unlabeled. In order to effectively use the unlabeled data in the learning process, we have seen that some assumptions on the unlabeled data must be satisfied, such as the cluster, smoothness, and manifold assumptions.

The most active area of research in the field of semi-supervised learning has been related to methods based on graphs or networks. A noticeable advantage of using networks for data analysis is the ability to reveal the topological structure of the data set. Once a graph is constructed, the goal is to propagate labels from labeled to unlabeled instances in accordance with a diffusive process. Many graph-based methods can be expressed in terms of a regularization framework, in which the goal is to minimize a cost or energy function that is composed of two terms: the loss and regularization functions. The loss function tends to penalize decisions that flip the labels of pre-labeled vertices. In contrast, the regularization function is responsible for modeling the cost of propagating labels to unlabeled vertices. Given that many algorithms rely on the smoothness assumption, this function must be smooth in dense regions of the network. A main drawback of these techniques is that a matrix inversion operation if often necessary in the propagation process. Hence, the application of these techniques in large-scale graphs is reduced.

In the case study of semi-supervised learning in Chap. 10, we present an alternative semi-supervised learning technique that relies on a competitive-cooperative process among several particles. These particles navigate in the network forming teams. The goal of each team is to conquer new vertices, while also defending their previously conquered vertices. The visiting process of the particles has an analogy with the label propagation process in these regularization frameworks. However, the particle navigation does not need matrix inversion, which enables us to use it in large-scale problems. In addition, we show that the particle competition algorithm can be used to detect and prevent error propagation. In this regard, the algorithm considers that the initial beliefs or pre-labeled instances are not totally reliable. In the learning process, the algorithm then rearranges these labels whenever it spots non-smoothness in the learning problem.

References

1. Baluja, S., Seth, R., Sivakumar, D., Jing, Y., Yagnik, J., Kumar, S., Ravichandran, D., Aly, M.: Video suggestion and discovery for youtube: taking random walks through the view graph. In: Proceedings of the 17th International Conference on World Wide Web, WWW '08, pp. 895–904. Association for Computing Machinery, New York, NY (2008)
2. Belkin, M., Matveeva, I., Niyogi, P.: Regularization and semi-supervised learning on large graphs. In: Shawe-Taylor, J., Singer, Y. (eds.) Learning Theory, Lecture Notes in Computer Science, vol. 3120, pp. 624–638. Springer, Berlin, Heidelberg (2004)
3. Belkin, M., Niyogi, P., Sindhwani, V.: On manifold regularization. In: Proceedings of the Tenth International Workshop on Artificial Intelligence and Statistics (AISTAT 2005), pp. 17–24. Society for Artificial Intelligence and Statistics, Cliffs, NJ (2005)
4. Belkin, M., Niyogi, P., Sindhwani, V.: Manifold regularization: a geometric framework for learning from labeled and unlabeled examples. J. Mach. Learn. Res. 7, 2399–2434 (2006)
5. Blum, A., Chawla, S.: Learning from labeled and unlabeled data using graph mincuts. In: Proceedings of the Eighteenth International Conference on Machine Learning, pp. 19–26. Morgan Kaufmann, San Francisco (2001)
6. Blum, A., Lafferty, J., Rwebangira, M.R., Reddy, R.: Semi-supervised learning using randomized mincuts. In: Proceedings of the Twenty-first International Conference on Machine Learning, p. 13. Association for Computing Machinery, New York, NY (2004)
7. Callut, J., Françoise, K., Saerens, M., Duppont, P.: Semi-supervised classification from discriminative random walks. European Conference on Machine Learning and Principles and Practice of Knowledge Discovery in Databases, Lecture Notes in Artificial Intelligence, vol. 5211, pp. 162–177 (2008)
8. Chapelle, O., Schölkopf, B., Zien, A. (eds.): Semi-supervised Learning. Adaptive Computation and Machine Learning. MIT Press, Cambridge, MA (2006)
9. Cupertino, T.H., Gueleri, R., Zhao, L.: A semi-supervised classification technique based on interacting forces. Neurocomputing 127, 43–51 (2014)
10. García, S., Fernández, A., Luengo, J., Herrera, F.: Advanced nonparametric tests for multiple comparisons in the design of experiments in computational intelligence and data mining: experimental analysis of power. Inf. Sci. 180(10), 2044–2064 (2010)
11. Gazi, V., Passino, K.M.: Stability analysis of swarms. IEEE Trans. Autom. Control 48, 692–697 (2003)

12. Grady, L.: Random walks for image segmentation. IEEE Trans. Pattern Anal. Mach. Intell. **28**(11), 1768–1783 (2006)
13. Joachims, T.: Transductive learning via spectral graph partitioning. In: Proceedings of International Conference on Machine Learning, pp. 290–297. Association for the Advancement of Artificial Intelligence Press, Palo Alto, CA (2003)
14. Karypis, G., Han, E.H., Kumar, V.: Chameleon: Hierarchical clustering using dynamic modeling. Computer **32**(8), 68–75 (1999)
15. Lafferty, J.D., McCallum, A., Pereira, F.C.N.: Conditional random fields: probabilistic models for segmenting and labeling sequence data. In: Proceedings of the Eighteenth International Conference on Machine Learning, ICML '01, pp. 282–289. Morgan Kaufmann Publishers Inc., San Francisco, CA (2001)
16. Liu, Y., Passino, K.M., Polycarpou, M.: Stability analysis of one-dimensional asynchronous swarms. IEEE Trans. Autom. Control **48**, 1848–1854 (2003)
17. Ng, A.Y., Jordan, M.I.: On discriminative vs. generative classifiers: a comparison of logistic regression and naive Bayes. In: Dietterich, T., Becker, S., Ghahramani, Z. (eds.) Advances in Neural Information Processing Systems, vol. 14, pp. 841–848. MIT Press, Cambridge, MA (2002)
18. Rabiner, L., Juang, B.H.: Fundamentals of Speech Recognition. Prentice-Hall, Englewood Cliffs (1993)
19. Sheskin, D.J.: Handbook of Parametric and Nonparametric Statistical Procedures. Chapman & Hall/CRC, Boca Raton (2007)
20. Silva, T.C., Zhao, L.: Network-based high level data classification. IEEE Trans. Neural Netw. Learn. Syst. **23**(6), 954–970 (2012)
21. Silva, T.C., Zhao, L.: Network-based stochastic semisupervised learning. IEEE Trans. Neural Netw. Learn. Syst. **23**(3), 451–466 (2012)
22. Silva, T.C., Zhao, L.: Semi-supervised learning guided by the modularity measure in complex networks. Neurocomputing **78**(1), 30–37 (2012)
23. Silva, T.C., Zhao, L.: Stochastic competitive learning in complex networks. IEEE Trans. Neural Netw. Learn. Syst. **23**(3), 385–398 (2012)
24. Silva, T.C., Zhao, L.: Uncovering overlapping cluster structures via stochastic competitive learning. Inf. Sci. **247**, 40–61 (2013)
25. Silva, T.C., Zhao, L.: High-level pattern-based classification via tourist walks in networks. Inf. Sci. **294**(0), 109–126 (2015). Innovative Applications of Artificial Neural Networks in Engineering
26. Singh, A., Nowak, R.D., Zhu, X.: Unlabeled data: now it helps, now it doesn't. In: The Conference on Neural Information Processing Systems NIPS, pp. 1513–1520 (2008)
27. Singla, P., Domingos, P.: Discriminative training of Markov logic networks. In: Proceedings of the 20th National Conference on Artificial Intelligence, AAAI'05, vol. 2, pp. 868–873. Association for the Advancement of Artificial Intelligence Press, Menlo Park, CA (2005)
28. Stevens, K.: Acoustic Phonetics. MIT Press, Cambridge, MA (2000)
29. Szummer, M., Jaakkola, T.: Partially labeled classification with Markov random walks. In: Advances in Neural Information Processing Systems, vol. 14, pp. 945–952 (2001)
30. Talukdar, P.P., Crammer, K.: New regularized algorithms for transductive learning. In: Proceedings of the European Conference on Machine Learning and Knowledge Discovery in Databases: Part II, ECML PKDD '09, pp. 442–457. Springer, Berlin, Heidelberg (2009)
31. Vidal, R., Tron, R., Hartley, R.: Multiframe motion segmentation with missing data using powerfactorization and GPCA. Int. J. Comput. Vis. **79**(1), 85–105 (2008)
32. Wang, F., Zhang, C.: Label propagation through linear neighborhoods. IEEE Trans. Knowl. Data Eng. **20**(1), 55–67 (2008)
33. Wang, F., Li, T., Wang, G., Zhang, C.: Semi-supervised classification using local and global regularization. In: AAAI'08: Proceedings of the 23rd National Conference on Artificial Intelligence, pp. 726–731. Association for the Advancement of Artificial Intelligence Press, Palo Alto, CA (2008)

34. Wu, M., Schölkopf, B.: Transductive classification via local learning regularization. In: 11th International Conference on Artificial Intelligence and Statistics, pp. 628–635. Microtome, Brookline, MA (2007)
35. Zhou, D., Bousquet, O., Lal, T.N., Weston, J., Schölkopf, B.: Learning with local and global consistency. In: Advances in Neural Information Processing Systems, vol. 16, pp. 321–328. MIT Press, Cambridge, MA (2004)
36. Zhu, X.: Semi-supervised learning literature survey. Tech. Rep. 1530, Computer Sciences, University of Wisconsin-Madison (2005)
37. Zhu, X.: Semi-supervised learning with graphs. Doctoral thesis, Carnegie Mellon University CMU-LTI-05-192 (2005)
38. Zhu, X., Ghahramani, Z.: Learning from labeled and unlabeled data with label propagation. Tech. Rep. CMU-CALD-02-107, Carnegie Mellon University, Pittsburgh (2002)
39. Zhu, X., Goldberg, A.B.: Introduction to Semi-Supervised Learning. Synthesis Lectures on Artificial Intelligence and Machine Learning. Morgan and Claypool Publishers, San Rafael, CA (2009)
40. Zhu, X., Ghahramani, Z., Lafferty, J.: Semi-supervised learning using gaussian fields and harmonic functions. In: International Conference on Machine Learning, pp. 912–919 (2003)

Chapter 8
Case Study of Network-Based Supervised Learning: High-Level Data Classification

Abstract The power of computers to generalize to unseen data is intriguing. Computers have been used successfully to accurately predict prices of non-catalogued houses, trends in financial time series, or even to classify whether cancer tumors are benign or malign. One thing that all these tasks have in common is that computers are put forward to output answers to which they have not been explicitly programmed. A natural computational solution to estimate unseen data is to rely on the knowledge bases to which computers have been exposed, effectively mimicking the past behaviors. This chapter deals with supervised learning from a new learning perspective: a hybrid classification framework is presented that combines the decisions of low- and high-level classifiers. The low-level classifier realizes the classification task considering physical features of the input data, such as geometrical or statistical characteristics. In contrast, the high-level classification process checks the compliance of new test instances with the characteristic patterns formed by each of the classes that composes the training data. Test instances are declared members of those classes whose formed patterns are maintained with the introduction of those test instances. For this end, the high-level classifier extracts suitable organizational and topological descriptors of the network constructed from the input data. Using these network-based descriptors in a convenient collective manner, the high-level term is expected to promote the detection of data patterns with semantic and global meanings. The way we extract the patterns using these descriptors gives rise to several strategies to build up the high-level framework. In this book, we show two forms of pattern extraction strategies: using classical network measurements and employing dynamic information that is generated by several *tourist walk processes*. The ability of discovering high-level features formed by the data relationships is investigated using several artificial and real-world data sets. Here, we focus in situations in which the high-level term is able to identify intrinsic data patterns, but the low-level term alone fails to do so. This provides a clear motivation for the employment of a dual classification procedure (low + high). The obtained results reveal that the hybrid classification technique is able to improve the already optimized performances of traditional classification techniques. Finally, the hybrid classification approach is applied to recognize handwritten digits images.

© Springer International Publishing Switzerland 2016
T.C. Silva, L. Zhao, *Machine Learning in Complex Networks*,
DOI 10.1007/978-3-319-17290-3_8

8.1 A Quick Overview of the Chapter

This chapter treats the issue of supervised data classification by using not only physical features of data items, but also their high-level characteristics. As examples of low-level features, we may highlight: distance between data instances, conditional distribution of the data, and composition of the data neighborhood. In contrast, high-level characteristics can be defined under several different perspectives. One can understand them as semantic relationships that extrapolate the classical raw end-to-end relationships of the data (such as the edges in a network context). In this regard, subsets of these raw relationships may give rise to new concepts of the data organization that are ultimately not seen by low-level visions. For instance, data members of the same class (subset of relationships) may share homogeneous visions to each other, such as in a well-behaved distribution. Meanwhile, they may also indicate heterogeneous organizations for different classes. Hence, here we consider high-level features as descriptors that summarize the organization of the data relationships in a structural sense.

Despite being an interesting problem, most methods in the literature ignore the high-level relations among the data, such as the formation of clear patterns in the data relationships. In view of this gap, a hybrid classification technique has been proposed that takes into account both types of learning [24]. In essence, the low-level classification is guided by the labels and the physical features of the data items. In practical terms, it can therefore be implemented by any traditional techniques in the literature. In contrast, the high-level classification uses, besides the data labels, structural or pattern information of the data relationships. Here, new forms of extracting high-level features of the data relationships are discussed using a network-based approach. In this respect, the pattern extraction is conducted by exploring the complex topological properties of the underlying network constructed from the input data. In this framework, the low- and high-level classifiers are joined together via a suitable convex linear combination, which is calibrated by what is called the *compliance term*. Basically, the compliance term adjusts the importance that is given for the low- and high-level decisions.

In this chapter, two different implementations of the high-level term are discussed, both relying on a networked representation of the data, as follows:

- The first one comprises a weighted combination of three classical network measurements, namely the assortativity, the clustering coefficient, and the average degree;
- The second one is composed of two quantities that are directly derived from the dynamics of tourist walks, which are the cycle and the transient lengths.

The compliance term plays a crucial role in the classification process. As such, several analyses are going to be provided in order to show the impact of the compliance term in data sets with different distributions and particularities that range from completely well-posed classes to highly overlapping classes. As a quick glimpse of the final results, we show that one must raise the decision influence of the high-level classifier as the joint distribution of the classes becomes more complex in the sense of the existence of overlapping regions.

Once the hybrid classification technique is properly presented, we explore the effectiveness of the model by delving into the real-world application of handwritten digits and letters recognition. Additionally, to illustrate the influence that the compliance term makes upon the final decision in this real-world problem, a small manuscript digits network sampled from the real data set is displayed. Such a network shows that the high-level term is really necessary in special occasions.

8.2 Motivation

Data items are not isolated points in the attribute space but instead tend to form certain patterns when looked in a collective manner. For example, in Fig. 8.1, the test instance represented by the "triangle" (purple) will probably be classified as a member of the "square" (blue) class if only distances among data instances are considered. In contrast, if we take into account the relationship and semantic meaning among the data items, we would intuitively classify the "triangle" item as a member of the "circle" (red) class, since a clear pattern of a "moon" contour is formed. The human (animal) brain has the ability to identify patterns according to the semantic meaning of the input data. But, this feature still stands as a hard task for computers. Supervised data classification that not only considers physical

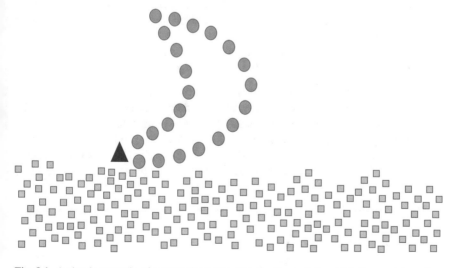

Fig. 8.1 A simple example of a supervised data classification task in which there exists a class with a clear pattern, the "*circle*" (*red*) class, and another without apparent structural organization, the "*square*" (*blue*) class. The goal is to classify the "*triangle*" (*purple*) data item. Traditional (low-level) classifiers would have trouble in classifying such item, since they derive their decisions merely based on physical measures

attributes of data items but also their pattern formations is here referred to as *high-level classification.*

The hybrid classification technique presents a way to classify data combining two semantically different views: the low-level view applies physical features of the data and the high-level view checks pattern formation of the data. In this sense, the co-training technique [4] is related to the hybrid classification technique. Co-training requires two independent views of the data. It first learns a separate classifier for each view using the labeled data items. Then, the most confident predictions of each classifier on the unlabeled data are then used to iteratively construct additional labeled training data. However, the "independent views" in co-training are generated by low-level classification techniques, i.e., the "independence" is at the physical feature level. On the other hand, the hybrid technique [24, 28] gives "independent views" from different levels ranging from physical features to semantic meaningful patterns. In the same sense, another related technique is the committee machine, which consists of an ensemble of classifiers [12]. In this case, each classifier makes a decision by itself and all these decisions are combined into a single response by a voting scheme. The combined response of the committee machine is supposed to be superior to those of its constituent experts. Again, all the involved techniques are low-level ones.

Another strong feature of the hybrid classification technique is that it is an across-network technique, i.e., it considers the network constructed from the input data as a whole and the global pattern of the network is taken into account. In the across-network approach, we take a set of network measures for each constructed data network, in such a way to characterize the global patterns formed by the underlying network via a measure vector. In this extraction process, we observe each network as if we were outside of it and hence each extracted measure represents a different view of the network.

In contrast, in the within-network approach, we look at the network inside it. In the within-network case, we basically have two objectives:

1. Making a probabilistic or deterministic inference to find out the best route from one vertex to another. For example, we may determine the class label of a test vertex using an inference process starting from an already labeled vertex.
2. Information transmission or diffusion. In this case, we propagate some kind of information to the entire or a portion of the network. For example, we may propagate labels from some vertices to the entire network in semi-supervised learning.

Figure 8.2 illustrates a schematic of the differences of the across- and within-network approaches.

The across-network feature of the hybrid framework classification contrasts with several other related works, such as the Semantic Web [1, 7, 23] and Statistical Relational Classification (SRC). Semantic Web uses ontologies to describe the semantics of the data. Even though it is a promising idea, it still presents several difficult challenges. A key challenge in creating Semantic Web is the semantic mapping among the ontologies, i.e., there are more than one ontology to describe

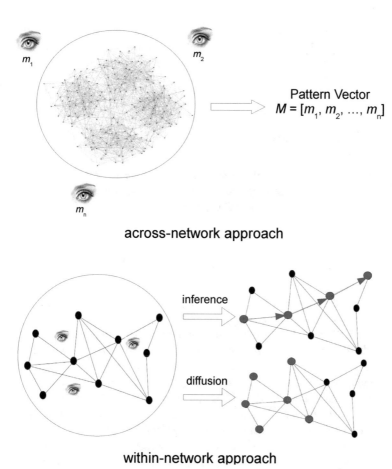

across-network approach

within-network approach

Fig. 8.2 Differences of across- and within-network approaches. In the across-network approach we look at the network as an "outsider" from different viewpoints. In the within-network, we either perform inference or label diffusion with processes that take place inside the network

the same data item. Another challenge is the one-to-many mapping of concepts in the Semantic Web, since most techniques are only able to induce one-to-one mapping, which does not correspond to real-world problems. In the case of statistical relational classification (SRC), it predicts the category of an object based on its attributes and its links and attributes of linked objects. Network-based SRC can be categorized into three main groups [9]: collective inference [9, 19, 29, 36–38, 40], network-based semi-supervised learning [5, 25–27, 39], and contextual classification techniques [2, 18, 20, 31, 32, 34, 35]. In all the cases, the labels are propagated from pre-labeled vertices to unlabeled vertices considering the local relationships or certain smoothness criteria. Therefore, those techniques are kind of within-network ones. On the other hand, the classification method introduced in high-level classification technique presents a different approach. It does not consider the paths within the network, instead, it characterizes the pattern formation of the whole network by exploiting its topological properties using a set of network measurements. In this case, each of these measures views the network as an outsider in different perspectives.

8.3   Model Description

In this section, the high-level classification technique is presented. Specifically, Sect. 8.3.1 supplies the general idea of the model and Sect. 8.3.2 deals with the methodology for building the hybrid classification framework.

8.3.1   Fundamental Ideas Behind the Model

Suppose that it is given a training set $\mathscr{X}_{\text{training}} = \{(x_1, y_1), \ldots, (x_L, y_L)\}$ with L labeled items, where the first component of the i-th tuple $x_i = (x_{i1}, \ldots, x_{iP})$ denotes the attributes of the P-dimensional i-th training instance and the second component $y_i \in \mathscr{Y}$ is the class label of x_i. Denote $Y = |\mathscr{Y}|$ as the number of classes in the classification problem. Recall that we have a binary classification problem when $Y = 2$ and a multiclass classification problem when $Y > 2$.

As usual, the goal of supervised learning is to learn a mapping $x \mapsto y$. Normally, the generalization power of the constructed classifier is checked against a test set of U items $\mathscr{X}_{\text{test}} = \{x_{L+1}, \ldots, x_{L+U}\}$ without label information.

The classification process consists of two phases: the *training phase* and the *classification phase*. In the training phase, the classifier is induced or trained by using the training instances (labeled data) in $\mathscr{X}_{\text{training}}$. In network-based models, the classifier's model is represented by a network that is formed from the input data and the associated labels. We term this output network from the training phase as the training network. In the classification phase, the labels of the test instances in $\mathscr{X}_{\text{test}}$ are predicted using the induced classifier. That is, we start off from the training network and make some modifications to accommodate the unseen test instances. Using this slightly modified network, we predict the label of the test instance. This modified network is referred to as the classification network.

8.3.1.1   Training Phase

In this phase, the training data is transformed to a network $\mathscr{G}$ using a network formation technique $g : \mathscr{X}_{\text{training}} \mapsto \mathscr{G} = \langle \mathscr{V}, \mathscr{E} \rangle$. Hence, we have $V = |\mathscr{V}|$ vertices and $E = |\mathscr{E}|$ edges in the training network. Each vertex in $\mathscr{V}$ represents a training instance in $\mathscr{X}_{\text{training}}$, so that $V = L$ holds.

The network is constructed using a combination of the ϵ-radius and k-nearest neighbors (k-NN) techniques. As shown in the network construction chapter, both approaches have their limitations,[1] i.e., these techniques, applied in the isolated form, may generate densely connected networks or may split the vertices into disconnected components.

For this reason, the combination of ϵ-radius and k-nearest neighbors techniques is used to construct the training network. The neighborhood of a training vertex x_i is given by:

$$\mathcal{N}_{\text{training}}(x_i) = \begin{cases} \epsilon\text{-radius}(x_i, y_i), & \text{if } |\epsilon\text{-radius}(x_i, y_i)| > k \\ k\text{-NN}(x_i, y_i), & \text{otherwise} \end{cases} \tag{8.1}$$

in which y_i denotes the class label of the training instance x_i, ϵ-radius(x_i, y_i) returns the set $\{x_j, j \in \mathcal{V} : d(x_i, x_j) <= \epsilon \wedge y_i = y_j\}$, and k-NN(x_i, y_i) returns, in principle, the set containing the k nearest vertices of the same class as x_i. There is a caveat, however, in this returned set of the k-NN technique. Suppose we rank all of data items in accordance with their similarities in relation to x_i. Let this sorted sequence be denoted by $\mathscr{S}(x_i) = \{x_i^{(1)}, \ldots, x_i^{(k-1)}, x_i^{(k)}, x_i^{(k+1)}, \ldots, x_i^{(Y(x_i)-1)}\}$, where $Y(x_i)$ is the number of data items of the same class as x_i. In this notation, $x_i^{(1)}$ and $x_i^{(Y(x_i)-1)}$ are the most and the least similar data items to x_i, respectively. As pointed out, we first try to connect x_i to its k most similar data items, i.e., $\{x^{(1)}, \ldots, x^{(k)}\}$. However, if we end up getting more than one graph component of the same class as x_i, we then drop the least similar data item among those k most similar data items, that is, we discard $x_i^{(k)}$, and attempt to connect x_i to the next most similar data item $x_i^{(k+1)}$. This process is recursively performed until we find connections of x_i to other data items in such a way to prevent the emergence of more than a network component of the same class.

Note that the ϵ-radius technique is used for dense regions ($|\epsilon$-radius$(x_i)| > k$), while the k-NN is employed for sparse regions. With this mechanism, it is expected that each class is represented by a unique and single component. Below, we present a simple contextual example showing the network formation technique.

Example 8.1. Consider the scatter plot in Fig. 8.3, where the task is to determine to which neighbors the central vertex of the red class (dark gray) connects. Consider that $k = 2$ and ϵ is the radius illustrated in the figure. As there are $3 > k$ vertices in the ϵ-neighborhood, the area depicted in the figure is considered as a dense region and the ϵ-radius technique is employed. In this way, the central red (dark gray) vertex is connected to the other three red (dark gray) vertices reached by this radius.

---

[1]Revisit Sect. 4.3 for a thorough analysis of the shortcomings and advantages of using k-NN and ϵ-radius network formation techniques.

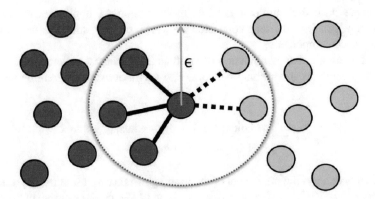

Fig. 8.3 Illustration of the network formation technique consisting in a combination of the k-nearest neighbor and the ϵ-radius techniques. In the depicted network, there are two classes: the *red* (*dark gray*) and the *blue* (*light gray*) classes. Since $k = 2$, the local region is considered as densely populated. Thus, the ϵ-radius technique is employed. As the centralized vertex belongs to the *red* (*dark gray*) class, it is only permitted to be linked to other *red* (*dark gray*) class vertices

For the sake of clarity, Fig. 8.4a shows a schematic of how the network looks like for a multiclass classification with $Y = 3$ when the training phase is completed. In this case, each class holds a representative component. In the figure, the surrounding circles denote these components: $\mathscr{G}_{C_1}$, $\mathscr{G}_{C_2}$, and $\mathscr{G}_{C_3}$.

8.3.1.2   Classification Phase

In the classification phase, the unlabeled data items (test instances) in the $\mathscr{X}_{\text{test}}$ are presented to the classifier one by one. The neighborhood of the test instance x_i is defined using the following rule:

$$\mathscr{N}_{\text{classification}}(x_i) = \begin{cases} \epsilon\text{-radius}(x_i), & \text{if } |\epsilon\text{-radius}(x_i)| > k. \\ k - \text{NN}(x_i), & \text{otherwise.} \end{cases} \tag{8.2}$$

Equation (8.2) means that the ϵ-radius connects every vertex within the radius ϵ, disregarding the class labels of the neighbor vertices. If the region is sparse enough, i.e., there are less than k vertices in this neighborhood, then the k-NN approach is employed. The modified network with the test instance is the classification network. In the high-level model, each class retains a network component. Once a test item is inserted, each component (class) calculates the changes that occur in its pattern formation with the insertion of this test instance by means of a set of complex network measures. If slight or no changes occur, then it is said that the test instance is in compliance with that class pattern. As a result, the high-level classifier yields a large membership value for that test instance on that class. Conversely, if these changes dramatically modify the class pattern, then the high-level classifier produces a small membership value on that class.

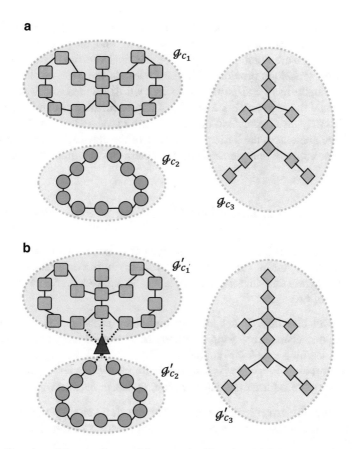

Fig. 8.4 Overview of the two phases of the supervised learning. (**a**) Schematic of the network in the training phase. (**b**) Schematic of how the inference in the classification phase is performed

Figure 8.4b shows a schematic of how the classification process is performed. The test instance (triangle-shaped) is inserted using the traditional ϵ-radius technique, resulting in the classification network. Due to its insertion, the class components become altered: $\mathscr{G}'_{C_1}$, $\mathscr{G}'_{C_2}$, and $\mathscr{G}'_{C_3}$, in which each of them is a component surrounded by a circle in Fig. 8.4b. It may occur that some class components do not share any links with this test instance. In the figure, this happens with $\mathscr{G}'_{C_3}$. In this case, the test instance does not comply with the pattern formation of the class component. For the components that share at least a link ($\mathscr{G}'_{C_1}$ and $\mathscr{G}'_{C_2}$), each of it calculates, in isolation, the impact on its pattern formation by virtue of the insertion of the test instance. For example, when we check the compliance of the test instance with the component $\mathscr{G}'_{C_1}$, the connections from the test instance to the component $\mathscr{G}'_{C_2}$ are ignored, and vice versa.

At the same time, a low-level classifier is also constructed to predict the membership of the test instance for each class. At the end, the predictions produced by both classifiers are combined to produce the final decision. Each of the low- and high-level technique provides a different view of the data set. The low-level techniques usually has good performance on well distributed and well separated data sets, while the high-level one has the ability to identify semantic meaning of the data. We may also understand the classification process in such a way that the low-level one guarantees the basic performance of the classification results and the high-level one explores complex and special patterns hidden in the data set.

8.3.2   Derivation of the Hybrid Classification Framework

The high-level classification complements the performance of the learning procedure by exactly capturing the formation of patterns in the data relationships. For this reason, a hybrid classification technique F is introduced. It consists of a convex combination of two terms:

1. A low-level classifier. It can be any traditional classification technique, for instance, a decision tree, Support Vector Machines (SVM), neural networks, Bayesian learning, or a k-NN classifier;
2. A high-level classifier, which is responsible for classifying test instances according to their pattern formation with the training data or network.

Specifically, the hybrid framework yields the decision $F_i^{(y)}$ representing the membership of the test instance $x_i \in \mathscr{X}_{\text{test}}$ with respect to the class $y \in \mathscr{Y}$ as follows:

$$F_i^{(y)} = (1 - \rho)L_i^{(y)} + \rho H_i^{(y)}, \tag{8.3}$$

in which $L_i^{(y)} \in [0, 1]$ and $H_i^{(y)} \in [0, 1]$ denote the low- and high-level predictions of test instance x_i's membership towards class y and $\rho \in [0, 1]$ is the *compliance term*, which plays the role of counterbalancing the classification decisions supplied by both classifiers. Whenever $L_i^{(y)} = 1$ and $H_i^{(y)} = 1$, x_i is very similar (low-level) and perfect complies (high-level) with class y. In contrast, whenever $L_i^{(y)} = 0$ and $H_i^{(y)} = 0$, x_i does not present any similarities (low-level) nor complies to the pattern formation (high-level) of class y. Values in-between these two extremes lead to natural uncertainty in the classification process and are found in the majority of times during a classification task. It is worth noting that (8.3) is a convex combination of two scalars, since the sum of the coefficients is unitary. Thus, as the domains of $L_i^{(y)}$ and $H_i^{(y)}$ range from 0 to 1, it is guaranteed that $F_i^{(y)} \in [0, 1]$. Therefore, (8.3) provides a fuzzy classification method. Moreover, it is valuable to mention that, when $\rho = 0$, (8.3) reduces to a common low-level classifier.

The test instance x_i receives the label from that class $y \in \mathscr{Y}$ that maximizes (8.3). Mathematically, the estimated label of the test instance x_i, $\hat{y}_i$, is computed as:

$$\hat{y}_i = \max_{y \in \mathscr{Y}} F_i^{(y)}. \tag{8.4}$$

Now, we proceed to a detailed analysis of the high-level classification term H. In the high-level classification, the inference of pattern formation within the data is processed using the generated training network. Due to the network formation process, the training network has the following structural constraints:

1. Each class $y \in \mathscr{Y}$ is an isolated network component; and
2. Each class $y \in \mathscr{Y}$ retains a representative and unique network component.

With these two network properties in mind, the pattern formation of the data is quantified via suitable combinations of network measurements. These measures are chosen in a way to cover relevant high-level aspects of the class component. In special, it is often desirable to capture strictly local, mixed, and global network characteristics, such as to cover structural network aspects in different perspectives.[2] The high-level classifier accepts an arbitrary number of network measurements. Suppose $m > 0$ network measures are selected to comprise the high-level classifier H. Mathematically, the high-level prediction on the membership of test instance $x_i \in \mathscr{X}_{\text{test}}$ with respect to class $y \in \mathscr{Y}$, here written as $H_i^{(y)}$, is given by:

$$H_i^{(y)} = \frac{\sum_{u=1}^{m} \alpha(u) \left[1 - f_i^{(y)}(u) \right]}{\sum_{g \in \mathscr{Y}} \sum_{u=1}^{m} \alpha(u) \left[1 - f_i^{(g)}(u) \right]}, \tag{8.5}$$

in which $\alpha(u) \in [0, 1], \forall u \in \{1, \ldots, m\}$, is a user-defined parameter that indicates the influence of each network measure in the classification process and $f_i^{(y)}(u)$ is a function that depends on the u-th network measure applied to the i-th data item with regard to the class y. This function is responsible for providing an answer whether or not the test instance x_i presents the same patterns or organizational features of the class y. The denominator in (8.5) has been introduced for normalization matters. Indeed, Eq. (8.5) is a valid convex combination of network measures if and only if:

$$\sum_{u=1}^{m} \alpha(u) = 1. \tag{8.6}$$

The functional form $f_i^{(y)}(u)$ is given by:

$$f_i^{(y)}(u) = \Delta G_i^{(y)}(u) p^{(y)}, \tag{8.7}$$

---

[2]Revisit Sect. 2.3.5 for definitions on the classification of network measurements.

in which $\Delta G_i^{(y)}(u) \in [0, 1]$ is the variation of the u-th network measure that occurs on the component representing class y if x_i joins it and $p^{(y)} \in [0, 1]$ is the proportion of data items pertaining to class y.

Remembering that each class has a single representative component, the strategy to check the pattern compliance of test instance x_i is to examine whether its insertion causes a great variation of the network measures in the class components. If for some class component the variation is small, then x_i is in compliance with all of the other training data items that comprise that class component, i.e., it follows the same pattern as the original members of that class. This case happens when the structural features of the network component are maintained with the introduction of x_i. Otherwise, if its insertion is responsible for a significant variation of the component's network measures, then x_i may not belong to that class in the structural sense. In this case, the structural properties of the class component are altered due to the insertion of x_i. These two behaviors are exactly captured by (8.5) and (8.7). To see that, note that a small variation of $f(u)$ causes a large membership value output by H; and vice versa. For didactic purposes, we show this concept in a simple example.

Example 8.2. Suppose a network in which there exists two equally sized classes, namely A and B. For simplicity, let us use a single network measure to quantify the pattern formation ($m = 1$). The goal is to classify a test instance x_i. Hypothetically, say that $\Delta G_i^{(A)}(1) = 0.7$ and $\Delta G_i^{(B)}(1) = 0.3$. In a pattern formation view, x_i has a bigger chance of belonging to class B, since its insertion causes a smaller impact on the pattern formation formed by class B than on the one formed by class A.

In general, a data set usually encompasses several classes of different sizes and many network measures are sensitive to the component size. In order to avoid the unbalanced problem, the term $p^{(y)}$ in Eq. (8.7) is introduced, which is the proportion of vertices that class y has. Formally, it is given by:

$$p^{(y)} = \frac{1}{V} \sum_{u=1}^{V} \mathbb{1}_{[y_u = j]}, \tag{8.8}$$

in which V is the number of vertices and $\mathbb{1}_{[.]}$ is the indicator function that yields 1 if the argument is logically true, or 0, otherwise.

The following simple example illustrates the effect of the term $p^{(y)}$.

Example 8.3. Consider a network in which there are two classes: A and B, but now A's size is ten times bigger than B's. From (8.8), $p^{(A)} = 10/11$ and $p^{(B)} = 1/11$. Without the term $p^{(y)}$, it is expected that variations of the network measures in the component A to be considerably smaller than those in component B, because of the size differences. This occurs even when the test instance x_i complies more with class B. By considering the term $p^{(y)}$, the larger value of $p^{(A)}$ over $p^{(B)}$ cancels out the effects of unbalanced classes in the calculation of the network measures. In this way, the component size modulates the variations of the topological descriptors when deciding on the compliance of new test instances.

8.4 Possible Ways of Composing High-Level Classifiers

Two network-based high-level classifications have been proposed [24, 28]. The first one makes use of a mixture of classical network measurements, namely the assortativity, the clustering coefficient, and the average degree measures. The second one uses the dynamical information generated by several tourist walks processes. In the following, the two techniques are discussed.

8.4.1 High-Level Classification Using a Mixture of Complex Network Measures

In this section, the first implementation of the high-level classification is introduced [24], which is composed of three complex network measures, which are: assortativity, clustering coefficient, and average degree. In spite of having chosen these measures, it is worth emphasizing that other network measures can be also plugged into the high-level classifier through Eq. (8.5). The reason these three measures have been chosen is as follows: the degree measure figures out strictly local or scalar information of each vertex in the network; the clustering coefficient of each vertex captures local structures by means of counting triangles formed by the current vertex and any of its two neighbors; the assortativity measure considers not only the current vertex and its neighbors, but also the second level of neighbors (neighbor of neighbor), the third level of neighbors, and so on. We perceive that the three measures characterize the network topological properties in a local to global fashion. In this way, the combination of these measures is expected to capture the pattern formations of the underlying network in a systematic manner. Below, we revisit these three measures and show how to incorporate them into the high-level classification.

8.4.1.1   First Network Measure: Assortativity

Assortativity is the preference of vertices in a network to link to others that are similar in term of vertices' degrees. This measure has been discussed in the chapter dealing with the fundamentals of Complex Networks (cf. Definition 2.36). We now derive $\Delta G_i^{(y)}(1)$ using the assortativity measure. Consider that the membership of the test instance x_i with respect to the class y is going to be determined. The actual assortativity measure of the component representing class y (before the insertion of x_i) is given by $r^{(y)}$ (step performed in the training phase). Then, we temporarily insert x_i into the component representing class y using the explained network formation technique (classification phase) and quantify the new component's assortativity measure, here denoted as $r'^{(y)}$. This procedure is performed for all of the classes $y \in \mathcal{Y}$. It may occur that some classes $u \in \mathcal{Y}$ do not share any connections with the test instance x_i. Using this approach, $r^{(u)} = r'^{(u)}$, which is undesirable, since this configuration would state that x_i complies perfectly with class u. In order to overcome this problem, a simple postprocessing is necessary: for all components $u \in \mathcal{Y}$ that do not share at least one link with x_i, we deliberately set $r^{(u)} = -1$ and $r'^{(u)} = 1$, i.e., the maximum possible difference. One may interpret this postprocessing as a way to state that x_i does not share any pattern formation with class u, since it is not even connected with it.

In view of this, we are able to calculate $\Delta G_i^{(y)}(1)$ for all $y \in \mathcal{Y}$ as follows:

$$\Delta G_i^{(y)}(1) = \frac{|r'^{(y)} - r^{(y)}|}{\sum_{u \in \mathcal{Y}} |r'^{(u)} - r^{(u)}|}, \tag{8.9}$$

in which the denominator is introduced only for normalization matters. According to (8.9), for components in which the insertion of x_i result in a considerable variation of the assortativity measure, $\Delta G_i^{(y)}(1)$ is large, and, consequently, by (8.7), $f_i^{(y)}(1)$ is also large. In light of this, the high-level classifier H produces a small membership value, as (8.5) reveals. Conversely, for insertions that do not cause a considerable variation of the assortativity, $\Delta G_i^{(y)}(1)$ is small, resulting in a small $f_i^{(y)}(1)$. As a consequence, the high-level classifier H produces a large membership value. In this way, the high-level classifier favors test instances that do not impact much the organizational and pattern features of a class.

8.4.1.2   Second Network Measure: Clustering Coefficient

Clustering coefficient is an indicator of the degree to which nodes in a network tend to cluster together in a triangular manner. This measure has also been investigated in the chapter dealing with the fundamentals of Complex Networks (cf. Definitions 2.46 and 2.47). We motivate the use of the clustering coefficient by the following facts: components with large clustering coefficient are found to have a modular structure with a high density of local connections, while components with small average clustering values tend to have many long-range connections, with the absence of local structures.

The derivation of $\Delta G_i^{(y)}(2)$ is rather analogous to the previous case, except for a simple detail: In this case, for all components $u \in \mathscr{Y}$ that do not share at least one link with the test instance x_i, we intentionally fix $CC^{(u)} = 0$ and $CC'^{(u)} = 1$, i.e., the maximum possible difference, since $CC^{(u)}$ ranges from $[0, 1]$. In this way, we are able to define $\Delta G_i^{(y)}(2)$ as:

$$\Delta G_i^{(y)}(2) = \frac{|CC'^{(y)} - CC^{(y)}|}{\sum_{u \in \mathscr{Y}} |CC'^{(u)} - CC^{(u)}|}. \tag{8.10}$$

where CC^u and CC'^u are clustering coefficients before and after the insertion of the test instance to class component u.

8.4.1.3 Third Network Measure: Average Degree or Connectivity

This measure has also been explored in the chapter related to the fundamentals of Complex Networks (cf. Definitions 2.10 and 2.12). The component connectivity is a relative simple measure, which statistically quantifies the average degree of the vertices of a component. This measure by itself is weak in terms of finding patterns in the network, since the mean value may not exactly quantify the degrees of the majority of vertices in a component. However, if it is jointly used with other measures, its recognition power significantly increases.

The derivation of $\Delta G_i^{(y)}(3)$ is similar to the previous case, except for a simple particularity: for all components $u \in \mathscr{Y}$ that do not share at least one link with test instance x_i, we purposefully assign:

$$\langle k'^{(u)} \rangle = \max \left(\langle k^{(u)} \rangle - \min_j \left(k_j^{(u)} \right), \max_j \left(k_j^{(u)} \right) - \langle k^{(u)} \rangle \right), \tag{8.11}$$

i.e., the maximum possible difference from the mean of the component. In this way, we are able to define $\Delta G_i^{(y)}(3)$ as:

$$\Delta G_i^{(y)}(3) = \frac{|\langle k'^{(y)} \rangle - \langle k^{(y)} \rangle|}{\sum_{u \in \mathscr{Y}} |\langle k'^{(y)} \rangle - \langle k^{(u)} \rangle|}. \tag{8.12}$$

Recall that $\langle k'^{(u)} \rangle$ and $\langle k^{(u)} \rangle$ represent the average degree of the component u before and after the test instance x_i is inserted into the class component u, respectively.

8.4.2   High-Level Classification Using Tourist Walks

Now we study the second approach of high-level classification [28]. Instead of using classical network measures, we use the dynamics generated by several tourist walk processes to extract high-level information from the network constructed from the input data. For the sake of clarity, we retrieve, in a synthetic manner, the main concepts that we utilize in this section.[3]

A tourist walk can be conceptualized as a walker (tourist) aiming at visiting sites (data items) in a P-dimensional map, representing the data set. At each step, the tourist follows a simple deterministic rule: it visits the nearest site that has not been visited in the previous μ steps. In other words, the walker performs partially self-avoiding deterministic walks over the data set, where the self-avoiding factor is limited to the memory window $\mu - 1$. This quantity can be understood as a repulsive force emanating from the sites in this memory window, which prevents the walker from visiting them in this interval (refractory time). Each tourist walk can be decomposed in two terms: (1) the initial *transient part* of length t and (2) a *cycle* (attractor) with period c. Since the tourist walker must respect the network topology, it may get in a dead end with no available neighboring vertex to go. In this case, we say that the cycle length is null. In spite of being a simple rule, it has been shown that this movement dynamic possesses complex behavior when $\mu > 1$ [15]. Moreover, the transient and cycle lengths are dependent on the choice of the memory length μ.

In the previous implementation of the high-level classifier, three different network measures have been employed: connectivity or average degree, clustering coefficient, and assortativity. An immediate question that arises is whether or not this set of selected measures is really sufficient to extract patterns from a network. In addition, in case they are sufficient, how one may come up with other sets of measures to construct new high-level classifiers? Therefore, a serious open problem of the previous approach is how one may choose other network measures in an intuitive way and also how one may define the learning weights that are associated with each of them. For instance, those three network measures have been chosen under a series of trial and error attempts against several well-known network measures. These issues are addressed when tourist walks are employed to construct high-level classifiers. Firstly, a unified measure to capture the patterns formed by the data is presented. In this way, one does not need to discover suitable and convenient sets of network measures to build up the high-level classifier, as occurs in the previous approach. In this new implementation, we show that the dynamical information generated by tourist walks processes is able to extract local-to-global organizational and complex features of the network by adjustments in the walker's memory length parameter. For example, when the memory window of the tourist is small, local structural features of the network are extracted. Conversely, as the

---

[3]Revisit Sect. 2.4.4 for a comprehensive review on tourist walks.

memory window grows larger, the walker is forced to venture far away from its starting point, permitting it to learn global features of the network. Secondly, the model selection procedure is simplified. As one can see in (8.5), several learning weights of the high-level classifier must be carefully fine-tuned. Because they are in large numbers, the model selection procedure may take a considerable amount of time to complete. In large-scale data sets, therefore, the application of the previous approach would be unfeasible. In this new approach, the learning weights are endogenously adjusted or fit from the training data. For this end, we utilize efficient statistical procedures to adjust the learning weights that run in linear time. As a result, the model selection effort is reduced to a large extent.

In addition to these advantages, the process of tourist walks presents some other interesting characteristics. One of them is the presence of class-dependent critical memory lengths. For a specific class $y \in \mathscr{Y}$, the critical memory length is defined as an emergent point in which larger memory length values make no changes in the behaviors of the transient and cycle lengths. This phenomenon is observed when the memory length assumes sufficient large values. We say that, when this happen, the walks have reached the "complexity saturation" of the class component. In this occasion, the global topological and organizational features of the network are said to be completely characterized in the sense of the tourist walks process. Moreover, this phenomenon can be related to phase transition in the context of complex networks.

Having in mind these concepts, the decision output of the high-level classifier based on tourist walks is given by:

$$H_i^{(y)} = K_H \sum_{\mu=0}^{\mu_c^{(y)}} w_{\text{inter}}^{(y)}(\mu) \left[w_{\text{intra}}^{(y)}(\mu) T_i^{(y)}(\mu) + (1 - w_{\text{intra}}^{(y)}(\mu)) C_i^{(y)}(\mu) \right], \qquad (8.13)$$

in which:

- $\mu_c^{(y)}$ is a critical value that indicates the maximum memory length of the tourist walks performed in the training phase for class y;
- $T_i^{(y)}(\mu)$ and $C_i^{(y)}(\mu)$ are functions that depend on the transient and cycle lengths, respectively, of the tourist walk applied to the i-th data item with regard to class y. These functions are responsible for providing an estimate of whether or not the data item i under analysis possesses the same patterns of component y;
- $w_{\text{inter}}^{(y)}(\mu)$ is the weight or influence that is given for the tourist walk with memory length μ on class y. Observe that we have used the subscript *inter* to make clear that this coefficient deals with the regulation of tourist walks with different μ;
- $w_{\text{intra}}^{(y)}(\mu)$ is the weight or influence of the transient length of a particular tourist walk with memory length μ on class y. The complementary value, i.e., $(1 - w_{\text{intra}}^{(y)}(\mu))$, records the same information but for the cycle length. Note that we have used the subscript *intra* to denote that such coefficient is modulating the dynamic generated within the same tourist walk;
- K_H is a normalization constant which ensures the fuzziness of the high-level classifier H.

8.4.2.1 Calculating the Variational Descriptors $T_i^{(y)}(\mu)$ and $C_i^{(y)}(\mu)$

The descriptors that characterize the structural variations on the component representing class $y \in \mathcal{Y}$ due to insertion of the test instance x_i are defined as:

$$T_i^{(y)}(\mu) = 1 - \Delta t_i^{(y)}(\mu) p^{(y)},$$
$$C_i^{(y)}(\mu) = 1 - \Delta c_i^{(y)}(\mu) p^{(y)},$$

$$(8.14)$$

in which $\Delta t_i^{(y)}(u), \Delta c_i^{(y)}(u) \in [0, 1]$ are the variations of the transient and cycle lengths on the component representing class y if test instance i joins it and $p^{(y)} \in [0, 1]$ is the proportion of training data items pertaining to class y.

In order to compute $\Delta t_i^{(y)}(\mu)$ and $\Delta c_i^{(y)}(\mu)$ that appear in (8.14), for a fixed μ, we perform tourist walks initiating from each of the vertices of component $y \in \mathcal{Y}$. In this way, we get the average transient and cycle lengths of component that represents class y, $\langle t^{(y)}(\mu) \rangle$ and $\langle c^{(y)}(\mu) \rangle$, respectively. For the purpose of estimating the variation of the component's network measures, consider that $x_i \in \mathcal{X}_{\text{test}}$ is a test instance. After x_i is inserted into an arbitrary class $y \in \mathcal{Y}$, we recalculate the average transient and cycle lengths of this component, denoted as $\langle t'^{(y)}_i(\mu) \rangle$ and $\langle c'^{(y)}_i(\mu) \rangle$, respectively. This procedure is performed for all classes $y \in \mathcal{Y}$. Again, if some classes $u \in \mathcal{Y}$ do not share any connections with the test instance x_i, we set a high value for $\langle t'^{(y)}_i(\mu) \rangle$ and $\langle c'^{(y)}_i(\mu) \rangle$.

Then, we can calculate $\Delta t_i^{(y)}(\mu)$ and $\Delta c_i^{(y)}(\mu)$, $\forall y \in \mathcal{Y}$, as follows:

$$\Delta t_i^{(y)}(\mu) = \frac{|\langle t'^{(y)}_i(\mu) \rangle - \langle t^{(y)}(\mu) \rangle|}{\sum_{u \in \mathcal{Y}} |\langle t'^{(u)}_i(\mu) \rangle - \langle t^{(u)}(\mu) \rangle|},$$

$$\Delta c_i^{(y)}(\mu) = \frac{|\langle c'^{(y)}_i(\mu) \rangle - \langle c^{(y)}(\mu) \rangle|}{\sum_{u \in \mathcal{Y}} |\langle c'^{(u)}_i(\mu) \rangle - \langle c^{(u)}(\mu) \rangle|},$$

$$(8.15)$$

in which the denominator is just for normalization matters. According to (8.15), large variations of a component's transient and cycle lengths, $\Delta t_i^{(y)}(\mu)$ and $\Delta c_i^{(y)}(\mu)$, yield small membership values $T_i^{(y)}(\mu)$ and $C_i^{(y)}(\mu)$ and small variations produce large membership values.

The memory length μ has a high influence on the classification result. According to (8.13), the above described procedure is performed by varying the memory length μ, ranging from 0 (memoryless) to a critical value μ_c. In this way, the descriptors can capture complex patterns of each of the representative class components in a local to global fashion. When μ is small, the walks tend to possess a small transient and cycle parts, so that the walker does not wander far away from the starting vertex. In this way, the walking mechanism is responsible for capturing the local structures of the class component. On the other hand, when μ increases, the walker is compelled to venture deep into the component, possibly very far away from its starting vertex. In this case, the walking process is responsible for capturing the global features of the component.

8.4.2.2 Determining the Intra-Modulation Parameters $w_{intra}^{(y)}(\mu)$

The idea to estimate the parameters $w_{intra}^{(y)}(\mu)$ is simple. Intuitively, the transient length and cycle length of each tourist walk can be considered as a peculiar or unique vision of the class component. For a fixed μ, if the variation of transient lengths (cycle lengths) is small, all the walks have a homogenous vision on a class component; otherwise, they have heterogeneous visions. In the event of an insertion of a new test instance, it is reliable to state that its impact on the organizational formation of the component is much stronger if its insertion causes a break in the homogenous vision rather than in the heterogeneous vision. That is, a greater influence should be intuitively given for the changes in homogeneous visions in detriment to changes in heterogeneous visions. With this idea in mind, we propose a calibration of $w_{intra}^{(y)}(\mu)$ using the variances of the transient and cycle lengths generated by the training set.

In the following, we formalize this idea. For a fixed μ and $y \in \mathcal{Y}$, let the variances of the tourist walk's transient and cycle lengths be $\sigma_c^{(y)}(\mu)$ and $\sigma_t^{(y)}(\mu)$, respectively. Then, $w_{intra}^{(y)}(\mu), \forall \mu \in \{0, \ldots, \mu_c\}, y \in \mathcal{Y}$, is given as follows:

$$w_{intra}^{(y)}(\mu) = \frac{\sigma_c^{(y)}(\mu) + 1}{\sigma_t^{(y)}(\mu) + \sigma_c^{(y)}(\mu) + 2}, \tag{8.16}$$

$$1 - w_{intra}^{(y)}(\mu) = \frac{\sigma_t^{(y)}(\mu) + 1}{\sigma_t^{(y)}(\mu) + \sigma_c^{(y)}(\mu) + 2}, \tag{8.17}$$

in which Eq. (8.16) refers to the influence of the transient length and (8.17) provides the influence of the cycle length when tourist walks with memory length μ are performed. Note that we have employed the Laplace Smoothing technique into the determination of such parameters, which adds 1 for the variances of each variable. This allows Eqs. (8.16) and (8.17) to be always defined for every $\sigma_c^{(y)}(\mu) \times \sigma_t^{(y)}(\mu) \in \mathbb{R}^2$.

8.4.2.3 Determining the Inter-Modulation Parameters $w_{inter}^{(y)}(\mu)$

In order to estimate the influence weights of tourist walks with different μ, the same strategy based on variances given in the previous subsection can be applied. i.e., a higher weight value is given to the walks that have less total variance (transient + cycle). This strategy is the same as to declare that we are favoring homogenous visions against heterogeneous visions.

The idea is formalized as follows. For a fixed μ and $y \in \mathcal{Y}$, say that the variances of the transient and cycle lengths of tourist walks with $\mu \in \{0, \ldots, \mu_c\}$ are given by $\sigma_c^{(y)}(\mu)$ and $\sigma_t^{(y)}(\mu)$, respectively. Then, $w_{inter}^{(y)}(\mu)$, $\mu \in \{0, \ldots, \mu_c\}$, is given as follows:

$$w_{inter}^{(y)}(\mu) = \frac{\sum_{\Delta=0, \Delta \neq \mu}^{\mu_c^{(y)}} \sigma_t^{(y)}(\Delta) + \sigma_c^{(y)}(\Delta)}{\sum_{\rho=0}^{\mu_c^{(y)}} \sum_{\Delta=0, \Delta \neq \rho}^{\mu_c^{(y)}} \sigma_t^{(y)}(\Delta) + \sigma_c^{(y)}(\Delta)}$$

$$= \frac{k_{inter}^{(y)} - (\sigma_t^{(y)}(\mu) + \sigma_c^{(y)}(\mu))}{\mu_c^{(y)} k_{inter}^{(y)}}, \qquad (8.18)$$

in which:

$$k_{inter}^{(y)} = \sum_{\Delta=0}^{\mu_c^{(y)}} \sigma_t^{(y)}(\Delta) + \sigma_c^{(y)}(\Delta). \qquad (8.19)$$

Note that, for a fixed μ, if $\sigma_t^{(y)}(\mu) + \sigma_c^{(y)}(\mu)$ is large, then the corresponding influence of that tourist walk, $w_{inter}^{(y)}(\mu)$, is small in relation to the others $w_{inter}^{(y)}(\Delta)$, $\Delta \neq \mu$. Conversely, if the sum of the descriptors of the tourist walk with fixed μ is small, we get a large $w_{inter}^{(y)}(\Delta)$, showing the relative importance of those tourist walks in the learning process.

8.5   Numerical Analysis of the High-Level Classification

In this section, we assess the performance of the high-level classifier composed by the dynamical information generated from the tourist walks. Section 8.5.1 reviews a problematic situation where the high-level of learning is welcomed and Sect. 8.5.2 supplies a parameter sensitivity analysis of the model.

8.5.1   An Illustrative Example

Figure 8.5 shows a segment of line representing the red or circular-shaped class (9 vertices) and also a condensed hollow circular class (torus) depicted by the blue or square-shaped class (1000 vertices). The network formation parameters are fixed as $k = 1$ and $\epsilon = 0.07^4$. The fuzzy SVM technique [16] with RBF kernel

---

[4](This radius covers, for any vertex in the straight line, two adjacent vertices, except for the vertices in each end).

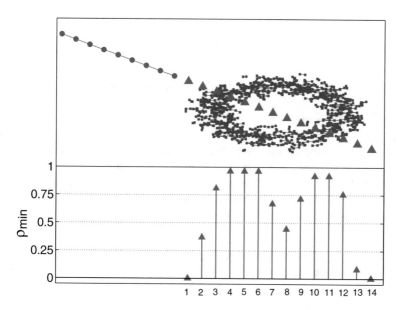

Fig. 8.5 Minimum value of the compliance term, ρ_{min}, that results in the classification of the *purple or triangle*-shaped test instances as members of the *red or circular*-shaped class. Reproduced from [28] with permission from Elsevier

($C = 2^2$ and $\gamma = 2^{-1}$) is employed as the traditional low-level classifier. The task is to classify the 14 test instances represented by the big triangle-shaped items from left to right. After a test instance is classified, it is incorporated to the training set with the corresponding predicted label (self-learning). The low part of the figure shows the minimum required value of ρ_{min} for which the triangle-shaped items are classified as members of the red or circular-shaped class. From this figure, we see that the big triangle-shaped items can be identified to form a line pattern with the red or circular elements, even if the straight line crosses the condensed region of the blue or square-shaped class. Another feature is that the required compliance term takes small values when the test instances stay long away from the blue or square-shaped class (simple cases of classification) but it takes large values when the straight line crosses the "torus" class (complex cases of classification). This means that the high-order of learning is very useful in complex situations of classification.

8.5.2   *Parameter Sensitivity Analysis*

In this section, several simulations are performed to show the influence of the model's parameters. We only provide a parameter sensitivity analysis of the high-level classifier that is based on tourist walk processes, because this approach self-adjusts the learning of the network descriptors. Therefore, model selection procedures become viable. In contrast, the model based on complex network measures has as many learning weights as there are complex network measures.

8.5.2.1 Influence of the Parameters Related to the Network Formation

The network formation step plays a crucial role in the learning process of the high-level classification. It is regulated by the parameters k and ϵ. For every data item (vertex), the algorithm determines if it is located in a sparse or dense region by counting the number of data items within a hyper-sphere with radius ϵ. If this number is smaller than k, the vertex is declared to be in a sparse region and the k-NN is used. Otherwise, the ϵ-radius is employed. If k is bounded by the number of training data items V, some mathematical consequences of this procedure are given below:

- If $k \to V \Rightarrow$ all vertices are declared to be in sparse regions $\Rightarrow$ always use the k-NN technique, regardless of ϵ;
- If $k \to 0 \Rightarrow$ all vertices are declared to be in dense regions $\Rightarrow$ always use the ϵ-radius technique;
- If $\epsilon \to \infty \Rightarrow$ all vertices are declared to be in dense regions $\Rightarrow$ always use the ϵ-radius technique, regardless of k;
- If $\epsilon \to 0 \Rightarrow$ all vertices are declared to be in sparse regions $\Rightarrow$ always use the k-NN technique.

The aforementioned behaviors imply that only for intermediate values of k and ϵ, both the k-NN and ϵ-radius techniques are enabled. This is because, if k is too large, the ϵ-radius technique is disabled. Conversely, if ϵ is too large, the k-NN technique is turned off.

8.5.2.2 Influence of the Critical Memory Length

In Ref. [28], the authors uncovered a phenomenon regarding the critical memory length $\mu_c^{(y)}$, called *complexity saturation*. In order to facilitate the understanding of such property, let us first investigate some simulation results on synthetic and real-world data sets.

For the experiments, we consider the Wine data set (unbalanced classes), which is a well-known data set from the UCI Machine Learning Repository [8]. Figure 8.6a and b portrays the transient and cycle lengths computed for each class reported in the Wine data set. With respect to the transient length behavior, we can see that the transient length increases as μ increases. However, when μ is sufficiently large, the components' transient lengths settle down in a flat region. In contrast, the cycle length behavior shows an interesting behavior, which can be roughly divided in three different regions: (1) for a small μ, the cycle length is directly proportional to μ; (2) for intermediate values of μ, the cycle length is inversely proportional to μ; and

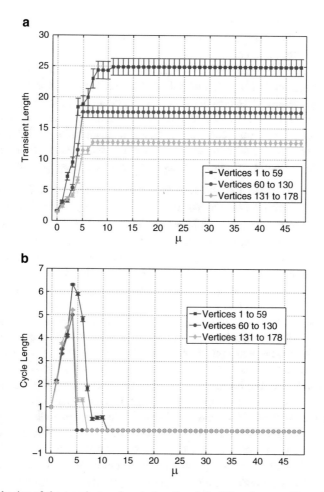

Fig. 8.6 Behavior of the transient and cycle lengths of the Wine data set. Network formation parameters: $k = 3$ and $\epsilon = 0.04$. Reproduced from [28] with permission from Elsevier. (a) Transient length vs. μ_c and (b) cycle length vs. μ_c

(3) for sufficiently large values of μ, the cycle length also settles down in a steady region . One can interpret these results as follows:

- When μ is small, it is very likely that the transient and cycle parts are also small, because the memory of the tourist is very limited. We can conceive this as a walk with almost no restrictions;
- When μ assumes an intermediate value, the transient length keeps increasing but the cycle length reaches a peak and starts to decrease afterwards. This peak characterizes the topological complexity of the component and varies from one to another. Hence, this is the most important region for capturing pattern formation of the class component by using the network topological structure;

- When μ is large, the tourist has a greater chance of getting trapped in a vertex of the graph. This happens when the entire neighborhood of the visited vertex is contained within the memory window μ. In this scenario, the transient length is expected to be very high and the cycle length, null. This phenomenon explains the steady regions in Fig. 8.6a and b. In this region, the tourist walks have already covered all the global aspects of the class component, and increasing the memory length μ will not capture any new topological features or pattern formation of the class components. In this scenario, it is said that the tourist walks have completely described the topological complexity of the class component (saturation). In view of this, the calculation of tourist walks by further increasing μ is redundant.

This analysis suggests that the accuracy of the high-level classification may not change given that we choose suitable $\mu_c^{(y)}$, $y \in \mathscr{Y}$, residing near these steady regions for each class in the problem. This means that larger values for $\mu_c^{(y)}$ only cause redundant computations as the accuracy rate does not enhance nor reduce.

This phenomenon of complexity saturation observed in these data sets can also be related to phase transition in networks. For example, when $\mu < \mu_c$, we can conceive the tourist walks in the network to be in an exploratory phase, where the dynamics of the transient and cycle lengths change as the parameter μ is modified. Therefore, in this initial exploratory phase, parameter μ is sensitive to the outcome of the tourist walks' dynamics. However, when $\mu = \mu_c$, the walk changes from the exploratory to the stationary phase, in which the lengths of the transient and cycle parts of the tourist walks performed on networks are not sensitive (independent) anymore. This holds true for all $\mu \geq \mu_c$. In this phase, the graph topology restrains the tourist walk such as to not change its dynamical information anymore. As the network becomes more dense, more different walks are probabilistic possible, and μ_c is expected to take on larger values. In this regard, in a complete graph, the μ_c of a networked topology would be exactly equal to a networkless (lattice) approach.

Based on these experiments, a heuristic for estimating the critical memory length $\mu_c^{(y)}$ is provided as follows. For a particular class, the dynamics of the tourist walks are calculated starting from $\mu = 0$. Once finished, μ is incremented and the same calculations are performed for the new μ. Say that $t^{(y)}(\mu)$ and $c^{(y)}(\mu)$ are the curves drawn from these calculations for the transient and cycle lengths, respectively. Once the derivatives of $t^{(y)}(\mu)$ and $c^{(y)}(\mu)$, i.e., $t'^{(y)}(\mu)$ and $c'^{(y)}(\mu)$ are zero, we store the $\mu_c^{(y)}$ in which this happened and start out a counter, which monitors how many iterations of μ the derivatives of these measures have not changed. This counter is incremented as μ increases. If $t'^{(y)}(\mu)$ and $c'^{(y)}(\mu)$ remain zero-valued in few iterations, the learning process is stopped and all calculations for which $\mu > \mu_c^{(y)}$ are discarded.

8.6   Application: Handwritten Digits Recognition

In this section, we show the appealing feature of high-level classification through a real pattern recognition application—handwritten digits recognition. We focus on the performance of the high-level classification that relies on a linear combination of tourist walks because the model selection procedure is simpler.

Section 8.6.1 motivates the importance of handwritten recognition in real-world applications and the challenges involved in this process. Section 8.6.2 describes the data set composed of handwritten digits that is employed in the automated recognition task. Section 8.6.3 presents a suitable image-based similarity measure that we use when constructing the training network. Section 8.6.4 lists a small set of low-level classification techniques that are plugged into the hybrid classification framework to test its robustness. Section 8.6.5 reports the results of the hybrid classifier that comprises a suitable combination of the low-level and the high-level classification. Section 8.6.6 illustrates how the training network of handwritten digits is and also shows how the high-level classifier can really help in classifying digits of a real-world data set.

8.6.1   Motivation

Handwritten recognition is the ability of computers to receive and interpret intelligible handwritten input from sources such as paper documents, photographs, touchscreens, data sets, and other devices [17, 30]. Ideally, the handwriting recognition systems should be able to read and understand any handwriting [3]. Handwriting recognition has been one of the most fascinating and challenging research areas in the field of image processing and pattern recognition in the recent years [22]. It contributes immensely to the advancement of an automation process and can improve the interface between man and machine in numerous applications [3, 30]. In general, handwritten recognition is classified into off-line or on-line. In the first case, the writing is obtained by an electronic device and the captured writing is completely available as an image to the handwritten recognition method. In the second case, the coordinates of successive points are available by means of a function dependent on time, i.e., the complete image is not given [22, 30]. In summary, several research works have been proposed [11, 21, 22] in an attempt to reduce the processing time of both off-line and on-line methods, while, at the same time, providing higher recognition accuracy. Due to the high complexity that this topic offers, it still has a wide range of problems to be addressed, such as the efficient recognition of images that are distorted or suffered a nonlinear transformation [3, 6, 30]. In view of these complexities, we attempt to utilize complex networks to help in the task of handwritten digits and letters recognition by taking advantage of the topological characteristics of the constructed network of patterns.

8.6.2   Description of the MNIST Data Set

Handwriting digits recognition is a well accepted benchmark for comparing pattern recognition methods. Here, the Modified NIST (National Institute of Standards and Technology) data set [14], MNIST for short, is used. It was created by "re-mixing" the samples from NIST's original data sets. While the NIST's training data set was taken from American Census Bureau employees, the test data set was taken from American high school students. In view of the different data distributions of the training and test sets, the NIST's complete data set was considered too hard.

The database contains 60,000 training images and 10,000 testing images. Half of the training set and half of the test set were taken from NIST's training data set, while the other half of the training set and the other half of the test set were taken from NIST's testing data set. This data set is almost balanced with regard to the size of the ten existing classes, each of which representing a digit. Similarly to [14], a pre-processing step is conducted. In this respect, the gray-level images (samples) are reduced to fit in a 20×20 pixel box, while preserving their aspect ratio.

8.6.3   A Suitable Similarity Measure for Images

In a network-based data representation, the images (data items) are represented by the vertices, while the relationships between them are given by the links. A link connecting two vertices (images) holds a weight that numerically translates the similarity between them. Each image can be represented by a "square" matrix $\eta \times \eta$. For rectangle images, a pre-processing is required to transform it into a square image. We conventionally set the pixels' values range to lie within the interval $[0, 1]$ by normalization. Thus, an arbitrary data item (image) x_i can be seen as a matrix with dimensions $\eta \times \eta$, where each pixel $x_i^{(u,j)} \in [0, 1]$, $\forall (u, j) \in \{1, \ldots, \eta\} \times \{1, \ldots, \eta\}$.

In order to construct the network, we are required to establish a similarity measure. The traditional pixel-per-pixel distance is rather insufficient in terms of reliably representing data, since such measure is very sensitive to rotations and scale modifications. With the purpose of overcoming this difficulty, we propose a measure based on the eigenvalues that each image inherently carries with it. First of all, we remove the mean associated to each data item (image), so that we have a common basis of comparison. After that, we calculate the ϕ greatest eigenvalues of the image. Efficient methods have been developed for finding the leading eigenvalues of real-valued asymmetric matrices [10, 33]. The magnitudes of the eigenvalues are related to the variations that the image possesses; hence, it is a natural carrier of information [13]. The greater its value, more information about the image it conveys. By virtue of that, a good choice is to only extract the greatest $\phi < \eta$ eigenvalues and drop the smaller values, since these do not transport too much information about the image. Also, in order to give more emphasis to the largest eigenvalues, a weight is associated to each one so that the larger an eigenvalue is, the larger is its associated weight.

Consider that we are to compare the similarity between two images, say x_i and x_j, in relation to the ϕ largest eigenvalues. We firstly sort the ϕ eigenvalues of each image as: $|\lambda_i^{(1)}| \geq |\lambda_i^{(2)}| \geq \ldots \geq |\lambda_i^{(\phi)}|$ and $|\lambda_j^{(1)}| \geq |\lambda_j^{(2)}| \geq \ldots \geq |\lambda_j^{(\phi)}|$, where $|\lambda_i^{(k)}|$ marks the k-th eigenvalue of the i-th data item. In this case, the dissimilarity $d(i,j)$ (or, equivalently, the similarity $s(i,j) = 1 - d(i,j)$) between image i and j is given by:

$$d(i,j) = \frac{1}{\rho_{max}} \sum_{k=1}^{\phi} \beta(k) \left[|\lambda_i^{(k)}| - |\lambda_j^{(k)}| \right]^2, \tag{8.20}$$

in which $\rho \in [0,1]$, $\rho_{max} > 0$ is a normalization constant, $\beta : \mathbb{N}^* \rightarrow (0, \infty)$ indicates a monotonically decreasing function that can be arbitrarily chosen by the user.

8.6.4 Configurations of the Low-Level Classification Techniques

Three low-level classification techniques are going to be used in the following computer simulations. For more details about the setup and architectural characteristics of the first two techniques, one can refer to [14].

- A *Perceptron neural network*: each input pixel value contributes to a weighted sum for each output neuron. The output neuron with the highest sum (including the contribution by virtue of the bias applied to that neuron) marks the class of the input character.
- A *k-nearest neighbors classifier*: we set the similarity as the reciprocal of the Euclidean distance and $k = 3$.
- A *network-based ϵ-radius classifier*: the ϵ-radius network formation technique is employed with a dissimilarity measure given by a weighted sum of the $\phi = 4$ greatest eigenvalues, as described in the previous section.

8.6.5 Experimental Results

Figure 8.7 shows the performance of the three low-level techniques acting together with the high-level classification in a networked environment. For example, the Perceptron alone reaches 88 % of accuracy rate, while a small increase in the compliance term is able to increase the overall model's accuracy rate to 91 %

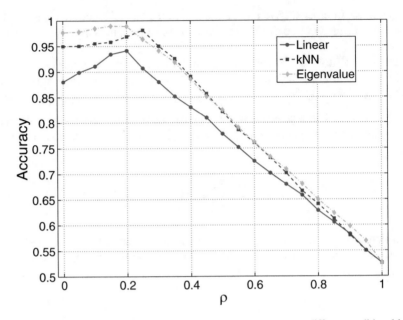

Fig. 8.7 A detailed analysis of the impact of the compliance term ρ on different traditional low-level techniques applied to the MNIST database. One can see that a mixture of traditional and high-level techniques does give a boost in the accuracy rate in this real-world data set. Reproduced from [28] with permission from Elsevier

($\rho = 0.2$). Regarding the k-nearest neighbor algorithm, for a pure traditional classifier, we obtain 95 % of accuracy rate, against 97.6 % when $\rho = 0.25$. For the weighted eigenvalue measure, we obtain 98 % of accuracy rate when $\rho = 0$, against 99.1 % when $\rho = 0.2$. It is worth noting that these enhancements are significant. Even in the third case, the improvement is quite welcomed, because it is a hard task to increase an already very high accuracy rate.

8.6.6 Illustrative Examples: High-Level Classification vs. Low-Level Classification

In this section, we provide illustrative examples to show the situations where the high-level classification works but the low-level term fails.

For simplification matters, we consider only two classes: digits "5" and "6". Figure 8.8a and b illustrates how the digit classification is carried out by using simple networks containing these small samples of digits.

Firstly, let us consider Fig. 8.8a, where the digits "5" and "6" surrounded by brown and blue boxes, respectively, represent the training set. The task is to classify the test instance represented by the digit in the red box. If only the low-level

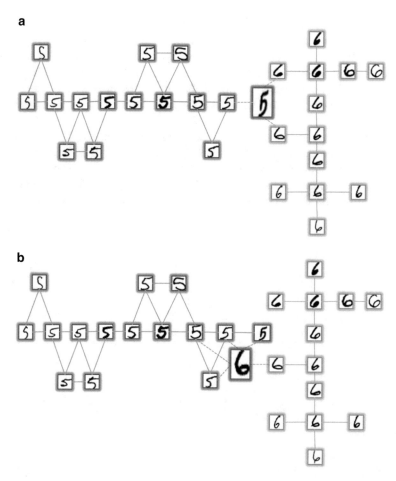

Fig. 8.8 Illustration of the pattern formation impact in a subset of samples extracted from the MNIST data set. The training instances are displayed by the *brown* (digit 5) and *blue* (digit 6) colors. The test instances are indicated by a *red color* (bigger sizes). Reproduced from [28] with permission from Elsevier. (**a**) Insertion of a digit 5 test instance and (**b**) insertion of a digit 6 test instance

classification is applied, the test digit will probably be classified as a digit "6", because there are more neighbors of digit "6" than that of "5" in the vicinity. On the other hand, if we also consider the class' geometrical disposition (high-level classification), it is more suggestive that the referred test instance is a member of the digit "5" class, because it complies more to the pattern formed by training digits of the class "5" than to the one formed by the digits of the class "6". In organizational terms, if the test digit is inserted into the class "5" as displayed, it will just extend the somewhat formed horizontal "line" pattern. As a consequence, the inclusion of the test digit in this class will disturb (change) the class organization in a small extent, i.e., its representative descriptors (transient and cycle length) will not vary significantly. However, if the test digit is inserted into the class "6", larger variations of the component measures will occur, since cycles are formed in the component. Taking into consideration that, before the insertion of the test instance, there were no cycles in the components, it is clear that the representative descriptors of the component representing the class "6" will vary considerably by virtue of this abrupt change.

Figure 8.9a and b exhibits the transient and cycle lengths, as well as their corresponding variations, as a function of μ, when the digit "5" test instance is inserted into the component represent the digit "5" cluster. As we expected, we see that the variations are very small in the class representing the digit "5", indicating and suggesting the strong compliance of the digit "5" test instance with the pattern

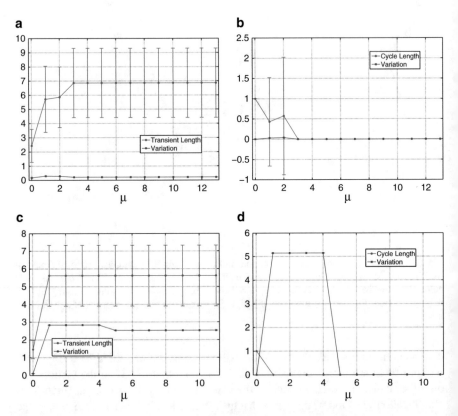

Fig. 8.9 Transient and cycle lengths of the graph components representing the training instances of the digits 5 (*brown class*) and 6 (*blue class*), as shown in Fig. 8.8a. In addition, the variation on these two measurements are reported due to the insertion of the digit 5 test instance (*red instance*). Reproduced from [28] with permission from Elsevier. (**a**) *Brown class* transient length, (**b**) *brown class* cycle length, (**c**) *blue class* transient length and (**d**) *blue class* cycle length

already formed by the representative graph component of the digit "5". On the other hand, Fig. 8.9c and d show the variations of the transient and cycle lengths of the component representing the digit "6" cluster when the digit "5" test instance is inserted. Here, we see that larger variations occur, which means that the digit "5" test instance does not comply with the pattern formed by the representative component of the digit "6" cluster.

Putting together these two observations, we conclude that the high-level classification will correctly classify the test instance as a digit "5". The same reasoning can be applied to the digit network shown in Fig. 8.8b. In this case, the transient and cycle lengths as well as the corresponding variations are shown in Fig. 8.10a–d, when the digit "6" test instance is inserted into the component of digit "5" or "6" cluster, respectively. Using the same arguments and aforementioned plots, in this situation, we can verify that the digit "6" test instance is correctly classified as a digit "6".

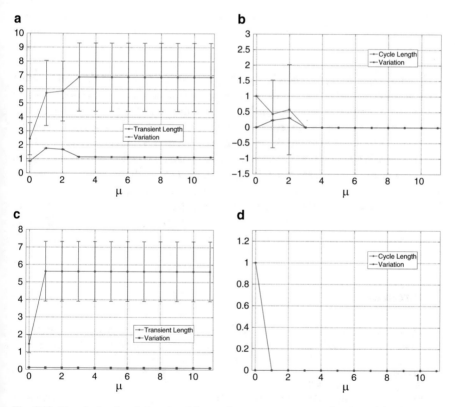

Fig. 8.10 Transient and cycle lengths of the graph components representing the training instances of the digits 5 (*brown class*) and 6 (*blue class*), as shown in Fig. 8.8b. In addition, the variation on these two measurements are reported due to the insertion of the digit 6 test instance (*red instance*). Reproduced from [28] with permission from Elsevier. (**a**) *Brown class* transient length, (**b**) *brown class* cycle length, (**c**) *blue class* transient length and (**d**) *blue class* cycle length

8.7   Chapter Remarks

In this chapter, we have studied a general classification framework that is composed of a novel combination of low- and high-level classifiers. The low-level term classifies test instances according to their physical features, while the second term measures how well new test instances comply with the existing patterns formed by the data relationships. This is performed by exploiting the complex topological properties of the network built from the training data.

In addition to the novel definition of this hybrid classification, two implementations for the high-level term are reviewed, both running in a networked environment. In the first one, the high-level classification is composed of three complex network measures: the average degree, clustering coefficient, and assortativity. In the second implementation, the complex dynamics generated by several tourist walks processes are employed. Specifically, the model is characterized by linear weighted combinations of transient and cycle lengths of different tourist walks. It stores tourist walks with varying values (up to a critical value) for the memory parameter. The motivation behind taking combinations of tourist walks with different memory values is that they can capture local (small memory values) to global (large memory values) aspects of the network.

Several experiments are conducted on synthetic and real-world data sets, so that we can better assess the performance of the hybrid classification framework. A quite interesting feature of this technique is that the influence of the high-level term has to be increased as the complexity of the classes increases. This suggests that the high-level term is specially useful in complex situations of classification.

The high-level classification techniques have been applied to handwritten digits recognition and we have seen that the hybrid model can really improve the accuracy rates of traditional classification techniques in certain conditions. It is worth noting that the employment of the high-level term in isolation generally does not perform very well. However, when utilized together with a suitable low-level classification technique, it can really boost the performance of the overall classification procedure.

References

1. Berners-Lee, T., Hendler, J., Lassila, O.: The semantic web. Sci. Am. **284**(5), 34–43 (2001)
2. Binaghi, E., Gallo, I., Pepe, M.: A cognitive pyramid for contextual classification of remote sensing images. IEEE Trans. Geosci. Remote Sens. **41**(12), 2906–2922 (2003)
3. Bishop, C.M.: Pattern Recognition and Machine Learning (Information Science and Statistics). Springer, New York (2007)
4. Blum, A., Mitchell, T.: Combining labeled and unlabeled data with co-training. In: Proceedings of the 11th Annual Conference on Computational Learning Theory, pp. 92–100 (1998)
5. Chapelle, O., Schölkopf, B., Zien, A. (eds.): Semi-Supervised Learning. Adaptive Computation and Machine Learning. MIT Press, Cambridge, MA (2006)
6. Duda, R.O., Hart, P.E., Stork, D.G.: Pattern Classification. Wiley, New York, NY (2001)

7. Feigenbaum, L., Herman, I., Hongsermeier, T., Neumann, E., Stephens, S.: The semantic web in action. Sci. Am. **297**(6), 90–97 (2007)
8. Lichman, M., UCI machine learning repository, University of California, Irvine, School of Information and Computer Sciences (2013)
9. Gallagher, B., Tong, H., Eliassi-rad, T., Faloutsos, C.: Using ghost edges for classification in sparsely labeled networks. In: Knowledge Discovery and Data Mining, pp. 256–264 (2008)
10. Goldhirsch, I., Orszag, S.A., Maulik, B.K.: An efficient method for computing leading eigenvalues and eigenvectors of large asymmetric matrices. J. Sci. Comput. **2**, 33–58 (1987)
11. Govindan, V.K., Shivaprasad, A.P.: Character recognition: a review. Pattern Recogn. **23**, 671–683 (1990)
12. Haykin, S.S.: Neural Networks and Learning Machines. Prentice Hall, Englewood Cliffs, NJ (2008)
13. Jolliffe, I.T.: Principal Component Analysis. Springer Series in Statistics, New York (2002)
14. LeCun, Y., Bottou, L., Bengio, Y., Haffner, P.: Gradient-based learning applied to document recognition. Proc. IEEE **86**(11), 2278–2324 (1998)
15. Lima, G.F., Martinez, A.S., Kinouchi, O.: Deterministic walks in random media. Phys. Rev. Lett. **87**(1), 010603 (2001)
16. Lin, C.F., Wang, S.D.: Fuzzy support vector machines. IEEE Trans. Neural Netw. **13**, 464–471 (2002)
17. Liu, C.L., Sako, H., Fujisawa, H.: Performance evaluation of pattern classifiers for handwritten character recognition. IJDAR **4**, 191–204 (2002)
18. Lu, D., Weng, Q.: Survey of image classification methods and techniques for improving classification performance. Int. J. Remote Sens. **28**(5), 823–870 (2007)
19. Macskassy, S.A., Provost, F.: Classification in networked data: a toolkit and a univariate case study. J. Mach. Learn. Res. **8**, 935–983 (2007)
20. Micheli, A.: Neural network for graphs: a contextual constructive approach. IEEE Trans. Neural Netw. **20**, 498–511(3) (2009)
21. Mori, S., Suen, C.Y., Yamamoto, K.: Historical review of OCR research and development. Proc. IEEE **80**, 1029–1058 (1992)
22. Pradeep, J., Srinivasan, E., Himavathi, S.: Diagonal based feature extraction for handwritten alphabets recognition system using neural network. Int. J. Comput. Sci. Inf. Technol. **3**, 27–38 (2011)
23. Shadbolt, N., Berners-Lee, T., Hall, W.: The semantic web revisited. IEEE Intell. Syst. **6**, 96–101 (2006)
24. Silva, T.C., Zhao, L.: Network-based high level data classification. IEEE Trans. Neural Netw. Learn. Syst. **23**(6), 954–970 (2012)
25. Silva, T.C., Zhao, L.: Network-based stochastic semisupervised learning. IEEE Trans. Neural Netw. Learn. Syst. **23**(3), 451–466 (2012)
26. Silva, T.C., Zhao, L.: Semi-supervised learning guided by the modularity measure in complex networks. Neurocomputing **78**(1), 30–37 (2012)
27. Silva, T.C., Zhao, L.: Stochastic competitive learning in complex networks. IEEE Trans. Neural Netw. Learn. Syst. **23**(3), 385–398 (2012)
28. Silva, T.C., Zhao, L.: High-level pattern-based classification via tourist walks in networks. Inform. Sci. **294**(0), 109–126 (2015). Innovative Applications of Artificial Neural Networks in Engineering
29. Skolidis, G., Sanguinetti, G.: Bayesian multitask classification with gaussian process priors. IEEE Trans. Neural Netw. **22**(12), 2011–2021 (2011)
30. Theodoridis, S., Koutroumbas, K.: Pattern Recognition. Academic, London (2008)
31. Tian, B., Azimi-Sadjadi, M.R., Haar, T.H.V., Reinke, D.: Temporal updating scheme for probabilistic neural network with application to satellite cloud classification. IEEE Trans. Neural Netw. **11**(4), 903–920 (2000)

32. Tian, Y., Yang, Q., Huang, T., Ling, C.X., Gao, W.: Learning contextual dependency network models for link-based classification. IEEE Trans. Data Knowl. Eng. **18**(11), 1482–1496 (2006)
33. Tsai, S.H., Lee, C.Y., Wu, Y.K.: Efficient calculation of critical eigenvalues in large power systems using the real variant of the Jacobi-Davidson QR method. IET Gener. Transm. Distrib. **4**, 467–478 (2010)
34. Tuia, D., Camps-Valls, G., Matasci, G., Kanevski, M.: Learning relevant image features with multiple-kernel classification. IEEE Trans. Geosci. Remote Sens. **48**(10), 3780–3791 (2010)
35. Williams, D., Liao, X., Xue, Y., Carin, L.: On classification with incomplete data. IEEE Trans. Pattern Anal. Mach. Intell. **29**(3), 427–436 (2007)
36. Zhang, D., Mao, R.: Classifying networked entities with modularity kernels. In: International Conference on Information and Knowledge Management, pp. 113–122 (2008)
37. Zhang, H., Liu, J., Ma, D., Wang, Z.: Data-core-based fuzzy min-max neural network for pattern classification. IEEE Trans. Neural Netw. **22**(12), 2339–2352 (2011)
38. Zhang, T., Popescul, A., Dom, B.: Linear prediction models with graph regularization for web-page categorization. In: Conference on Knowledge Discovery and Data Mining, pp. 821–826. Association for Computing Machinery, New York (2006)
39. Zhu, X.: Semi-supervised learning literature survey. Tech. Rep. 1530, Computer Sciences, University of Wisconsin-Madison (2005)
40. Zhu, S., Yu, K., Chi, Y., Gong, Y.: Combining content and link for classification using matrix factorization. In: Special Interest Group on Information Retrieval, pp. 487–494. Association for Computing Machinery, New York (2007)

Chapter 9
Case Study of Network-Based Unsupervised Learning: Stochastic Competitive Learning in Networks

Abstract Many business and day-to-day problems that arise in our lives must be dealt with under several constraints, such as the prohibition of external interventions of human beings. This may be due to high operational costs or physical or economical impossibilities that are inherently involved in the process. The unsupervised learning—one of the existing machine learning paradigms—can be employed to address these issues and is the main topic discussed in this chapter. For instance, a possible unsupervised task would be to discover communities in social networks, find out groups of proteins with the same biological functions, among many others. In this chapter, the unsupervised learning is investigated with a focus on methods relying on the complex networks theory. In special, a type of competitive learning mechanism based on a stochastic nonlinear dynamical system is discussed. This model possesses interesting properties, runs roughly in linear time for sparse networks, and also has good performance on artificial and real-world networks. In the initial setup, a set of particles is released into vertices of a network in a random manner. As time progresses, they move across the network in accordance with a convex stochastic combination of random and preferential walks, which are related to the offensive and defensive behaviors of the particles, respectively. The competitive walking process reaches a dynamic equilibrium when each community or data cluster is dominated by a single particle. Straightforward applications are in community detection and data clustering. In essence, data clustering can be considered as a community detection problem once a network is constructed from the original data set. In this case, each vertex corresponds to a data item and pairwise connections are established using a suitable network formation process.

9.1 A Quick Overview of the Chapter

Competition is a natural process observed in nature and in many social systems that have limited resources, such as water, food, mates, territory, recognition, etc. Competitive learning is an important machine learning approach that is widely employed in artificial neural networks to realize unsupervised learning. Early developments include the famous self-organizing map (SOM—*Self-organizing Map*) [19], differential competitive learning [20], and adaptive resonance theory (ART—*Adaptive Resonance Theory*) [6, 14]. From then on, many competitive

© Springer International Publishing Switzerland 2016
T.C. Silva, L. Zhao, *Machine Learning in Complex Networks*,
DOI 10.1007/978-3-319-17290-3_9

learning neural networks have been proposed [1–3, 16, 17, 24, 25, 28, 31, 39] and a wide range of applications has been considered. Some of these application include data clustering, data visualization, pattern recognition, and image processing [4, 7, 9, 10, 22, 41]. Without a doubt, competitive learning represents one of the main successes of the unsupervised learning development.

The network-based unsupervised learning technique that we present here is one type of competitive learning process. In essence, the model relies on a competitive mechanism of multiple homogeneous particles originally proposed in [32]. Thereafter, the particle competition technique has been enhanced and formally modeled by a stochastic nonlinear dynamical system and applied to data clustering tasks in [35]. In this chapter, we explore the particle competition algorithm by providing several empirical and analytical analyses. In this investigation, we attempt to show the potentialities and shortcomings of the particle competition technique. Given that the models of interactive walking processes correspond to many natural and artificial systems, and due to the relative lack of theory for such systems, the analytical analysis of this model is an important step to understanding such systems.

Once the fundamental idea and the model definition are properly presented, several applications that use the particle competition model are discussed in various interesting problems indicated in the literature. One of these problems is the creation of efficient evaluation indices for estimating the most likely number of clusters or communities in data sets. We show that these indices explore dynamic variables that are constructed from the competitive behavior of the particles inside the network. In this way, the evaluation of these indices is embedded within the mechanics of the particle competition process. As a result, if one takes into account that the number of clusters is far less than the quantity of data items, the process of determining the most likely number of clusters does not increase the model's time complexity order. Since the determination of the actual number of clusters is an important issue in data clustering [38, 40], the particle competition model also presents a contribution to this topic.

Following the same line, an index for detecting overlapping cluster structures is also discussed, which, under some assumptions, may also not increase the model's time complexity order due to its embedded nature within the competitive process.

With all these tools at hand, the chapter is finalized by investigating how the model behaves in an application of handwritten digits and letters clustering. Therein, we see that the competitive model is able to satisfactorily cluster several variations and distortions of the same handwritten digits and letters into their corresponding clusters.

9.2   Description of the Stochastic Competitive Model

In this section, the competitive dynamical system consisting of multiple particles [35] is discussed.

Section 9.2.1 provides the intuition behind the mechanics of the model. Section 9.2.2 builds on the caveats for constructing the transition matrix of the stochastic dynamical system that the particle competition model relies on. Section 9.2.3 formally defines the corresponding dynamical system. Section 9.2.4 explores the application of estimating the most likely number of communities or groups in a data set. Section 9.2.5 introduces another application of detecting overlapping vertices and communities. Section 9.2.6 supplies a parameter sensitivity analysis of the model's parameters. Finally, Sect. 9.2.7 analyzes convergence issues of the particle competition algorithm.

9.2.1 Intuition of the Model

Consider a network $\mathscr{G} = \langle \mathscr{V}, \mathscr{E} \rangle$, where $\mathscr{V}$ is the set of vertices and $\mathscr{E} \subset \mathscr{V} \times \mathscr{V}$ is the set of links (or edges). There are $V = |\mathscr{V}|$ vertices and $E = |\mathscr{E}|$ edges in the network. In the competitive learning model, a set of particles $\mathscr{K} = \{1, \ldots, K\}$ is inserted into the vertices of the network in a random manner. Essentially, each particle can be conceived as a flag carrier whose goal is to conquer new vertices, while defending its current dominated vertices. Given that we have a finite number of vertices, competition among particles naturally occurs. Note that the vertices play the role as valuable resources in this competition process. When a particle visits an arbitrary vertex, it strengthens its own domination level on that vertex and, at the same time, weakens the domination levels of all of the other rival particles on the same vertex. Finally, it is expected that each particle will be confined within a subnetwork corresponding a community. In this way, the communities are uncovered. Figure 9.1a, b portray a possible initial condition, in which particles are randomly inserted into network vertices, and the expected long-run dynamic of the particle competition system for an artificial clustered network with three well-defined communities.

Due to the competition effect, a particle is either in the *active* or in the *exhausted* state. Whenever the particle is active, it navigates in the network guided by a combination of two orthogonal walking rules: the random and the preferential movements. The random walking term permits particles to randomly visit neighboring vertices regardless of their current conditions and the neighborhood. Therefore, the random walking term is an unconditional rule that depends only on the immutable network topology and hence is responsible for the particle's exploratory behavior. On the contrary, the preferential walking term accounts for the defensive behavior of the particles by favoring particles to revisit and reinforce their dominated territory rather than to visit non-dominated vertices. This walking term is a conditional rule that depends on the particles' domination levels on the neighborhood. Therefore, while the movement distribution that models the exploratory behavior is fixed, that distribution that describes the defensive behavior is mutable, being dependent both on the particles and the time dimension.

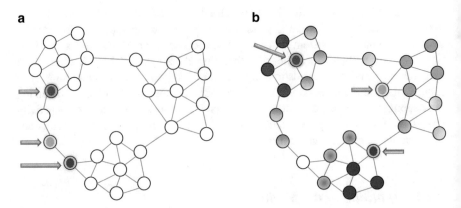

Fig. 9.1 Illustration of the initial conditions and long-run dynamic of the particle competition model. (**a**) Possible initial setup, (**b**) expected long-run dynamic

In the particle competition process, each particle carries a time-dependent energy variable that reflects its instantaneous exploration ability. The energy variable increases when the particle is visiting a vertex that it dominates, and decreases whenever it visits a vertex dominated by a rival particle. If the energy variable drops below a minimum threshold, the particle becomes exhausted and it is brought back to one of the vertices that it dominates. With this mechanism, the network always has a constant number of particles and frequent intrusions of particles to regions dominated by rival particles can be avoided. The exhaustion of particles in the learning process can be related to the smoothness assumption of unsupervised learning algorithms, because this process delimits the community borders that each particle dominates.

9.2.2   Derivation of the Transition Matrix

During the competition process, each particle $k \in \mathcal{K}$ performs two distinct types of movements:

- a *random movement term*, modeled by the matrix $\mathbf{P}_{\text{rand}}^{(k)}$, which allows the particle to venture throughout the network, without accounting for the defense of the previously dominated vertices; and
- a *preferential movement term*, modeled by the matrix $\mathbf{P}_{\text{pref}}^{(k)}$, which is responsible for inducing the particle to reinforce vertices that the particle dominates, effectively creating a preferential visiting rule to dominated vertices rather than random ones.

Consider the random vector $p(t) = [p^{(1)}(t), p^{(2)}(t), \dots, p^{(K)}(t)]$, which denotes the location of the set of K particles presented to the network. Its k-th entry, $p^{(k)}(t)$, indicates the location of particle k in the network at time t, i.e., $p^{(k)}(t) \in \mathscr{V}, \forall k \in \mathscr{K}$. With the intent of keeping track of the current states of all particles, we introduce the random vector $S(t) = [S^{(1)}(t), \dots, S^{(K)}(t)]$, where the k-th entry, $S^{(k)}(t) \in \{0, 1\}$, indicates whether particle k is active ($S^{(k)}(t) = 0$) or exhausted ($S^{(k)}(t) = 1$) at time t. When a particle is active, it performs the combined random-preferential movements; when it is exhausted, the particle switches its movement policy to a new transition matrix, here referred to as $\mathbf{P}_{\mathrm{rean}}^{(k)}(t)$. This matrix is responsible for taking the particle back to its dominated territory, in order to reanimate the corresponding particle by recharging its energy. This sequence of steps is called the *reanimation procedure*. After the particle's energy has been properly recharged, it again walks in the network. With these notations at hand, we can define a transition matrix that governs the probability distribution of the movement of the particles to the immediate future state $p(t + 1) = [p^{(1)}(t + 1), p^{(2)}(t + 1), \dots, p^{(K)}(t + 1)]$ as follows:

$$\mathbf{P}_{\mathrm{transition}}^{(k)}(t) \triangleq (1 - S^{(k)}(t)) \left[\lambda \mathbf{P}_{\mathrm{pref}}^{(k)}(t) + (1 - \lambda) \mathbf{P}_{\mathrm{rand}}^{(k)} \right] + S^{(k)}(t) \mathbf{P}_{\mathrm{rean}}^{(k)}(t), \quad (9.1)$$

in which k is the particle index, $\lambda \in [0, 1]$ modulates the desired fraction of preferential and random movements. Larger values of λ favor preferential walks in detriment to random walks. The entry $\mathbf{P}_{\mathrm{transition}}^{(k)}(i, j, t)$ indicates the probability that particle k performs a transition from vertex i to j at time t. Now we define the random and the preferential movement matrices.

Each entry $(i, j) \in \mathscr{V} \times \mathscr{V}$ of the random movement matrix is given by:

$$\mathbf{P}_{\mathrm{rand}}^{(k)}(i, j) \triangleq \frac{\mathbf{A}_{ij}}{\sum_{u \in \mathscr{V}} \mathbf{A}_{iu}}, \quad (9.2)$$

in which $\mathbf{A}_{ij}$ denotes the (i, j)-th entry of the adjacency matrix $\mathbf{A}$ of the network. It means that the probability of an adjacent neighbor j to be visited from vertex i is proportional to the edge weight linking these two vertices. The matrix is time-invariant and it is the same for every particle in the network; therefore, whenever the context makes it clear, we drop the superscript k for convenience.

In order to derive the preferential movement matrix, $\mathbf{P}_{\mathrm{pref}}^{(k)}(t)$, we introduce the following random vector:

$$\mathbf{N}_i(t) \triangleq [\mathbf{N}_i^{(1)}(t), \mathbf{N}_i^{(2)}(t), \dots, \mathbf{N}_i^{(K)}(t)]^T, \quad (9.3)$$

in which $\dim(\mathbf{N}_i(t)) = K \times 1$, T denotes the transpose operator, and $\mathbf{N}_i(t)$ registers the number of visits received by vertex i up to time t by each of the particles in the network. Specifically, the k-th entry, $\mathbf{N}_i^{(k)}(t)$, indicates the number of visits made by particle k to vertex i up to time t. Then, the matrix that contains the number of visits made by each particle in the network to all the vertices is defined as:

$$\mathbf{N}(t) \triangleq [\mathbf{N}_1(t), \mathbf{N}_2(t), \dots, \mathbf{N}_V(t)]^T, \tag{9.4}$$

in which $\dim(\mathbf{N}(t)) = V \times K$. Let us also formally define the domination level vector of vertex i, $\bar{\mathbf{N}}_i(t)$, according to the following random vector:

$$\bar{\mathbf{N}}_i(t) \triangleq [\bar{\mathbf{N}}_i^{(1)}(t), \bar{\mathbf{N}}_i^{(2)}(t), \dots, \bar{\mathbf{N}}_i^{(K)}(t)]^T, \tag{9.5}$$

in which $\dim(\bar{\mathbf{N}}_i(t)) = K \times 1$ and $\bar{\mathbf{N}}_i(t)$ denotes the relative frequency of visits of all particles in the network to vertex i at time t. In particular, the k-th entry, $\bar{\mathbf{N}}_i^{(k)}(t)$, indicates the relative frequency of visits performed by particle k to vertex i at time t. Then, the matrix form of the domination level of all vertices is defined as:

$$\bar{\mathbf{N}}(t) \triangleq [\bar{\mathbf{N}}_1(t), \bar{\mathbf{N}}_2(t), \dots, \bar{\mathbf{N}}_V(t)]^T, \tag{9.6}$$

in which $\dim(\bar{\mathbf{N}}(t)) = V \times K$. Mathematically, each entry of $\bar{\mathbf{N}}_i^{(k)}(t)$ is defined as:

$$\bar{\mathbf{N}}_i^{(k)}(t) \triangleq \frac{\mathbf{N}_i^{(k)}(t)}{\sum_{u \in \mathcal{K}} \mathbf{N}_i^{(u)}(t)}. \tag{9.7}$$

With these notations at hand, the preferential movement rule can be defined as:

$$\mathbf{P}_{\text{pref}}^{(k)}(i, j, t) \triangleq \frac{\mathbf{A}_{ij}\bar{\mathbf{N}}_j^{(k)}(t)}{\sum_{u \in \mathcal{V}} \mathbf{A}_{iu}\bar{\mathbf{N}}_u^{(k)}(t)}. \tag{9.8}$$

Equation (9.8) defines the probability of a single particle k to perform a transition from vertex i to j at time t, using solely the preferential movement term. It can be observed that each particle has a different transition matrix associated to its preferential movement. Moreover, each matrix is time-varying with dependence on the domination levels of all of the vertices ($\bar{\mathbf{N}}(t)$) at time t. Since the preferential movement term of particles directly depends on their visiting frequency to a specific vertex, as more visits are performed by a particle to a determined vertex, the higher is the chance for that particle to repeatedly visit the same vertex. Furthermore, if the domination level of the visiting particle on a vertex is strengthened, the domination levels of all other particles on the same vertex are consequently weakened. This feature occurs on account of the normalization process in (9.7): if one domination level increases, all of the others must go down, so that the overall sum still produces 1.

For didactic purposes, we now summarize and consolidate the key concepts introduced so far in a simple example given in the following.

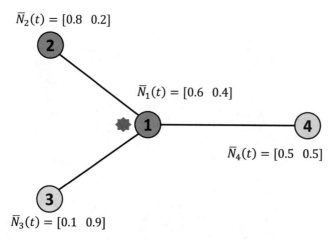

$$\bar{N}_2(t) = [0.8 \quad 0.2]$$

$$\bar{N}_1(t) = [0.6 \quad 0.4]$$

$$\bar{N}_4(t) = [0.5 \quad 0.5]$$

$$\bar{N}_3(t) = [0.1 \quad 0.9]$$

Fig. 9.2 A typical situation where the *red (dark gray)* particle, presently located at vertex 1, has to choose a neighbor to visit in the next iteration. For illustration purposes, the domination level vector for each vertex is displayed, in which the two entries represent the domination levels of the *red (dark gray)* and the *blue (gray)* particles, in that order. In this example, there are two particles, *red (dark gray)* and *blue (gray)*. The beige *(light gray) color* denotes vertices that are not being dominated by any particles in the system at time t

Example 9.1. Consider the network portrayed in Fig. 9.2, where there are two particles, namely red (dark gray) and blue (gray), and four vertices. For illustrative purposes, we only depict the location of the red (dark gray) particle, which is currently visiting vertex 1. In this example, the role that the domination level plays in determining the resulting transition probability matrix is presented. Within the figure, we also didactically supply the domination level vector of each vertex at time t. Note that the ownership of the vertex (in the figure, the color of the vertex) is set according to the particle that is imposing the highest domination level on that specific vertex. For instance, in vertex 1, the red (dark gray) particle is imposing a domination of 60 %, and the blue (gray) particle, of only 40 %. The goal here is to derive the transition matrix of the red particle in agreement with (9.1). Suppose at time t the red particle is *active*, therefore, $S^{(\text{red})}(t) = 0$. Consequently, the second term of the convex combination in (9.1) vanishes. On the basis of (9.2), the random movement term of the red particle is given by:

$$\mathbf{P}^{(\text{red})}_{\text{rand}} = \begin{bmatrix} 0 & 1/3 & 1/3 & 1/3 \\ 1 & 0 & 0 & 0 \\ 1 & 0 & 0 & 0 \\ 1 & 0 & 0 & 0 \end{bmatrix}, \tag{9.9}$$

(continued)

Example 9.1 (continued)
and the preferential movement matrix at the immediate posterior time $t + 1$, according to (9.8), is given by:

$$
\mathbf{P}_{\text{pref}}^{(\text{red})}(t + 1) = \begin{bmatrix} 0 & 0.57 & 0.07 & 0.36 \\ 1 & 0 & 0 & 0 \\ 1 & 0 & 0 & 0 \\ 1 & 0 & 0 & 0 \end{bmatrix}. \tag{9.10}
$$

Finally, the transition matrix associated to the red particle is determined by a weighted combination of the random (time-invariant) and the preferential matrices at time $t + 1$, given that the particle is active (see (9.1)). If $\lambda = 0.8$, then such matrix is given by:

$$
\mathbf{P}_{\text{transition}}^{(\text{red})}(t + 1) = 0.2 \begin{bmatrix} 0 & 1/3 & 1/3 & 1/3 \\ 1 & 0 & 0 & 0 \\ 1 & 0 & 0 & 0 \\ 1 & 0 & 0 & 0 \end{bmatrix} + 0.8 \begin{bmatrix} 0 & 0.57 & 0.07 & 0.36 \\ 1 & 0 & 0 & 0 \\ 1 & 0 & 0 & 0 \\ 1 & 0 & 0 & 0 \end{bmatrix}
$$

$$
= \begin{bmatrix} 0 & 0.52 & 0.12 & 0.36 \\ 1 & 0 & 0 & 0 \\ 1 & 0 & 0 & 0 \\ 1 & 0 & 0 & 0 \end{bmatrix}. \tag{9.11}
$$

Therefore, the red particle, which is currently in vertex 1, has a higher chance to visit vertex 2 (52 % chance of visiting) than the others. This behavior can be controlled by adjusting the λ parameter. A large value of λ induces particles to perform mostly preferential movements, i.e., particles keep visiting their dominated vertices in a frequent manner. A small value of λ, in contrast, provides a larger weight to the random movement term, making particles resemble traditional Markovian walkers as $\lambda \to 0$ [8]. In the extreme case, i.e., when $\lambda = 0$, the mechanism of competition is turned off and the model reduces to multiple non-interactive random walks. In this way, we can see that the particle competition model generalizes the dynamical system of multiple random walks, according to the parameter λ.

Now we define $\mathbf{P}_{\text{rean}}^{(k)}(t)$ matrix that is responsible for transporting an exhausted particle $k \in \mathscr{K}$ back to its dominated territory, with the purpose of recharging its energy (reanimation process). Suppose particle k is visiting vertex i when its energy is completely depleted. In this situation, the particle must regress to an arbitrary vertex j of its possession at time t, according to the following expression:

$$\mathbf{P}_{\text{rean}}^{(k)}(i,j,t) \triangleq \frac{\mathbb{1}\left[\arg\max_{m\in\mathscr{K}}\left(\bar{\mathbf{N}}_{j}^{(m)}(t)\right)=k\right]}{\sum_{u\in\mathscr{V}}\mathbb{1}\left[\arg\max_{m\in\mathscr{K}}\left(\bar{\mathbf{N}}_{u}^{(m)}(t)\right)=k\right]}, \tag{9.12}$$

in which $\mathbb{1}_{[.]}$ is the indicator function that yields 1 if the argument is logically true and 0, otherwise. The operator $\arg\max_{m\in\mathbb{K}}(.)$ returns an index M, where $\bar{\mathbf{N}}_{u}^{(M)}(t)$ is the maximal value among all $\bar{\mathbf{N}}_{u}^{(m)}(t)$ for $m = 1, 2, \ldots, K$. We note that (9.12) reduces to a uniform distribution when we take the subset of vertices that are dominated by particle k. For all of the non-dominated vertices, the transition probability is zero. Observe also that the transition probability is independent of the network topology. If no vertex is being dominated by particle k at time t, we put it in any vertex of the network in a random manner (uniform distribution on the whole network).

Example 9.2. Figure 9.3 illustrates how the reanimation scheme takes place. Consider that the red (dark gray) particle is exhausted possibly because it has visited several non-dominated vertices, which led to the depletion of its energy. The reanimation procedure consists in transporting back that particle to one of its dominated vertices, regardless of the network topology. The intuition of this procedure is that, with a relatively high probability, its energy will be renewed in the next iterations, for the neighborhood is expected to be dominated by the same particle.

Let also the random vector $E(t) = [E^{(1)}(t), \ldots, E^{(K)}(t)]$ represent the energy that each particle holds. In special, its k-th entry, $E^{(k)}(t)$ denotes the energy level of particle k at time t. In view of these definitions, the energy update rule is given by:

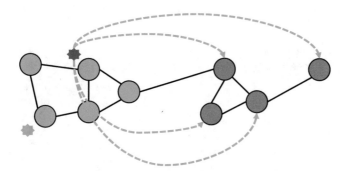

Fig. 9.3 Illustration of the reanimation scheme. The *red (dark gray)* particle is exhausted and is forced to be transported back to its dominated territory. The transition probability follows a uniform distribution on the dominated vertices

$$E^{(k)}(t) = \begin{cases} \min(\omega_{max}, E^{(k)}(t-1) + \Delta), & \text{if owner}(k,t) \\ \max(\omega_{min}, E^{(k)}(t-1) - \Delta), & \text{if } \neg\text{owner}(k,t) \end{cases} \qquad (9.13)$$

in which the parameters ω_{min} and ω_{max} characterize the minimum and maximum energy levels, respectively, that a particle may possess. Therefore, $E^{(k)}(t) \in [\omega_{min}, \omega_{max}]$. The owner(k, t) is defined as:

$$\text{owner}(k,t) = \left(\arg\max_{m \in \mathscr{K}} \left(\bar{\mathbf{N}}^{(m)}_{p^{(k)}(t)}(t) \right) = k \right) \qquad (9.14)$$

is a logical expression that essentially yields true if the vertex that particle k visits at time t (i.e., vertex $p^{(k)}(t)$) is being dominated by that same visiting particle, and results in a logical false otherwise; $\dim(E(t)) = 1 \times K$; $\Delta > 0$ symbolizes the increment or decrement of energy that each particle receives at time t. The first expression in (9.13) represents the increment of the particle's energy and occurs whenever particle k visits a vertex $p^{(k)}(t)$ that it dominates, i.e., $\arg\max_{m \in \mathbb{K}} \left(\bar{\mathbf{N}}^{(m)}_{p^{(k)}(t)}(t) \right) = k$. Similarly, the second expression in (9.13) indicates the decrement of the particle's energy that happens when it visits a vertex dominated by rival particles. Therefore, in this model, particles are given a penalty if they are wandering in rival territory, so as to minimize aimless navigation trajectories in the network.

The term $S(t)$ is responsible for determining the movement policy of each particle at each time t. It is really a switching function and defined as follows:

$$S^{(k)}(t) = \mathbb{1}_{\left[E^{(k)}(t) = \omega_{min}\right]}, \qquad (9.15)$$

in which $\dim(S(t)) = 1 \times K$. Specifically, $S^{(k)}(t) = 1$ if $E^{(k)}(t) = \omega_{min}$ and 0, otherwise.

In the following, we apply the concepts introduced so far in a concise and simple example.

Example 9.3. Consider the network depicted in Fig. 9.4. Suppose there are two particles, namely, red (dark gray) and blue (gray), each of which located at vertices 13 and 1, respectively. As both particles are visiting vertices whose owners are rival particles, their energy levels drop. Consider, in this case, that both particles have reached the minimum allowed energy, i.e., ω_{min}, at time t. Therefore, according to (9.15), both particles are *exhausted*. Consequently, $S^{(red)}(t) = 1$ and $S^{(blue)}(t) = 1$, and the transition matrix associated to each particle reduces to the second term in the convex combination of (9.1). According to the mechanism of the dynamical system, these particles are

(continued)

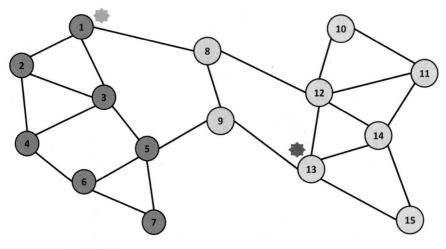

Fig. 9.4 Illustration of the reanimation procedure in a typical situation. There are two particles, namely *red* (*dark gray*) and *blue* (*gray*), located at vertices 13 and 1, respectively, at time *t*. The network encompasses 15 vertices. As both particles are visiting vertices whose owners are rival particles, their energy levels drop. In this example, suppose both energy levels of the particles reach the minimum possible value, $\omega_{\min}$. The vertex *color* represents the particle that is imposing the highest domination level at time *t*. The beige (*light gray*) denotes a non-dominated vertex

Example 9.3 (continued)
transported back to their dominated territory to recharge their energy levels. The transition occurs regardless of the network topology. This mechanism follows the distribution in (9.12). In view of that, the transition matrix for the red (dark gray) particle at time *t* is:

$$\mathbf{P}^{(\text{red})}_{\text{transition}}(i,j,t) = \frac{1}{7}, \forall i \in \mathcal{V}, j \in \{v_1, v_2, \dots, v_7\}, \tag{9.16}$$

$$\mathbf{P}^{(\text{red})}_{\text{transition}}(i,j,t) = 0, \forall i \in \mathcal{V}, j \in \mathcal{V} \setminus \{v_1, v_2, \dots, v_7\}, \tag{9.17}$$

and the transition matrix associated to the blue (gray) particle at time *t* is written as:

$$\mathbf{P}^{(\text{blue})}_{\text{transition}}(i,j,t) = \frac{1}{6}, \forall i \in \mathcal{V}, j \in \{v_{10}, v_{11}, \dots, v_{15}\}, \tag{9.18}$$

$$\mathbf{P}^{(\text{blue})}_{\text{transition}}(i,j,t) = 0, \forall i \in \mathcal{V}, j \in \mathcal{V} \setminus \{v_{10}, v_{11}, \dots, v_{15}\}. \tag{9.19}$$

One can verify that exhausted particles are transported back to their territory (set of dominated vertices) regardless of the network topology. The determination of which of the dominated vertices to visit follows a uniform distribution. In this way, vertices are equally probable to receive only those particles that dominate them.

Looking at (9.1), we see that each particle has a representative transition matrix. For compactness, we can join all of them together into a single representative transition matrix that we refer to $\mathbf{P}_{transition}(t)$, which models the transition of the random vector $p(t)$ to $p(t+1)$. This global matrix will prove useful in Sect. 9.3. Given the current system's state at time t, one can see that $p^{(k)}(t+1)$ and $p^{(u)}(t+1)$ are independent for every pair $(k, u) \in \mathcal{K} \times \mathcal{K}, k \neq u$. Another way of looking at this fact is that, given the immediate past position of each particle, it is clear that, via (9.1), the next particle's location is only dependent on the topology of the network (random term) and the domination levels of the neighborhood in the previous step (preferential term). In this way, $\mathbf{P}_{transition}(t)$ can be written as:

$$\mathbf{P}_{transition}(t) = \mathbf{P}^{(1)}_{transition}(t) \otimes \ldots \otimes \mathbf{P}^{(k)}_{transition}(t), \qquad (9.20)$$

in which $\otimes$ denotes the Kronecker tensor product operator. In this way, Eq. (9.20) completely specifies the transition distribution matrix for *all* of the particles in the network.

Essentially, when $K \geq 2$, $p(t)$ is a vector and we would no longer be able to conventionally define the row $p(t)$ of matrix $\mathbf{P}_{transition}(t)$. Owing to this, we define an invertible mapping $f : \mathcal{V}^K \mapsto \mathbb{N}$. The function f simply maps the input vector to a scalar number that reflects the natural ordering of the tuples in the input vector. For example, $p(t) = [1, 1, \ldots, 1, 1]$ (all particles at vertex 1) denotes the first state; $p(t) = [1, 1, \ldots, 1, 2]$ (all particles at vertex 1, except the last particle, which is at vertex 2) is the second state; and so on, up to the scalar state V^K. Therefore, with this tool, we can fully manipulate the matrix $\mathbf{P}_{transition}(t)$.

Remark 9.1. The matrix $\mathbf{P}_{transition}(t)$ in (9.20) possesses dimensions $V^K \times V^K$, which are undesirably high. In order to save up space, one can use the individual transition matrices associated to each particle (therefore, we maintain a collection of K matrices), as shown in (9.1), each of which with dimensions $V \times V$, to model the dynamic of the particles' transition with no loss of generality, by using the following method: once every transition of the collection of K matrices has been performed, one could concatenate the new particle positions to assemble the random vector that denotes the particles' localization, $p(t+1)$, in an ordered manner. With this technique, the spatial complexity would not surpass $\mathcal{O}(KV)$, provided that we implement the matrices in a sparse mode.

9.2.3   Definition of the Stochastic Nonlinear Dynamical System

We can stack up all of the dynamic variables that have been introduced in the previous section to make up the dynamical system's state $\mathbf{X}(t)$ as follows:

$$\mathbf{X}(t) = \begin{bmatrix} p(t) \\ \mathbf{N}(t) \\ E(t) \\ S(t) \end{bmatrix}, \qquad (9.21)$$

and the dynamical system that governs the particle competition model is given by:

$$
\phi : \begin{cases}
p^{(k)}(t+1) = j, \quad j \sim \mathbf{P}^{(k)}_{\text{transition}}(t) \\
\mathbf{N}^{(k)}_i(t+1) = \mathbf{N}^{(k)}_i(t) + \mathbb{1}_{[p^{(k)}(t+1)=i]} \\
E^{(k)}(t+1) = \begin{cases} \min(\omega_{\max}, E^{(k)}(t) + \Delta), \text{ if owner}(k,t) \\ \max(\omega_{\min}, E^{(k)}(t) - \Delta), \text{ if } \neg\text{owner}(k,t) \end{cases} \\
S^{(k)}(t+1) = \mathbb{1}_{[E^{(k)}(t+1)=\omega_{\min}]}
\end{cases} \tag{9.22}
$$

The first equation of system ϕ addresses the transition rules from i to a neighbor j, in which j is determined according to the time-varying transition matrix in (9.1). In other words, the acquisition of $p(t+1)$ is performed by generating random numbers following the distribution of the transition matrix $\mathbf{P}^{(k)}_{\text{transition}}(t)$. The second equation updates the number of visits that vertex i has received by particle k up to time t. The third equation is used to update the energy levels of all of the particles inserted in the network. Finally, the fourth equation indicates whether the particle is active or exhausted, depending on its actual energy level. Note that system ϕ is nonlinear. This occurs on account of the indicator function, which is nonlinear.

Observe that system ϕ can also be written in matrix form as:

$$
\phi : \begin{cases}
p(t+1) = \mathbf{f}_p(p(t)), \quad \mathbf{f}_p(p(t)) \sim \mathbf{P}_{\text{transition}}(t) \\
\mathbf{N}(t+1) = \mathbf{f}_\mathbf{N}(\mathbf{N}(t), p(t+1)) \\
E(t+1) = f_E(\mathbf{N}(t+1), p(t+1)) \\
S(t+1) = f_S(E(t+1))
\end{cases} , \tag{9.23}
$$

in which $\mathbf{f}_p(.)$, $\mathbf{f}_\mathbf{N}(.)$, $f_E(.)$, and $f_S(.)$ are suitable random matrix functions, whose entries have been defined in (9.22). An important characteristic of system ϕ is its Markovian property (see Proposition 9.1).

Now we discuss how to settle the initial condition of the dynamical system's state $\mathbf{X}(0)$. Firstly, the particles are randomly inserted into the network, i.e., the values of $p(0)$ are randomly set. A desirable and interesting feature of the particle competition method is that the initial positions of the particles do not affect the community detection or data clustering results, due to the competition nature. This behavior occurs even when particles are put together at the beginning of the process.

Each entry of matrix $\mathbf{N}(0)$ is initialized according to the following expression:

$$
\mathbf{N}^{(k)}_i(0) = \begin{cases} 2, & \text{if particle } k \text{ is generated at vertex } i. \\ 1, & \text{otherwise.} \end{cases} \tag{9.24}
$$

Remark 9.2. The initialization of $\mathbf{N}(0)$ may be awkward, but there is a mathematical reason behind it. The domination level matrix, $\bar{\mathbf{N}}(0)$, is a row-normalization of $\mathbf{N}(0)$. Therefore, if all entries of a same row are zero, then (9.8) is undefined. In order to overcome this problem, all entries of matrix $\mathbf{N}(0)$ are evenly set to 1, with exception of those in which the particles are initially spawned, whose starting values are 2. In this setup, a consistent initial configuration for the competitive scheme is provided.

Since a fair competition among the particles is desired, all particles $k \in \mathcal{K}$ start out with the same energy level:

$$E^{(k)}(0) = \omega_{min} + \left(\frac{\omega_{max} - \omega_{min}}{K} \right). \tag{9.25}$$

Lastly, all particles are active in the beginning of the competitive process, i.e.:

$$S^{(k)}(0) = 0. \tag{9.26}$$

9.2.4   Method for Estimating the Number of Communities

The particle competition algorithm described by the dynamical system ϕ produces a large quantity of useful information. Some of these dynamical variables can be used to solve other kinds of problems beyond community detection. In this section, we review the method for determining the most likely number of communities or clusters in a data set presented in [35]. In order to do so, an efficient evaluator index called average maximum domination level $\langle R(t) \rangle \in [0, 1]$ that monitors the information generated by the competitive model itself is constructed. Mathematically, this index is given by:

$$\langle R(t) \rangle = \frac{1}{V} \sum_{u \in \mathcal{V}} \max_{m \in \mathcal{K}} \left(\bar{N}_u^{(m)}(t) \right), \tag{9.27}$$

in which $\bar{N}_u^{(m)}(t)$ indicates the domination level that particle m is imposing on vertex u at time t (see (9.7)) and $\max_{m \in \mathcal{K}} \left(\bar{N}_u^{(m)}(t) \right)$ yields the maximum domination level imposed on vertex u at time t.

The basic idea can be described as follows. For a given network with K real communities, if we put exactly K particles in the network, each of them will dominate a community. Thus, one particle will not interfere much in the acting region of the other particles. As a consequence, $\langle R(t) \rangle$ will be large. In the extreme case when each vertex is completely dominated by a single particle, $\langle R(t) \rangle$ reaches 1. However, if we add more than K particles in the network, inevitably more than one particle will share the same community. Consequently, they will dispute the same group of vertices. In this case, one particle will lower the domination levels imposed by the other particles, and vice versa. As a result, $\langle R(t) \rangle$ will be small. Conversely, if we insert in the network a quantity of particles less than the number of real communities K, some particles will attempt to dominate more than one community. Again, $\langle R(t) \rangle$ will be small. In this way, the actual number of communities or clusters can be effectively estimated by checking for each K the index $\langle R(t) \rangle$ is maximized.

As it turns out, the optimal number of particles K to be inserted into a network is exactly the number of real communities that it has. In this way, the index $\langle R(t) \rangle$ is employed both to estimate the actual number of communities or clusters and also the number of particles K. The last point will be made precise in Sect. 9.2.6, where we study the sensitivity of the parameters that compose the particle competition model.

9.2.5 Method for Detecting Overlapping Structures

A measure that detects overlapping structures or vertices in a given network has been proposed in [36]. For this purpose, the domination level matrix $\bar{N}(t)$ generated by the particle competition process is employed. The intuition is as follows. When the maximum domination level imposed by an arbitrary particle k on a specific vertex i is much larger than the second maximum domination level imposed by another particle on the same vertex, then we can conclude that this vertex is being strongly dominated by particle k and no other particle is influencing it in a relevant manner. Therefore, the overlapping nature of such vertex is minimal. In contrast, when these two quantities are similar, then we can infer that the vertex in question holds an inherently overlapping characteristic. In light of these considerations, we can model this behavior as follows: let $M_i(x, t)$ denote the xth greatest domination level value imposed on vertex i at time t. In this way, the overlapping index of vertex i, $O_i(t) \in [0, 1]$, is given by:

$$O_i(t) = 1 - (M_i(1, t) - M_i(2, t)), \tag{9.28}$$

i.e., the overlapping index $O_i(t)$ measures the difference between the two greatest domination levels imposed by any pair of particles in the network on vertex i.

9.2.6 Parameter Sensitivity Analysis

The particle competition model requires a set of parameters to work. In special, we need to set the number of particles (K), the desired fraction of preferential movement (λ), the energy that each particle gains or loses (Δ), and a stopping factor (ϵ). In this section, we give the intuition on how to choose all of these parameters based on the type of data set we are dealing with.

In this section, we also discuss candidates as termination criteria.

9.2.6.1 Impact of the Parameter λ

Parameter λ is responsible for counterweighting the proportion of preferential and random walks performed by all particles in the network. Recall that the preferential

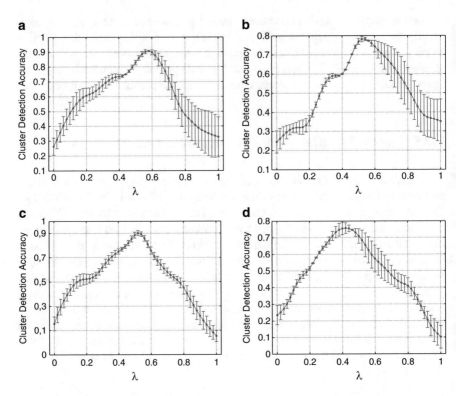

Fig. 9.5 Community detection rate vs. parameter λ. We fix $\Delta = 0.15$. Taking into account the steep peek that is verified and the large negative derivatives that surround it, one can see that the parameter λ is sensible to the overall model's performance. Results are averaged over 30 simulations. Reproduced from [36] with permission from Elsevier. (**a**) $\gamma = 2$ and $\beta = 1$, (**b**) $\gamma = 2$ and $\beta = 2$, (**c**) $\gamma = 3$ and $\beta = 1$, (**d**) $\gamma = 3$ and $\beta = 2$

term is related to the defensive behavior of the particles, while the random term is associated to the exploratory behavior. If we have small values for λ, we favor randomness over preferential visiting. As we increase λ, the tendency is to prefer reinforcing dominated territories instead of exploring new vertices. The two terms serve different and important roles in the community detection task and, in this section, we show that a combination of randomness and preferential behavior can really improve the performance of community detection tasks.

To study the role of λ in the learning process, we use artificial clustered networks that are generated following the benchmark of Lancichinatti et al. [21], which we have introduced in Sect. 6.2.4. We fix $V = 10,000$ vertices and the average network degree or network connectivity as $\bar{k} = 15$. Recall that the benchmark consists in varying the mixing parameter μ while evaluating the attained community detection rates.

Figure 9.5a–d portray the community detection rate of the particle competition model as a function of λ. We vary the counterweighting parameter λ from 0 (pure random walks) to 1 (pure preferential walks) for different values of γ and β, which are the exponents of the power-law degree and community size distributions.

We can observe that the community detection rate of the particle competition algorithm is very sensitive to parameter λ. Though choosing several different values for γ and β, we can see a very clear behavior from these pictures: when $\lambda = 0$ or $\lambda = 1$, the particle competition algorithm does not produce satisfactory results. These values correspond to walks with only random or preferential terms, respectively. This observation suggests that a mixture of these two terms can improve the algorithm's performance to a significant extent. One reason for that is because each of these terms serve a different role in the community detection process: while the random term expands community borders, the preferential term guarantees that community cores stay strongly dominated. The tradeoff between the speed of expanding community borders and guaranteeing the control of the subset of dominated vertices is performed by tuning parameter λ. By our results, we see that the particle competition algorithm provides good results when we have a sustainable increase and defense of the community borders, which happens when we select intermediate values for λ.

As a rule-of-thumb, the model gives good community detection rates in networks with well-defined communities when $0.2 \leq \lambda \leq 0.8$.

9.2.6.2 Impact of the Parameter Δ

Parameter Δ is responsible for updating the particles' energy levels as described in (9.13). We use the same type of artificial clustered networks as in the previous analysis. Figure 9.6a–d display the community detection rate of the particle competition model as a function of Δ. Again, we see that the competitive model does not behave well for extremal values of parameter Δ. The intuition for that is as follows. When Δ is very small, particles are not penalized enough and hence they do not get exhausted often. Consequently, particles are expected to frequently visit vertices that should belong to rival particles, possibly getting into the core of other communities. Therefore, all of the vertices in the network will be in constant competition and no community borders will be established and consolidated. As such, the algorithm's performance is expected to be poor. On the other extreme, when Δ is very large, particles are expected to be constantly exhausted once they visit vertices dominated by rival particles, thus frequently returning to their community core. In this setup, the initial positions of the particles become sensitive to the competitive model. Once we randomly put the particles inside the network at $t = 0$, they are expected to not venture far away from their initial positions due to the reanimation procedure. As such, whenever we put particles near each other, the community detection rate will be poor. In this way, it is unattainable for the particles to switch the ownership of already conquered vertices. We can conceive this phenomenon as an artificial "hard labeling."

Another interesting characteristic that can be extracted from the sensitivity curves in Fig. 9.6a–d is that the competitive model becomes robust against variations of Δ

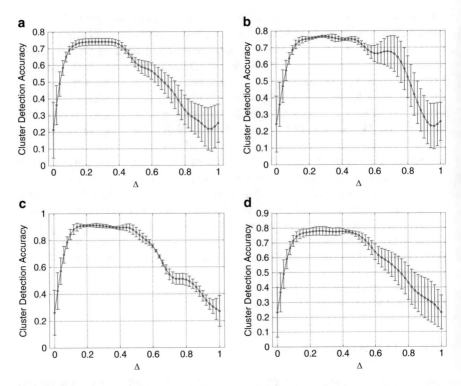

Fig. 9.6 Community detection rate vs. parameter Δ. We fix $\lambda = 0.6$. Taking into account the large steady region around $0.1 \le \Delta \le 0.4$, one can see that the parameter Δ is conditionally not sensible to the overall model's performance. Results are averaged over 30 simulations. Reproduced from [36] with permission from Elsevier. (**a**) $\gamma = 2$ and $\beta = 1$, (**b**) $\gamma = 2$ and $\beta = 2$, (**c**) $\gamma = 3$ and $\beta = 1$, (**d**) $\gamma = 3$ and $\beta = 2$

when we are in the region $0.1 \le \Delta \le 0.4$, i.e., it is conditionally not sensitive to Δ in this relatively wide interval. In this way, as a rule-of-thumb, we should choose intermediate but small values of Δ.

9.2.6.3   Impact of the Parameter K

Parameter K quantifies the number of particles that are inserted into the network to perform the community detection process. In comparison to all of the other parameters of the particle competition algorithm, K is the most sensitive parameter for the model's performance. Hence, the correct determination or at least estimation of K stands as an important problem when employing the particle competition model. Considering that, in the long-run dynamic, each particle dominates a single community, the heuristic presented for estimating the actual number of clusters or communities in Sect. 9.2.4 is a perfect candidate for estimating the proper value for the K parameter. That is, we estimate the number of particles K as the estimated number of communities in the data set using the index $\langle R(t) \rangle$. Mathematically, we choose a candidate K, $K_{\text{candidate}}$, such as to maximize the measure $\langle R(t) \rangle$, $\langle R(t)_{\max} \rangle$, as follows:

$$K = \{K_{\text{candidate}} \in \mathbb{N} : \langle R(t) \rangle = \langle R_{\max}(t) \rangle \}, \tag{9.29}$$

in which $\langle R(t)_{\max} \rangle$ is given by:

$$\langle R(t)_{\max} \rangle = \arg \max_{|\mathscr{K}| \in \mathbb{N}} \frac{1}{V} \sum_{u \in \mathscr{V}} \max_{m \in \mathscr{K}} \left(\bar{\mathbf{N}}_u^{(m)}(t) \right)$$

$$\propto \arg \max_{|\mathscr{K}| \in \mathbb{N}} \sum_{u \in \mathscr{V}} \max_{m \in \mathscr{K}} \left(\bar{\mathbf{N}}_u^{(m)}(t) \right). \tag{9.30}$$

In computational terms, we iterate the particle competition algorithm using $K = 2$ up to a small positive number, while maintaining the best K associated to the maximum achieved $\langle R(t)_{\max} \rangle$. We do not need to try large values for K because the number of communities is often far less than the number of data items.

9.2.7 Convergence Analysis

In this section, we present two possible stopping criteria for the particle competition model. Both of them assume that the particle competition converges. The termination criteria stands as an important issue as we are dealing with a dynamical system that can evolve indefinitely. In essence, we investigate the properties of the indices $\langle R(t) \rangle$ and $|\bar{\mathbf{N}}(t+1) - \bar{\mathbf{N}}(t)|_\infty$ when employed as stopping criteria. We inspect their behavior as a function of time and conclude for the convergence of the dynamical system using an empirical analysis. Based on convergence issues, we give evidences favoring $|\bar{\mathbf{N}}(t+1) - \bar{\mathbf{N}}(t)|_\infty$ in detriment to $\langle R(t) \rangle$.

In our analysis in this section, we use the synthetic data sets shown in Fig. 9.7a–c, which is composed of two communities: the red or "circle" and the green or "square" communities. The two groups in Fig. 9.7a are well-posed as their distributions are distinct and do not overlap. Figure 9.7b portrays an intermediate situation, in which the two groups slightly overlap. Finally, Fig. 9.7c depicts an ill-posed situation, in which the groups largely overlap. In the latter, the clustering task is extremely difficult since the smoothness and cluster assumptions do not hold.

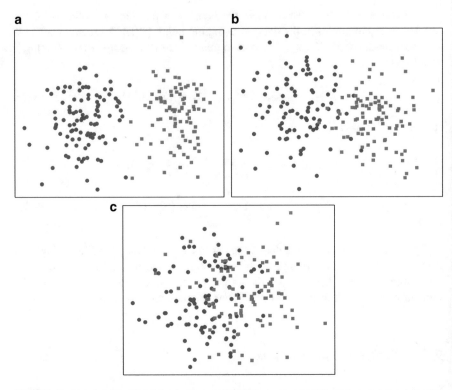

Fig. 9.7 Scatter plot of artificial databases constituted by two groups. The data is constructed using two bi-dimensional Gaussian distributions with varying mean and unitary covariance. Reproduced from [36] with permission from Elsevier. (**a**) Well-posed groups, (**b**) somewhat well-posed groups, (**c**) not well-posed groups

Now, we investigate how the index $\langle R(t) \rangle$, which has been introduced in Sect. 9.2.4, behaves as the competitive dynamical system progresses in time. The simulation results with regard to the synthetic data sets displayed in Fig. 9.7a–c are depicted in Fig. 9.8a–c, respectively. In all of these plots, we have explicitly indicated two important dynamical properties: (1) t_s, which is the time to reach the "almost-stationary" state of the model and (2) the diameter of the region in which the almost-stationary state is confined within. Note that, since the competition is always taking place, the model never reaches a perfect stationary state. Rather, the dynamic variables float around quasi-stationary states because of the constant visits that particles perform on vertices of the network. These fluctuations are expected, since the random walk behavior of the particles, which is denoted by the second term in Eq. (9.1), compels particles to visit vertices that they do not dominate. This behavior creates oscillations in the domination levels between rival particles. However, if we conduct walks with no random behavior, i.e., with only preferential movements, these fluctuations are expected to be eliminated, since the exploratory behavior of the particles would cease to exist. In this case, only the defensive behavior would

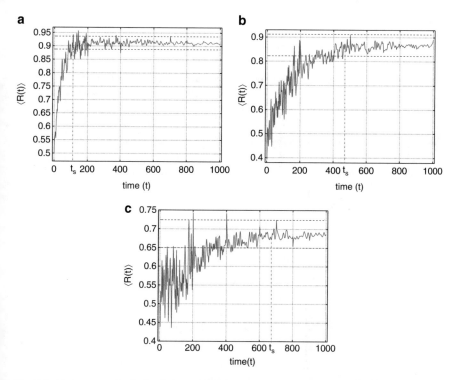

Fig. 9.8 Convergence analysis of the particle competition algorithm when $\langle R(t) \rangle$ is used. The algorithm is run against the binary artificial databases in Fig. 9.7. Here, we inspect how the index $\langle R(t) \rangle$ varies as a function of time. Reproduced from [36] with permission from Elsevier. (**a**) $\langle R(t) \rangle$ for Fig. 9.7a, (**b**) $\langle R(t) \rangle$ for Fig. 9.7b, (**c**) $\langle R(t) \rangle$ for Fig. 9.7c

be used by particles. However, each of the two kinds of movements (random and preferential) has its role in the competition process, in a such a way that disabling one or another would drastically affect the community or cluster detection. As such, good values for λ must reside between 0 and 1 and not in the extremes.

From Fig. 9.8a–c, we see that the time to reach the almost-stationary state t_s lingers to be established as the overlapping region of the groups gets larger. In this respect, t_s is roughly $150, 430, 650$ for Fig. 9.8a–c, respectively. This is because competition in the community border regions gets stronger as the overlap width increases. As a consequence, the dominance of each particle takes longer to be established. Another interesting phenomenon is that of the diameter of the confinement region of $\langle R(t) \rangle$, which grows larger as the overlap width increases. In these simulations, the diameters of such regions are roughly $0.06, 0.07, 0.08$. This is expected by the same reasons stated before.

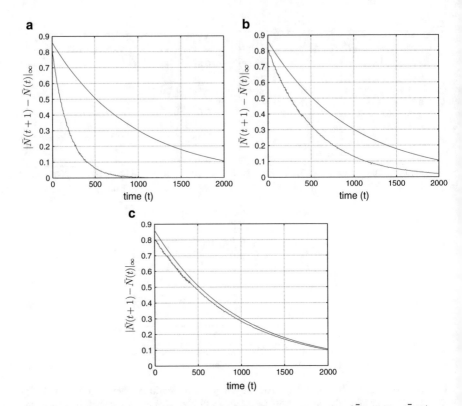

Fig. 9.9 Convergence analysis of the particle competition algorithm when $|\bar{\mathbf{N}}(t+1) - \bar{\mathbf{N}}(t)|_\infty$ is used. The upper theoretical limit is shown in the *blue curve* when $c = K$. The algorithm is run against the binary artificial databases in Fig. 9.7. Here, we inspect how $|\bar{\mathbf{N}}(t+1) - \bar{\mathbf{N}}(t)|_\infty$ varies as a function of time. Reproduced from [36] with permission from Elsevier. (a) $|\bar{\mathbf{N}}(t+1) - \bar{\mathbf{N}}(t)|_\infty$ for Fig. 9.7b, (b) $|\bar{\mathbf{N}}(t+1) - \bar{\mathbf{N}}(t)|_\infty$ for Fig. 9.7b, (c) $|\bar{\mathbf{N}}(t+1) - \bar{\mathbf{N}}(t)|_\infty$ for Fig. 9.7b

Figure 9.9a–c shows the variation term $|\bar{\mathbf{N}}(t+1) - \bar{\mathbf{N}}(t)|_\infty$ as a function of time when applied to the data sets in Fig. 9.7a–c. We see that the variation of $\bar{\mathbf{N}}(t+1)$ in relation to $\bar{\mathbf{N}}(t)$ reduces as time evolves. This happens because the total number of visits performed by particles always increases, since each particle must visit at least a vertex in any given time. Looking at Eq. (9.7), we see that the denominator always increases faster than the numerator. Therefore, it provides an upper limit for $\bar{\mathbf{N}}(t)$. In view of this, the variations from one iteration to another, i.e., $|\bar{\mathbf{N}}(t+1) - \bar{\mathbf{N}}(t)|_\infty$, tend to vary less and less. Analytically, we can verify that, when the particles start to walk, i.e., when $t = 1$, the maximum variation of $|\bar{\mathbf{N}}(t+1) - \bar{\mathbf{N}}(t)|_\infty$ is given by:

$$| \bar{\mathbf{N}}(1) - \bar{\mathbf{N}}(0) |_\infty \leq c \left(\frac{2}{V} - \frac{1}{V+1} \right), \tag{9.31}$$

in which c is a real positive constant which depends on the level of competitiveness, which in turn is directly proportional to λ. Such upper limit expression translates the maximum variation that occurs from time $t = 0$ to $t = 1$, which happens when, at $t = 0$, a vertex is not receiving any visits, but, at time $t = 1$, it is being visited by exactly one particle. Generalizing this equation for an arbitrary t, $|\bar{\mathbf{N}}(t+1) - \bar{\mathbf{N}}(t)|_\infty$ is always bounded by the following expression:

$$
| \bar{\mathbf{N}}(t+1) - \bar{\mathbf{N}}(t) |_\infty \leq c \left(\frac{t+2}{V+t} - \frac{t}{V+t+1} \right),
\tag{9.32}
$$

which demonstrates that, as the previous case, for any $t \leq \infty$, the model presents fluctuations around a quasi-stationary state. From this analysis, it is clear that the $|\bar{\mathbf{N}}(t+1) - \bar{\mathbf{N}}(t)|_\infty$ can be used as a termination criterion.

In summary, we find that the particle competition algorithm does not converge to a fixed point, but the dynamics of the system gets confined within a small finite sub-region in the space. The intuition behind that is, though in the long-run the communities have already being established, the competition among particles is always occurring. In this way, the domination levels of vertices keep changing, though with less magnitude as time progresses due to the accumulative effect that the number of visits plays in establishing the vertices' domination levels.

9.3 Theoretical Analysis of the Model

In this section, a mathematical analysis of the competitive system is supplied. Also, we show that the competitive system reviewed in the previous section reduces to multiple independent random walks when a special situation occurs. Some of the results have been presented in [37] and others are new results. In this book, we present the mathematical analysis in a self-contained manner.

9.3.1 Mathematical Analysis

To estimate the long-run dynamic of the stochastic competitive learning model, we first need to derive the transition probabilities between the different states in the dynamical system. Let the transition probability function of system ϕ be $P(\mathbf{X}(t+1) \mid \mathbf{X}(t))$. Observe that the marginal probability of the system's state $P(\mathbf{X}(t))$ can be written in terms of the joint probability of each of the components of the system's state, meaning $P(\mathbf{X}(t)) = P(\mathbf{N}(t), p(t), E(t), S(t))$. Thus, applying the product rule on the transition probability function, we have:

$P(\mathbf{X}(t+1) \mid \mathbf{X}(t))$

$$
\begin{aligned}
&= P(\mathbf{N}(t+1), p(t+1), E(t+1), S(t+1) \mid \mathbf{N}(t), p(t), E(t), S(t)) \\
&= P(S(t+1) \mid \mathbf{N}(t+1), p(t+1), E(t+1), \mathbf{N}(t), p(t), E(t), S(t)) \\
&\quad \times P(\mathbf{N}(t+1), p(t+1), E(t+1) \mid \mathbf{N}(t), p(t), E(t), S(t)) \\
&= P_{S(t+1)} P(E(t+1) \mid \mathbf{N}(t+1), p(t+1), \mathbf{N}(t), p(t), E(t), S(t)) \quad\quad (9.33) \\
&\quad \times P(\mathbf{N}(t+1), p(t+1) \mid \mathbf{N}(t), p(t), E(t), S(t)) \\
&= P_{S(t+1)} P_{E(t+1)} P(\mathbf{N}(t+1) \mid p(t+1), \mathbf{N}(t), p(t), E(t), S(t)) \\
&\quad \times P(p(t+1) \mid \mathbf{N}(t), p(t), E(t), S(t)) \\
&= P_{S(t+1)} P_{E(t+1)} P_{\mathbf{N}(t+1)} P_{p(t+1)}.
\end{aligned}
$$

in which:

$$
P_{S(t+1)} = P(S(t+1) \mid \mathbf{N}(t+1), p(t+1), E(t+1), \mathbf{X}(t)), \quad\quad (9.34)
$$

$$
P_{E(t+1)} = P(E(t+1) \mid \mathbf{N}(t+1), p(t+1), \mathbf{X}(t)), \quad\quad (9.35)
$$

$$
P_{\mathbf{N}(t+1)} = P(\mathbf{N}(t+1) \mid p(t+1), \mathbf{X}(t)), \quad\quad (9.36)
$$

$$
P_{p(t+1)} = P(p(t+1) \mid \mathbf{X}(t)). \quad\quad (9.37)
$$

Next, the algebraic derivations of these four quantities are explored.

9.3.1.1 Discovering the Factor $P_{p(t+1)}$

Observing that the random vector $p(t+1)$ is directly evaluated from $\mathbf{P}_{\text{transition}}(t)$ given in (9.20), which in turn only requires $p(t)$ and $\mathbf{N}(t)$ to be constructed ($\mathbf{X}(t)$ is given), then the following equivalence holds:

$$
P_{p(t+1)} = P(p(t+1) \mid \mathbf{X}(t)) = \mathbf{P}_{\text{transition}}(\mathbf{N}(t), p(t)). \quad\quad (9.38)
$$

Here, we have used $\mathbf{P}_{\text{transition}}(\mathbf{N}(t), p(t))$ to emphasize the dependence of the transition matrix on $\mathbf{N}(t)$ and $p(t)$.

9.3.1.2 Discovering the Factor $P_{\mathbf{N}(t+1)}$

In this case, taking a close look at $P_{\mathbf{N}(t+1)} = P(\mathbf{N}(t+1) \mid p(t+1), \mathbf{X}(t))$, we can verify that, besides the previous state $\mathbf{X}(t)$, we also know the value of the random vector $p(t+1)$. By a quick analysis of the update rule given in the second expression of system ϕ, it is possible to completely determine $\mathbf{N}(t+1)$, since $p(t+1)$ and $\mathbf{N}(t)$ are known. Owing to that, the following equation holds:

$$P_{\mathbf{N}(t+1)} = P(\mathbf{N}(t+1) \mid p(t+1), \mathbf{X}(t))$$
$$= \mathbb{1}_{[\mathbf{N}(t+1)=\mathbf{N}(t)+\mathbf{Q}_N(p(t+1))]}, \tag{9.39}$$

in which $\mathbf{Q}_N(p(t+1))$ is a matrix with $\dim(\mathbf{Q}_N) = V \times K$ and dependent on $p(t+1)$. The (i,j)-th entry of $\mathbf{Q}_N(p(t+1))$ is given by:

$$\mathbf{Q}_N(p(t+1))(i,k) = \mathbb{1}_{[p^{(k)}(t+1)=i]}, \tag{9.40}$$

The argument in the indicator function shown in (9.39) is essentially the first expression of system ϕ, but in a matrix notation. In brief, Eq. (9.39) results in 1 if the computation of $\mathbf{N}(t+1)$ is correct, given $p(t+1)$ and $\mathbf{N}(t)$, i.e., it is in compliance with the dynamical system rules; and 0, otherwise.

9.3.1.3 Discovering the Factor $P_{E(t+1)}$

For the third term, $P_{E(t+1)}$, we have knowledge of the previous state $\mathbf{X}(t)$, as well as of $p(t+1)$ and $\mathbf{N}(t+1)$. By (9.7), we see that $\bar{\mathbf{N}}(t+1)$ can be directly calculated from $\mathbf{N}(t+1)$, i.e., having knowledge of $\mathbf{N}(t+1)$ permits us to evaluate $\bar{\mathbf{N}}(t+1)$, which, probabilistically speaking, is also a given information. In light of this, together with (9.13), one can see that $E(t+1)$ can be evaluated if we have information of $E(t)$, $p(t+1)$, and $\bar{\mathbf{N}}(t+1)$, which we actually do. On account of that, $P_{E(t+1)}$ can be surely determined and, analogously to the calculation of the $P_{\mathbf{N}(t+1)}$, is given by:

$$P_{E(t+1)} = P(E(t+1) \mid \mathbf{N}(t+1), p(t+1), \mathbf{X}(t))$$
$$= \mathbb{1}_{[E(t+1)=E(t)+\Delta \times Q_E(p(t+1),\mathbf{N}(t+1))]}, \tag{9.41}$$

in which $Q_E(p(t+1), \mathbf{N}(t+1))$ is a random vector with $\dim(Q_E) = 1 \times K$ and dependence on $\mathbf{N}(t+1)$ and $p(t+1)$. The k-th entry, $k \in \mathcal{K}$, of such matrix is calculated as:

$$Q_E^{(k)}(p(t+1), \mathbf{N}(t+1)) = \mathbb{1}_{[\text{owner}(k,t+1)]} - \mathbb{1}_{[\neg\text{owner}(k,t+1)]}. \tag{9.42}$$

Note that the argument of the indicator function in (9.42) is essentially (9.13) in a compact matrix form. Indicator functions were employed to describe the two types of behavior that this variable can have: an increment or decrement of the particle's energy. Suppose that particle $k \in \mathcal{K}$ is visiting a vertex that it dominates, then only the first indicator function in (9.42) is enabled; hence, $Q_E^{(k)}(p(t+1), \mathbf{N}(t+1)) = 1$. Similarly, if particle k is visiting a vertex that is being dominated by a rival particle, then the second indicator function is enabled, yielding $Q_E^{(k)}(p(t+1), \mathbf{N}(t+1)) = -1$. This behavior together with (9.41) is exactly the expression given by (9.13) in a compact matrix form.

9.3.1.4   Discovering the Factor $P_{S(t+1)}$

Lastly, for the fourth term, $P_{S(t+1)}$, we have knowledge of $E(t+1), \mathbf{N}(t+1), p(t+1)$, and the previous internal state $\mathbf{X}(t)$. By a quick analysis of (9.15), one can verify that the calculation of the k-th entry of $S(t+1)$ is completely characterized once $E(t+1)$ is known. In this way, one can surely evaluate $P_{S(t+1)}$ in this scenario as follows:

$$P_{S(t+1)} = P(S(t+1) \mid E(t+1), \mathbf{N}(t+1), p(t+1), \mathbf{X}(t))$$
$$= \mathbb{1}_{[S(t+1)=Q_S(E(t+1))]}, \tag{9.43}$$

in which $Q_S(E(t+1))$ is a matrix with $\dim(Q_S) = 1 \times K$ and has dependence on $E(t+1)$. The k-th entry, $k \in \mathscr{K}$, of such matrix is calculated as:

$$Q_S^{(k)}(E(t+1)) = \mathbb{1}_{[E^{(k)}(t+1)=\omega_{min}]}. \tag{9.44}$$

9.3.1.5   The Transition Probability Function

Substituting (9.38), (9.39), (9.41), and (9.43) into (9.33), we are able to encounter the transition probability function of the competitive dynamical system:

$$
\begin{aligned}
P(\mathbf{X}(t+1) \mid \mathbf{X}(t)) &= \mathbb{1}_{[\mathbf{N}(t+1)=\mathbf{N}(t)+\mathbf{Q}_N(p(t+1))]} \\
&\times \mathbb{1}_{[S(t+1)=Q_S(E(t+1))]} \\
&\times \mathbb{1}_{[E(t+1)=E(t)+\Delta Q_E(p(t+1),\mathbf{N}(t+1))]} \\
&\times \mathbf{P}_{\text{transition}}(\mathbf{N}(t), p(t)) \\
&= \mathbb{1}_{[\text{Compliance}(t)]}\mathbf{P}_{\text{transition}}(\mathbf{N}(t), p(t)),
\end{aligned}
\tag{9.45}
$$

in which Compliance(t) is a logical expression given by:

$$
\begin{aligned}
\text{Compliance}(t) &= [\mathbf{N}(t+1) = \mathbf{N}(t) + \mathbf{Q}_N(p(t+1))] \\
&\wedge [S(t+1) = Q_S(E(t+1))] \wedge [E(t+1) \\
&= E(t) + \Delta Q_E(p(t+1), \mathbf{N}(t+1))],
\end{aligned}
\tag{9.46}
$$

i.e., Compliance(t) encompasses all the rules that have to be satisfied in order to all the indicator functions in (9.45) produce 1. If all the values provided to (9.45) are in compliance with the dynamic of the system, then Compliance(t) = true and the indicator function $\mathbb{1}_{[\text{Compliance}(t)]}$ yields 1; otherwise, if there is at least one measure that does not satisfy the system, then, from (9.46), the chain of logical-AND produces false. As a consequence, Compliance(t) = false and the indicator function $\mathbb{1}_{[\text{Compliance}(t)]}$ in (9.45) yields 0, resulting in a zero-valued transition probability.

9.3.1.6 Discovering the Distribution $P(\mathbf{N}(t))$

With the transition probability function derived in the previous section, we now turn our attention to determining the marginal distribution $P(\mathbf{N}(t))$ for a sufficiently large t. First, the Markovian property of system ϕ is demonstrated as follows.

Proposition 9.1. $\{\mathbf{X}(t) : t \geq 0\}$ *is a Markovian process.*

Proof. We seek to infer that system ϕ is completely characterized by only the acquaintance of the present state, i.e., it is independent of all the past states. Having that in mind, the probability expression to make a transition to a specific event X_{t+1} (a set with an element representing an arbitrary next state) in time $t + 1$, given the complete history of the state trajectory, is denoted by:

$$P\left(\mathbf{X}(t + 1) \in X_{t+1} \mid \mathbf{X}(t), \ldots, \mathbf{X}(0)\right)$$

$$= P\left(p_{t+1} : \begin{bmatrix} f_N(\mathbf{N}(t), p_{t+1}) \\ f_E(\mathbf{N}(t + 1), p_{t+1}) \\ f_S(E(t + 1)) \end{bmatrix} \in X_{t+1} \mid \mathbf{X}(t), \ldots, \mathbf{X}(0)\right). \tag{9.47}$$

Noting that the determination of p_{t+1} only depends on $\mathbf{N}(t)$ and $p(t)$, then:

$$P\left(p_{t+1} : \begin{bmatrix} f_N(\mathbf{N}(t), p_{t+1}) \\ f_E(\mathbf{N}(t + 1), p_{t+1}) \\ f_S(E(t + 1)) \end{bmatrix} \in X_{t+1} \mid \mathbf{X}(t), \ldots, \mathbf{X}(0)\right)$$

$$= P\left(p_{t+1} : \begin{bmatrix} f_N(\mathbf{N}(t), p_{t+1}) \\ f_E(\mathbf{N}(t + 1), p_{t+1}) \\ f_S(E(t + 1)) \end{bmatrix} \in X_{t+1} \mid \mathbf{X}(t)\right)$$

$$= P\left(\mathbf{X}(t + 1) \in X_{t+1} \mid \mathbf{X}(t)\right). \tag{9.48}$$

Therefore, in view of (9.48), $\{\mathbf{X}(t) : t \geq 0\}$ is a Markovian process, since it only depends on the present state to specify the next state and, hence, the past history of the system's trajectory is irrelevant. ∎

The strategy to calculate the distribution $P(\mathbf{N}(t))$ is to marginalize the joint distribution of the system's states, i.e., $P(\mathbf{X}(0), \ldots, \mathbf{X}(t))$, with respect to $\mathbf{N}(t)$ (a component of $\mathbf{X}(t)$). Mathematically, using Proposition 1 on this joint distribution $P(\mathbf{X}(0), \ldots, \mathbf{X}(t))$, we get:

$$P(\mathbf{X}(0), \ldots, \mathbf{X}(t)) = P(\mathbf{X}(t) \mid \mathbf{X}(t - 1))$$

$$\times P(\mathbf{X}(t - 1) \mid \mathbf{X}(t - 2))$$

$$\times \ldots \times P(\mathbf{X}(1) \mid \mathbf{X}(0))P(\mathbf{X}(0)). \tag{9.49}$$

Using the transition function that governs system ϕ, as illustrated in (9.45), to each shifted term in (9.49), we get:

$$P(\mathbf{X}(0),\ldots,\mathbf{X}(t)) = P(\mathbf{X}(0)) \prod_{u=1}^{t-1} \left[\mathbb{1}_{[\text{Compliance}(u)]} \mathbf{P}_{\text{transition}}(\mathbf{N}(u),p(u)) \right], \quad (9.50)$$

in which $P(\mathbf{X}(0)) = P(\mathbf{N}(0),p(0),E(0),S(0))$. But, we are interested in knowing the marginal distribution $\mathbf{N}(t)$ as $t \rightarrow \infty$. We can obtain it from the joint distribution calculated in (9.50), summing over all the possible values of random variables with no relevance in the analysis, i.e., $\mathbf{N}(t-1),\ldots,\mathbf{N}(0),p(t),\ldots,p(0),E(t),\ldots,E(0),S(t),\ldots,S(0)$. In doing so, it is worth studying the limits of $\mathbf{N}(t)$ for an arbitrary t, because the domain that an entry of $\mathbf{N}(t)$ can take is $[1,\infty)$. With this study, we expect to find the reachable values of every entry of matrix $\mathbf{N}(t)$ for any t. In this way, values which exceed these limits are guaranteed to happen with probability 0. Lemma 9.1 precisely supplies this analysis.

Lemma 9.1. *The maximum reachable value of* $\mathbf{N}_i^{(k)}(t)$, $\forall (i,k) \in \mathcal{V} \times \mathcal{K}, t \in \mathbb{N}$, *is:*

$$\mathbf{N}_{i_{\max}}^{(k)}(t) = \begin{cases} \lceil \frac{t+1}{2} \rceil + 1, & \text{if } t \geq 0 \text{ and } a_{ii} = 0 \\ t+2, & \text{if } t \geq 0 \text{ and } a_{ii} > 0 \end{cases}. \quad (9.51)$$

Proof. The proof is based on encountering the particle's trajectory that increases $\mathbf{N}_i^{(k)}(t)$ in the quickest manner. In this situation, we suppose particle k is generated in vertex i; otherwise, the maximum theoretical value would never be reached in view of the second expression in (9.24). For the sake of clarity, consider two specific cases, both depicted in Fig. 9.10: (1) networks without self-loops and (2) networks with self-loops.

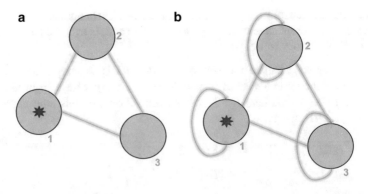

Fig. 9.10 An arbitrary network constructed with the purpose of obtaining the largest feasible entry of $\mathbf{N}(t)$ for a given t. (**a**) A network without the presence of self-loops; (**b**) a network with self-loops

For the first case, $\forall i \in \mathcal{V} : a_{ii} = 0$. By hypothesis, particle k starts in vertex i at $t = 0$. The quickest manner to increase $\mathbf{N}_i^{(k)}(t)$ happens when particle k visits a neighbor of vertex i and immediately returns to vertex i. Repeating this trajectory until time t, $\mathbf{N}_i^{(k)}(t)$ precisely matches the first expression in (9.51).

For the second case, $\exists i \in \mathcal{V} : a_{ii} > 0$. By hypothesis, particle k starts in vertex i at $t = 0$. It is clear that the quickest manner to increase $\mathbf{N}_i^{(k)}(t)$ occurs when particle k always travels through the self-loop for all t. In view of this, the maximum value that $\mathbf{N}_i^{(k)}(t)$ can reach is given by the second expression in (9.51). The "+2" factor appears because the particle initially spawns at vertex i, according to the second expression in (9.24). ∎

In what follows, we analyze the properties of the random vector $E(t)$. The upper limit of the k-th entry of $E(t)$ is always ω_{max}. Thus, provided that $\omega_{max} < \infty$, the upper limit is always well-defined. However, this entry does not only accept integer values in-between ω_{min} and ω_{max}. Lemma 9.2 provides all reachable values of $E(t)$ within this interval.

Lemma 9.2. *The reachable domain of $E^{(k)}(t), \forall k \in \mathcal{K}, t \in \mathbb{N}$, is:*

$$\mathcal{D}_E \triangleq \left\{ \omega_{min} + \frac{\omega_{max} - \omega_{min}}{K} + n\Delta, n = \{-\lfloor n_i \rfloor, \ldots, \lfloor n_m \rfloor\} \right\}$$
$$\cup \left\{ \omega_{min} + n\Delta, n = \left\{ 1, 2, \ldots, \left\lfloor \frac{\omega_{max} - \omega_{min}}{\Delta} \right\rfloor \right\} \right\}$$
$$\cup \left\{ \omega_{max} - n\Delta, n = \left\{ 1, 2, \ldots, \left\lfloor \frac{\omega_{max} - \omega_{min}}{\Delta} \right\rfloor \right\} \right\}, \tag{9.52}$$

in which $n_i = \frac{\omega_{max} - \omega_{min}}{K\Delta} \geq 0$ and $n_m = \frac{\omega_{max} - \omega_{min}}{\Delta} \left(1 - \frac{1}{K} \right) \geq 0$.

Proof. We divide this proof in three main steps, namely the three sets that appear in the expression in the *caput* of this lemma. The first set accounts for supplying all values that are multiples of Δ with the offset $E^{(k)}(0) = \omega_{min} + \left(\frac{\omega_{max} - \omega_{min}}{K} \right), \forall k \in \mathcal{K}$ (see (9.25)). The minimum reachable value is given when $n = n_i$:

$$n_i = \frac{\left(\omega_{min} + \frac{\omega_{max} - \omega_{min}}{K} \right) - \omega_{min}}{\Delta} = \frac{\omega_{max} - \omega_{min}}{K\Delta}, \tag{9.53}$$

whereas the maximum reachable value is given when $n = n_m$:

$$n_m = \frac{\omega_{max} - \left(\omega_{min} + \frac{\omega_{max} - \omega_{min}}{K} \right)}{\Delta} = \frac{\omega_{max} - \omega_{min}}{\Delta} \left(1 - \frac{1}{K} \right). \tag{9.54}$$

After some time, the particle k might reach one of the two possible extremes of energy value: ω_{min} or ω_{max}. On account of the max(.) operator in (9.13), it is also necessary to list all multiple numbers of Δ with these two offsets. The second and third sets precisely fulfill this aspect when the offsets are ω_{min} and ω_{max}, respectively. Once the particle enters one of these sets, it never leaves them. Hence, all values have been properly mapped. ∎

Lastly, the upper limit of an arbitrary entry of $S(t)$ is 1, since it is a boolean-valued variable. Observing now that $P(\mathbf{X}(0), \ldots, \mathbf{X}(t)) = P(\mathbf{N}(0), p(0), E(0), S(0), \ldots, \mathbf{N}(t), p(t), E(t), S(t))$, we marginalize this joint distribution with respect to $\mathbf{N}(t)$ as follows:

$$P(\mathbf{N}(t)) = \sum_{\sim \mathbf{N}(t)} P(\mathbf{X}(0), \ldots, \mathbf{X}(t)), \tag{9.55}$$

in which $\sim \mathbf{N}(t)$ means that we sum over all the possible values of $\mathbf{X}(0), \ldots, \mathbf{X}(t)$, except for $\mathbf{N}(t)$ which is inside $\mathbf{X}(t) = [\mathbf{N}(t)\, p(t)\, E(t)\, S(t)]^T$. Using (9.50) in (9.55), we are able to obtain $P(\mathbf{N}(t))$ as follows:

$$P(\mathbf{N}(t)) = \sum_{\sim \mathbf{N}(t)} \left\{ P(\mathbf{X}(0)) \prod_{u=1}^{t-1} \left[\mathbb{1}_{[\text{Compliance}(u)]} \mathbf{P}_{\text{trans}}(\mathbf{N}(u), p(u)) \right] \right\}. \tag{9.56}$$

Expanding (9.56) using Lemmas 9.1 and 9.2, we have:

$$P(\mathbf{N}(t)) = \sum_{p^{(1)}(0)\in\mathscr{V}} \sum_{p^{(2)}(0)\in\mathscr{V}} \cdots \sum_{p^{(K)}(0)\in\mathscr{V}} \cdots \sum_{p^{(K)}(t)\in\mathscr{V}}$$

$$\times \sum_{N_1^{(1)}(0)=1}^{N_{1\max}^{(1)}(0)} \sum_{N_1^{(2)}(0)=1}^{N_{1\max}^{(2)}(0)} \cdots \sum_{N_V^{(K)}(0)=1}^{N_{V\max}^{(K)}(0)} \cdots \sum_{N_V^{(K)}(t-1)=1}^{N_{V\max}^{(K)}(t-1)}$$

$$\times \sum_{E^{(1)}(0)\in\mathscr{D}_E} \sum_{E^{(2)}(0)\in\mathscr{D}_E} \cdots \sum_{E^{(K)}(0)\in\mathscr{D}_E} \cdots \sum_{E^{(K)}(t)\in\mathscr{D}_E}$$

$$\times \sum_{S^{(1)}(0)=0}^{1} \sum_{S^{(2)}(0)=0}^{1} \cdots \sum_{S^{(K)}(0)=0}^{1} \cdots \sum_{S^{(K)}(t)=0}^{1}$$

$$\left\{ P(\mathbf{X}(0)) \prod_{u=1}^{t-1} \left[\mathbb{1}_{[\text{Compliance}(u)]} \mathbf{P}_{\text{trans}}(\mathbf{N}(u), p(u)) \right] \right\}. \tag{9.57}$$

The summations in the first line of (9.57) account for passing through all possible values of $p(0), \ldots, p(t)$. The summations in the second line are responsible for passing through all reachable values of $\mathbf{N}(0), \ldots, \mathbf{N}(t-1)$, where the upper limit is set with the aid of Lemma 9.1. The third line supplies the summation over all possible values of $E(0), \ldots, E(t)$, in which it is utilized the set $\mathscr{D}_E$ defined in Lemma 9.2. Lastly, the fourth line summations cover all the values of $S(0), \ldots, S(t)$. Note that the logical expression Compliance(u) and the transition matrix inside the product are built up from all these summation indices previously fixed.

Remark 9.3. An interesting feature added by this theoretical analysis is that the particle competition model can also accept uncertainty revolving around its initial state, i.e., $P(\mathbf{X}(0)) = P(\mathbf{N}(0), p(0), E(0), S(0))$. In other terms, the particles' initial locations can be conceptualized as a true distribution itself.

9.3.1.7 Discovering the Distribution of the Domination Level Matrix $P(\mathbf{N}(t))$

The distribution of the domination level matrix $\bar{\mathbf{N}}(t)$ is the fundamental information needed to group up the vertices. First, one can observe that positive integer multiples of $\mathbf{N}(t)$ compose the same $\bar{\mathbf{N}}(t)$. Therefore, the mapping $\mathbf{N}(t) \rightarrow \bar{\mathbf{N}}(t)$ is not injective; hence, not invertible. Below, an illustrative example shows this property.

Example 9.4. Consider a network with 3 particles and 2 vertices. At time t, suppose that the random process is able to produce two distinct configurations for $\mathbf{N}(t)$, as follows:

$$\mathbf{N}(t) = \begin{bmatrix} 1 & 1 & 1 \\ 1 & 2 & 3 \end{bmatrix},$$

$$\mathbf{N}'(t) = \begin{bmatrix} 2 & 2 & 2 \\ 2 & 4 & 6 \end{bmatrix}. \tag{9.58}$$

Then, the setups in (9.58) applied to (9.7) make clear that both configurations yield the same $\bar{\mathbf{N}}(t)$ given by:

$$\bar{\mathbf{N}}(t) = \begin{bmatrix} 1/3 & 1/3 & 1/3 \\ 1/6 & 1/3 & 1/2 \end{bmatrix}. \tag{9.59}$$

In view of this, the mapping $\mathbf{N}(t) \rightarrow \bar{\mathbf{N}}(t)$ cannot be injective nor invertible.

Before proceeding further with the deduction of how to calculate $\bar{\mathbf{N}}(t)$ from $\mathbf{N}(t)$, let us present some helpful auxiliary results.

Lemma 9.3. *For any given time t, the following assertions hold* $\forall (i, k) \in \mathcal{V} \times \mathcal{K}$:

(a) *The minimum value of* $\bar{\mathbf{N}}_i^{(k)}(t)$ *is:*

$$\bar{\mathbf{N}}_{i_{\min}}^{(k)}(t) = \frac{1}{1 + \sum_{u \in \mathcal{K} \setminus \{k\}} \mathbf{N}_{i_{\max}}^{(u)}(t)}. \tag{9.60}$$

(b) The maximum value of $\bar{\mathbf{N}}_i^{(k)}(t)$ *is:*

$$\bar{\mathbf{N}}_{i_{\max}}^{(k)}(t) = \frac{\mathbf{N}_{i_{\max}}^{(k)}(t)}{\mathbf{N}_{i_{\max}}^{(k)}(t) + (K-1)}.$$ (9.61)

Proof. (a) According to (9.7), the minimum value occurs if three conditions are met: (i) particle k is not initially spawned at vertex i; (ii) particle k never visits vertex i; and (iii) all other $K-1$ particles $u \in \mathcal{K} \setminus \{k\}$ visit vertex i in the quickest possible manner, i.e., they follow the trajectory given in Lemma 9.1. In this way, vertex i is visited $\sum_{u \in \mathcal{K} \setminus \{k\}} \mathbf{N}_{i_{\max}}^{(u)}(t)$ times by other particles. However, having in mind the initial condition of $\mathbf{N}(0)$ shown in the second expression of (9.24), we must add 1 to the total number of visits received by vertex i. By virtue of that, it is expected that the total number of visits to be $1 + \sum_{u \in \mathcal{K} \setminus \{k\}} \mathbf{N}_{i_{\max}}^{(u)}(t)$. In view of this scenario, applying (9.7) to this configuration yields (9.60).

(b) The maximum value happens if three conditions are satisfied: (i) particle k initially spawns at vertex i; (ii) particle k visits i in the quickest possible manner; and (iii) all of the other particles $u \in \mathcal{K} \setminus \{k\}$ never visit i. In this scenario, vertex i receives $\mathbf{N}_{i_{\max}}^{(k)}(t) + (K-1)$ visits, where the second term in the summation is due to the initialization of $\mathbf{N}(0)$, as the second expression in (9.24) reveals. This information, together with (9.7), implies (9.61). ∎

Remark 9.4. If the network does not have self-loops, then (9.60) reduces to:

$$\bar{\mathbf{N}}_{i_{\min}}^{(k)}(t) = \frac{1}{1 + (K-1)\mathbf{N}_{i_{\max}}^{(k)}(t)}.$$ (9.62)

The following Lemma provides all reachable elements that an arbitrary entry of $\bar{\mathbf{N}}(t)$ can have.

Lemma 9.4. *Denote* num/den *as an arbitrary irreducible fraction. Consider that the set $\mathscr{I}_t$ retains all the reachable values of $\bar{\mathbf{N}}_i^{(k)}(t)$, $\forall (i,k) \in \mathscr{V} \times \mathscr{K}$, for a fixed t. Then, the elements of $\mathscr{I}_t$ are composed of all elements satisfying the following constraints:*

(a) The minimum element is given by the expression in (9.60).
(b) The maximum element is given by the expression in (9.61).
(c) All the irreducible fractions within the interval delimited by (a) and (b) such that:

    I.   num, den $\in \mathbb{N}^*$;
    II. num $\leq \mathbf{N}_{i_{\max}}^{(k)}(t)$;
    III. den $\leq \sum_{u \in \mathcal{K}} \mathbf{N}_{i_{\max}}^{(u)}(t)$.

Proof. (a) and (b) Straightforward from Lemma 9.3; (c) Firstly, we need to remember that $\mathbf{N}_i^{(k)}(t)$ may only take integer values. According to (9.7), $\bar{\mathbf{N}}_i^{(k)}(t) = $ num/den is a ratio of integer numbers. As a consequence, num and den must be integers and clause I is demonstrated. In view of (9.7), num only registers visits performed by a single particle. Therefore, the upper bound of it is established by Lemma 9.1, i.e., $\mathbf{N}_{i_{\max}}^{(k)}(t)$. Hence, clause II is proved. Looking at the same expression, observe that den registers the number of visits performed by all particles. Again, using Lemma 9.1 proves clause III. ∎

Another interesting feature of the set $\mathscr{I}_t$ is elucidated in the following Lemma.

Lemma 9.5. *Given $t \leq \infty$, the set $\mathscr{I}_t$ indicated in Lemma 9.4 is always finite.*

Proof. In order to demonstrate this lemma, it is enough to verify that each set appearing in the *caput* of Lemma 9.4 is finite.

(a) are (b) are scalars, hence, they are finite sets. (c) Clause I indicates a lower bound for the numerator and the denominator. Clauses II and III reveal upper bounds for the numerator and denominator, respectively. Also from clause I, it can be inferred that the interval delimited by the lower and upper bounds is discrete. Therefore, the number of irreducible fractions that can be made from these two limits is finite.

As all the sets are finite for any t, since $\mathscr{I}_t$ is the union of all these subsets, it follows that $\mathscr{I}_t$ is also finite for any t. ∎

Lemma 9.4 supplies the reachable values of an arbitrary entry of $\bar{\mathbf{N}}(t)$ by means of the definition of the set $\mathscr{I}_t$. Next, we simply extend this notion to the space spawned by the matrices $\bar{\mathbf{N}}(t)$ with dimensions $V \times K$, in such a way that each entry of it must be an element of $\mathscr{I}_t$ as follows:

$$\mathscr{M}_t \triangleq \{\bar{\mathbf{N}} : \bar{\mathbf{N}}_i^{(k)}(t) \in \mathscr{I}_t, \forall (i,k) \in \mathscr{V} \times \mathscr{K}\}. \tag{9.63}$$

In light of all these previous consideration, we can now provide a compact way of determining the distribution of $\bar{\mathbf{N}}(t)$. Following the aforementioned strategy, $P(\bar{\mathbf{N}}(t))$ can be calculated by summing over all multiples of $u\mathbf{N}(t)$, $u \in \{1, \ldots, t\}$ such that $f(u\mathbf{N}(t)) = \bar{\mathbf{N}}(t)$, where f is the normalization function defined in (9.7). On account of this, we have:

$$P\big(\bar{\mathbf{N}}(t) = \mathbf{U} : \mathbf{U} \in \mathscr{M}_t\big) = \sum_{u=1}^{t} P\big(f(u\mathbf{N}(t)) = \mathbf{U}\big), \tag{9.64}$$

in which the upper limit provided in summation of (9.64) is taken using a conservative approach. Indeed, the probability for events such that $\mathbf{N}_i^{(k)}(t) > \mathbf{N}_{i_{\max}}^{(k)}(t)$ is zero. By virtue of that, we can stop summing whenever any entry of matrix $u\mathbf{N}(t)$ exceeds this value. We have omitted this observation from (9.64) for the sake of clarity.

As $t \to \infty$, $P(\bar{\mathbf{N}}(t))$ provides enough information for grouping the vertices. In this case, they are grouped accordingly to the particle that is imposing the highest domination level. Since the domination level is a continuous random variable, the output of this model is fuzzy.

9.3.2   Linking the Particle Competition Model and the Classical Multiple Independent Random Walks System

Multiple random walks are modeled as dynamical systems and have been extensively studied by the literature [8]. In these systems, particles cannot communicate with each other. In effect, the model of multiple random walks can be understood as a system with multiple single random walks stacked up. The particle competition model that we have explored in this chapter, however, permits communication between different particles. The communication is modeled via the domination level matrix, which encodes the fraction of visits each vertex has received from particles in the network. This happens because the fraction of visits is computed by a normalization procedure that effectively entangles the walking dynamic of all particles with one another.[1]

The interaction or communication between particles, nonetheless, can be turned off when $\lambda = 0$ and $\Delta = 0$. This is equivalent to saying that the particle competitive model investigated in this chapter is a generalization of the classical dynamical system of multiple independent random walkers. Whenever $\lambda > 0$, the competitive mechanism is enabled and the combination of random-preferential interacting walks occurs. In this case, the reanimation feature is presented depending on the choice of Δ.

We now prove the assertion that when $\lambda = 0$ and $\Delta = 0$ holds, the particle competition model produces the same dynamics of multiple independent random walks.

Proposition 9.2. *If $\lambda = 0$ and $\Delta = 0$, then system ϕ reduces to the case of multiple independent random walks.*

Proof. First, note that, when $\lambda = 0$, the influence of the transition matrix that encodes the preferential movement, $\mathbf{P}_{\mathrm{pref}}(t)$, is taken away. Indeed, when $\lambda = 0$, the coupling between $\mathbf{N}(t)$ and $p(t)$ ceases to exist, because the calculation step of $P_{\mathrm{pref}}(t)$ (responsible for the coupling) can be skipped. Moreover, if $\Delta = 0$, then the particles can never get exhausted. In view of these characteristics, the dynamical system ϕ can be easily described by a traditional Markovian process given by:

$$p(t+1) = p(t)\mathbf{P}_{\mathrm{transition}}, \tag{9.65}$$

---

[1] Recall the evaluation of each entry of the domination matrix in (9.7).

in which $\mathbf{P}_{\text{transition}} = \mathbf{P}_{\text{rand}} \otimes \mathbf{P}_{\text{rand}} \otimes \cdots \otimes \mathbf{P}_{\text{rand}}$ and $p(t)$ is an enumerated state encompassing all the particles, as described before. Here, the independence among the particles is demonstrated by showing that the generated $\mathbf{N}(t)$ by system ϕ is exactly the same as the one produced by the potential matrix of the Markov chains theory as introduced in Definition 2.68. In other words, $\mathbf{N}(t)$ can be implicitly calculated from the stochastic process $\{p(t) : t \geq 0\}$.

We now find a closed expression for $\mathbf{N}(t)$ in terms of $\mathbf{N}(0)$. This can be easily done if we iterate the matrix equation $\mathbf{N}(t + 1) = \mathbf{N}(t) + Q$, where Q is as given in (9.40). In doing so, we get:

$$
\mathbf{N}(t) = \begin{bmatrix} 1 & \cdots & 1 \\ 1 & \cdots & 1 \\ \vdots & \ddots & \vdots \\ 1 & \cdots & 1 \end{bmatrix} + \sum_{i=0}^{t} \begin{bmatrix} \mathbb{1}_{[p^{(1)}(i)=1]} & \cdots & \mathbb{1}_{[p^{(K)}(i)=1]} \\ \mathbb{1}_{[p^{(1)}(i)=2]} & \cdots & \mathbb{1}_{[p^{(K)}(i)=2]} \\ \vdots & \ddots & \vdots \\ \mathbb{1}_{[p^{(1)}(i)=V]} & \cdots & \mathbb{1}_{[p^{(K)}(i)=V]} \end{bmatrix}. \tag{9.66}
$$

Since this process is stochastic, it is worth determining the expectation of the number of visits $\mathbf{N}(t)$ given the particle's initial location $p(0)$. Noting that $\mathbb{E}[\mathbb{1}_{[A]}] = P(A)$, we have:

$$
\mathbb{E}[\mathbf{N}(t) \mid p(0)] = \begin{bmatrix} 1 & \cdots & 1 \\ 1 & \cdots & 1 \\ \vdots & \ddots & \vdots \\ 1 & \cdots & 1 \end{bmatrix} + \sum_{i=0}^{t} \begin{bmatrix} \mathbf{P}^i(p_1(0), 1) & \cdots & \mathbf{P}^i(p_K(0), 1) \\ \mathbf{P}^i(p_1(0), 2) & \cdots & \mathbf{P}^i(p_K(0), 2) \\ \vdots & \ddots & \vdots \\ \mathbf{P}^i(p_1(0), V) & \cdots & \mathbf{P}^i(p_K(0), V) \end{bmatrix}, \tag{9.67}
$$

in which $\mathbf{P}^i(p_j(0), 1)$ denotes the $(p_j(0), 1)$-entry of $\mathbf{P}_{\text{transition}}$ to the i-th power. But, from the Markov chains theory, we have that the so-called truncated potential matrix [8] is given by:

$$
R_t(v, k) \triangleq \sum_{i=0}^{t} \mathbf{P}_{\text{transition}}^i(v, k). \tag{9.68}
$$

By virtue of (9.68), each entry of the matrix equation in (9.67) can be rewritten as:

$$
\mathbb{E}[\mathbf{N}_i^{(j)}(t) \mid p(0)] = 1 + R_t(p_j(0), i). \tag{9.69}
$$

From (9.69), we can infer that each particle does perform an independent random walk according to a Markov Chain. Thus, we are able to conclude that, for $\lambda = 0$ and $\Delta = 0$, all the states of system ϕ follow a traditional Markov Chain process, except for a constant, as demonstrated in (9.69). ∎

Proposition 9.2 states that system ϕ reduces to the case of multiple random walks when $\lambda = 0$ and $\Delta = 0$, i.e., we could think that there is a blind competition among the participants. Alternatively, when $0 < \lambda \leq 1$, some orientation is given to the participants, in the sense of defending their territory and not only keep adventuring through the network with no strategy at all. In either case, the reanimation procedure is enabled depending on the choice of Δ.

9.3.3 A Numerical Example

For the sake of clarity, we provide an example showing how to use the theoretical results supplied in the previous section. We limit the demonstration for a single iteration, which is the transition from $t = 0$ to $t = 1$. The simple example is composed of a trivial 3-vertex regular network, identical to the one in Fig. 9.10a. For the referred example, suppose there are $K = 2$ particles into the network, i.e., $\mathcal{K} = \{1, 2\}$. Let particle 1 be spawned at vertex 1 and particle 2 at vertex 2, i.e., we have certainty about the initial locations of the particles at $t = 0$:

$$P(\mathbf{X}(0)) = P\left(\mathbf{N}(0) = \begin{bmatrix} 2 & 1 \\ 1 & 2 \\ 1 & 1 \end{bmatrix}, p(0) = [1 \ 2], E(0), S(0)\right) = 1, \qquad (9.70)$$

i.e., there is 100 % chance (certainty) that particles 1 and 2 are generated at vertices 1 and 2, respectively. Observe that $\mathbf{N}(0)$, $E(0)$, and $S(0)$ are chosen such as to satisfy (9.24), (9.25), and (9.26), respectively. Otherwise, the probability should be 0, in view of (9.45). It is worth emphasizing that the competitive model accepts uncertainty about the initial location of the particles, in a way that we could specify different probabilities to each particle to spawn at different locations. This characteristic is not present in [32], in which it must be fixed a certain position for each particle.

From Fig. 9.10a we can deduce the adjacency matrix $\mathbf{A}$ of the network and, therefore, determine the transition matrix associated to the random movement term for a single particle. Recall that the random matrix is the same for all of the particles. Then, applying (9.2) on $\mathbf{A}$, we get:

$$\mathbf{P}_{\text{rand}} = \begin{bmatrix} 0 & 0.50 & 0.50 \\ 0.50 & 0 & 0.50 \\ 0.50 & 0.50 & 0 \end{bmatrix}. \qquad (9.71)$$

Given $\mathbf{N}(0)$, we can readily establish $\bar{\mathbf{N}}(0)$ with the aid of (9.7):

$$\bar{\mathbf{N}}(0) = \begin{bmatrix} 0.67 & 0.33 \\ 0.33 & 0.67 \\ 0.50 & 0.50 \end{bmatrix}. \qquad (9.72)$$

Using (9.8) we are able to calculate the matrices associated to the preferential movement policy for each particle in the network as:

$$\mathbf{P}_{\text{pref}}^{(1)}(0) = \begin{bmatrix} 0 & 0.40 & 0.60 \\ 0.57 & 0 & 0.43 \\ 0.67 & 0.33 & 0 \end{bmatrix}, \tag{9.73}$$

$$\mathbf{P}_{\text{pref}}^{(2)}(0) = \begin{bmatrix} 0 & 0.57 & 0.43 \\ 0.40 & 0 & 0.60 \\ 0.33 & 0.67 & 0 \end{bmatrix}. \tag{9.74}$$

In order to ease the calculations, let us assume that $\lambda = 1$, so that (9.20) reduces to $\mathbf{P}_{\text{transition}}(0) = \mathbf{P}_{\text{pref}}^{(1)}(0) \otimes \mathbf{P}_{\text{pref}}^{(2)}(0)$ at time $t = 0$,[2] which is a matrix with dimensions 9×9 that is given by:

$$\mathbf{P}_{\text{transition}}(0) = \begin{bmatrix} 0 & 0 & 0 & 0 & 0.228 & 0.172 & 0 & 0.342 & 0.258 \\ 0 & 0 & 0 & 0.160 & 0 & 0.240 & 0.240 & 0 & 0.360 \\ 0 & 0 & 0 & 0.132 & 0.268 & 0 & 0.198 & 0.402 & 0 \\ 0 & 0.325 & 0.245 & 0 & 0 & 0 & 0 & 0.245 & 0.185 \\ 0.228 & 0 & 0.342 & 0 & 0 & 0 & 0.172 & 0 & 0.258 \\ 0.188 & 0.382 & 0 & 0 & 0 & 0 & 0.142 & 0.288 & 0 \\ 0 & 0.382 & 0.288 & 0 & 0.188 & 0.142 & 0 & 0 & 0 \\ 0.268 & 0 & 0.402 & 0.132 & 0 & 0.198 & 0 & 0 & 0 \\ 0.221 & 0.449 & 0 & 0.109 & 0.221 & 0 & 0 & 0 & 0 \end{bmatrix}. \tag{9.75}$$

Since in the initial condition depicted in (9.70) particles 1 and 2 start out at vertices 1 and 2, respectively, the enumerated scalar state for the matters of calculating $p(t + 1)$ is $(1, 2) \rightarrow 2$. Hence, we turn our attention to the second row of $\mathbf{P}_{\text{transition}}(0)$, which completely characterizes the transition probabilities for the next state of the dynamical system. A quick analysis of the second row in (9.75) shows that, out of the 9 possible "next states" of the system, only 4 are plausible. (The remaining states have probability 0 to be reached.) In this way:

$$P\left(\mathbf{N}(1) = \begin{bmatrix} 2 & 2 \\ 2 & 2 \\ 1 & 1 \end{bmatrix}, \quad p(1) = \begin{bmatrix} 2 & 1 \end{bmatrix}, E(1), S(1) \mid \mathbf{X}(0) \right) = 0.160, \tag{9.76}$$

$$P\left(\mathbf{N}(1) = \begin{bmatrix} 2 & 1 \\ 2 & 2 \\ 1 & 2 \end{bmatrix}, \quad p(1) = \begin{bmatrix} 2 & 3 \end{bmatrix}, E(1), S(1) \mid \mathbf{X}(0) \right) = 0.240, \tag{9.77}$$

---

[2]Recall that all particles are active at the initial state in view of (9.26).

$$P\left(\mathbf{N}(1) = \begin{bmatrix} 2 & 2 \\ 1 & 2 \\ 2 & 1 \end{bmatrix}, \quad p(1) = [\,3\ 1\,], E(1), S(1) \mid \mathbf{X}(0)\right) = 0.240, \qquad (9.78)$$

$$P\left(\mathbf{N}(1) = \begin{bmatrix} 2 & 1 \\ 1 & 2 \\ 2 & 2 \end{bmatrix}, \quad p(1) = [\,3\ 3\,], E(1), S(1) \mid \mathbf{X}(0)\right) = 0.360, \qquad (9.79)$$

in which $\mathbf{X}(0)$ is as given in (9.70). Equations (9.76)–(9.79) match our intuition if we take a careful look at Fig. 9.10a: self-looping is not allowed, so the state space that is probabilistically possible of happening can only be these previous 4 states. In other terms, starting from vertex 1, there are only two different choices that the particle can make: either visit vertex 2 or 3. The same reasoning can be applied when we start at vertex 2. Since it is a joint distribution, we multiply these factors, which totalizes 4 different states. Furthermore, as we have fixed $\lambda = 1$, it is expected that the transition probabilities will be heavily dependent on the domination levels imposed on the neighboring vertices. In this case, strongly dominated vertices constitute repulsive forces that act against rival particles. In this regard, the preferential or defensive behavior of these particles prevents particles from visiting these type of vertices. This is exactly symbolized in (9.79), which denotes the transition probability $(1, 2) \rightarrow (3, 3)$ and also possesses the highest transition probability, in account of the neutrality of vertex 3, as opposed to the remaining two vertices.

Remark 9.5. Alternatively, we could have used the collection of two matrices 3×3, as given in (9.73) and (9.74) with no loss of generality. Here, we clarify this concept by calculating a single entry of $\mathbf{P}_{\text{transition}}(0)$ using this methodology. Consider we are to calculate the probability according to (9.76), i.e., particle 1 performs a transition from vertex 1 to vertex 2 and particle 2 executes a transition from vertex 2 to vertex 1. For the former case, according to the particle 1's transition matrix (see (9.73)) we have $\mathbf{P}_{\text{pref}}^{(1)}(0)(1, 2) = 0.40$. Likewise, for the last case (see (9.74)), we have $\mathbf{P}_{\text{pref}}^{(2)}(0)(2, 1) = 0.40$. Remembering that $p(0) = [1\ 2]$ in a scalar form corresponds to the second state of $\mathbf{P}_{\text{transition}}$ and $p(1) = [2\ 1]$ corresponds to the fourth state, then $\mathbf{P}_{\text{transition}}(0)(2, 4) = \mathbf{P}_{\text{pref}}^{(1)}(0)(1, 2) \times \mathbf{P}_{\text{pref}}^{(2)}(0)(2, 1) = 0.40 \times 0.40 = 0.16$, which is equal to the corresponding entry of the matrix in (9.75).

Before doing the calculation of the marginal distribution $P(\mathbf{N}(1))$, we are required to fix an upper limit for an arbitrary entry of the matrix $\mathbf{N}(1)$. This is readily evaluated from (9.51), which results in $\mathbf{N}_{i_{\max}}^{(k)}(1) = \mathbf{N}_{i_{\max}}^{(k)}(1) = 2, \forall (i, k) \in \mathcal{V} \times \mathcal{K}$, implying that we are only needed to take all numerical combinations for the matrix $\mathbf{N}(0)$ such that each entry may only take the values $\{1, 2\}$, since larger values yield probability 0 according to Lemma 9.1. Moreover, we need to iterate through every feasible value of every entry of $E(0)$ and $E(1)$. In order to do so, we fix $\Delta = 0.25$, $\omega_{min} = 0$, and $\omega_{max} = 1$. With that, we are able to make use of Lemma 9.2, which yields $E(t) \in \{0, 0.25, 0.5, 0.75, 1\}$. The limits of the remaining system variables $S(0)$ and $S(1)$ are straightforward. In the present conditions, we have enough information to calculate the marginal distribution $P(\mathbf{N}(1))$, according to (9.57):

$$P\left(\mathbf{N}(1) = \begin{bmatrix} 2 & 2 \\ 2 & 2 \\ 1 & 1 \end{bmatrix}\right) = 1 \times 0.160 = 0.160, \tag{9.80}$$

$$P\left(\mathbf{N}(1) = \begin{bmatrix} 2 & 1 \\ 2 & 2 \\ 1 & 2 \end{bmatrix}\right) = 1 \times 0.240 = 0.240, \tag{9.81}$$

$$P\left(\mathbf{N}(1) = \begin{bmatrix} 2 & 2 \\ 1 & 2 \\ 2 & 1 \end{bmatrix}\right) = 1 \times 0.240 = 0.240, \tag{9.82}$$

$$P\left(\mathbf{N}(1) = \begin{bmatrix} 2 & 1 \\ 1 & 2 \\ 2 & 2 \end{bmatrix}\right) = 1 \times 0.360 = 0.360. \tag{9.83}$$

As the last goal, our task is to determine the distribution $P(\bar{\mathbf{N}}(1))$. According to the specified steps in the previous section, we need to find all irreducible fractions that lie within the interval $[0, 1]$ with the constraints derived in the previous section. This means that we only have to consider entries of matrix $\bar{\mathbf{N}}(t)$ that contain elements of $\mathscr{I}_1$; the remainder $\bar{\mathbf{N}}(t)$ are infeasible and, thus, occur with probability 0. In view of the constraints previously enumerated, $\mathscr{I}_1 = \{1/4, 1/3, 1/2, 2/3, 3/4\}$. Observing that we have the complete distribution of $\mathbf{N}(1)$, it is an easy task to apply (9.64), as follows:

$$P\left(\bar{\mathbf{N}}(1) = \begin{bmatrix} 1/2 & 1/2 \\ 1/3 & 2/3 \\ 1/2 & 1/2 \end{bmatrix}\right) = 0.160, \tag{9.84}$$

$$P\left(\bar{\mathbf{N}}(1) = \begin{bmatrix} 2/3 & 1/3 \\ 1/2 & 1/2 \\ 1/3 & 2/3 \end{bmatrix}\right) = 0.240, \tag{9.85}$$

$$P\left(\bar{\mathbf{N}}(1) = \begin{bmatrix} 1/2 & 1/2 \\ 1/3 & 2/3 \\ 2/3 & 1/3 \end{bmatrix}\right) = 0.240, \tag{9.86}$$

$$P\left(\bar{\mathbf{N}}(1) = \begin{bmatrix} 2/3 \ 1/3 \\ 1/3 \ 2/3 \\ 1/2 \ 1/2 \end{bmatrix}\right) = 0.360. \tag{9.87}$$

It is noteworthy to reinforce that the mapping between the probabilities of $\mathbf{N}(t)$ and $\bar{\mathbf{N}}(t)$ is not bijective: in this special simple case that we are studying, we did not have distinct $\mathbf{N}(t)$ that could generate the same $\bar{\mathbf{N}}(t)$, but as t increases, this is likely to happen quite frequently. This process is repeated until a sufficiently large t or until the system converges to a quasi-stationary state of $\bar{\mathbf{N}}(t)$, as discussed in the empirical section.

9.4   Numerical Analysis of the Detection of Overlapping Vertices and Communities

In this section, some simulation results are presented with the purpose of assessing the effectiveness of the particle competition technique on detecting overlapping vertices and communities. Note that the index that estimates the overlapping nature of each network vertex is computed using (9.28). The obtained results are also compared to classical overlap vertex measures [11, 12].

9.4.1   Zachary's Karate Club Network

First, the particle competition technique is applied to detect fuzzy community structure in the Zachary's "karate club" network [42]. This is a well-known network from the social science literature, which has become a benchmark test for community detection algorithms. This network exhibits the pattern of friendship among 34 members of a club at an American University in the 1970s. The members are represented by vertices and an edge exists between two members if they know each other. Shortly after the formation of the network, the club dismembered in two as the consequence of an internal dispute, making it an interesting problem for detecting communities. Figure 9.11 shows the outcome of the community detection task. The red (dark gray) and blue (gray) colors denote the communities detected by the algorithm. Only vertex 3 (the yellow or light gray vertex) is incorrectly grouped as a member of the red (dark gray) community. In the literature, vertices 3 (e.g., see [13]) and 10 (e.g., see [29]) are often misclassified by many community detection algorithms. This happens because the number of edges that they share between the two communities is the same, i.e., they are inherently overlapping, making their clustering a hard problem. We apply the overlapping index of the particle competition model and report the results in Fig. 9.12. We see

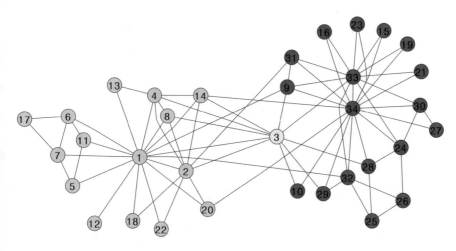

Fig. 9.11 Community detection result in the Zachary's karate club network when we apply the particle competition method. The *red* (*dark gray*) and *blue* (*gray*) *colors* denote the detected communities. Only the *yellow* or *light gray* vertex (vertex 3 in the original database) is incorrectly grouped. Reproduced from [36] with permission from Elsevier

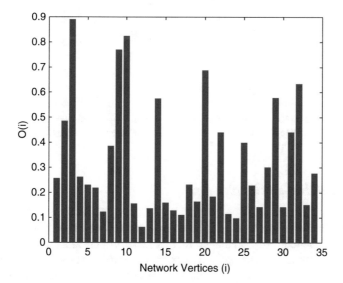

Fig. 9.12 Result of the calculation of the overlapping index for all vertices in the Zachary's "karate club" network. Reproduced from [36] with permission from Elsevier

that vertices 3 and 10 show the highest overlapping indices, confirming the previous analysis. Moreover, vertices 9, 14, 20, 29, and 32 also present a significant level of overlapping characteristics, since these are placed in the borders of each community.

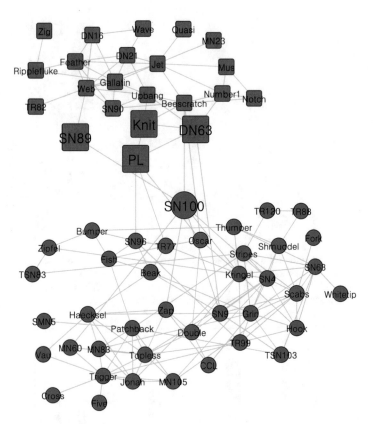

Fig. 9.13 Dolphin Social Network observed by Lusseau. $K = 2$ and $\lambda = 0.6$. The five vertices with the highest overlapping structure are displayed with larger sizes. Reproduced from [36] with permission from Elsevier

9.4.2 Dolphin Social Network

The Dolphin Social Network [26] is composed of 62 bottlenose dolphins living in Doubtful Sound, New Zealand. In this case, we consider that the dolphins represent the vertices, whereas edges between dolphin pairs are established by observation when there is a statistically significant frequent association between them. Figure 9.13 indicates the community detection outcome of the particle competition technique. The five most overlapping vertices are displayed in larger sizes. In this case, the number of particles that maximizes $\langle R(t) \rangle$ is $K = 2$, which corresponds to the division of the real problem indicated by Lusseau. The split into two communities seems to match the known division of the dolphin community, except for the dolphin "PL," which is a member of the blue (gray) community. Based on a 2-year research period, Lusseau reported that the bottlenose dolphins segregated into two communities, apparently by virtue of the disappearance of the dolphins located at the boundaries of each of the communities. When some of these dolphins later reappeared, the two halves of the network joined together once more. Surprisingly, these border dolphins are the ones that the particle competition algorithm captures as the vertices with the most overlapping nature, as we can verify in Fig. 9.13 from the larger vertices, i.e., "DN63," "Knit," "PL," "SN89," and "SN100."

9.4.3 Les misérables Novel Network

Les Misérables is an interaction network between major characters comprising the Victor Hugo's sprawling novel of crime and redemption in post-restoration France. Using the list of 77 character appearances by scene, compiled by Knuth [18], the network was constructed in a way that vertices represent characters and an edge between two vertices represents co-appearance of the corresponding characters in one or more scenes [30]. In this case, the quantity $\langle R(t) \rangle$ is maximized when $K = 6$. Figure 9.14 shows the outcome of the particle competition technique, along with the 10 most overlapping vertices portrayed in larger sizes. The communities clearly reflect the subplot structure of the book. As one can expect, the protagonist Jean Valjean and his nemesis, the police officer Javert, are captured as being the 2 most overlapping vertices of the network, since they are central to the Hugo's play and form the hubs of communities composed of their respective adherents.

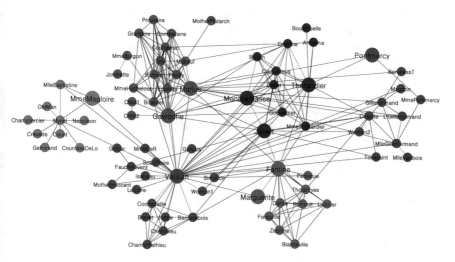

Fig. 9.14 Hugo's sprawling novel of crime and redemption in post-restoration France entitled Les Misérables. $K = 6$ and $\lambda = 0.6$. The 10 vertices with the highest overlapping structure are depicted with larger sizes. Reproduced from [36] with permission from Elsevier

9.5   Application: Handwritten Digits and Letters Clustering

In order to give a concrete vision of the particle competition model, we here review the application in data clustering for the particle competitive method presented in [36]. We employ three real-world data sets that are composed of handwritten digits and letters, which are the USPS, the MNIST, and the Letter Recognition data sets.

Section 9.5.1 briefly discusses on the employed data sets. Section 9.5.2 applies the index for estimating the most likely number of groups in these three data sets. Finally, Sect. 9.5.3 provides the data clustering results.

9.5.1   Brief Information of the Handwritten Digits and Letters Data Sets

The data sets in which the particle competition model is tested against are given in the following:

- *USPS data set*: Comprised of 9298 images of handwritten digits. The digits 0 to 9 have 1553, 1269, 929, 824, 852, 716, 834, 792, 708, and 821 samples respectively. The US Postal Service (USPS) digits data were gathered at the Center of Excellence in Document Analysis and Recognition (CEDAR) at SUNY Buffalo, as part of a project sponsored by the US Postal Service. For more details about this data set, refer to [15]. Each image has dimensions of 16×16 pixels, with 256 grey levels per pixel. We employ the weighted eigenvalue similarity measure as defined in Sect. 8.6.3. Instead of using 16 eigenvalues, we only work with the four greatest ones. In this case, we use the following β function: $\beta(x) = 16 \exp(\frac{x}{3})$.
- *MNIST data set*: Originally composed of images with dimensions 28×28. We only use the public set composed of 10,000 vertices. Moreover, we make use of the dissimilarity measure based on the first 4 eigenvalues of each image out of 28 eigenvalues. The same β function employed in the USPS data set is used here. More information is given in Sect. 8.6.2.
- *Letter Recognition data set*: Composed of characteristic vectors with 16 entries. There are 20,000 vertices.

Since none of these data sets are in a network form, the methodology is divided into two general steps: the network formation and the data clustering tasks. In the first, we use the k-nearest neighbor network formation technique with $k = 3$ after we apply a preprocessing step. In this preprocessing, we standardize the data such as to have zero mean and unitary standard deviation. As for the distance measure, we either use the weighted eigenvalue dissimilarity (for the first two data sets above) or the Euclidean distance (for the last one). The reason behind not using the weighted eigenvalue on the third data set is because the samples are not provided as images, but as image descriptors. Since the latter is formed by merely scalars, we cannot apply the dissimilarity measure. In the second step, the data clustering algorithm based on particle competition is applied. As we are dealing with unsupervised learning, we do not use any external information, such as labels or exogenous knowledge. Instead, we limit ourselves to discovering explicit or implicit relationships among the data by the mechanism of particle competition.

9.5.2 Determining the Optimal Number of Particles and Clusters

Figure 9.15a–c show the determination of the optimal K for the USPS, MNIST, and Letter Recognition data sets, respectively. One can verify that $\langle R(t) \rangle$ is maximized exactly when the number of particles is equal to the number of clusters in the network, confirming the effectiveness of such heuristic.

9.5.3 Handwritten Data Clustering

Here, we show "digit" and "letter" cluster detection results using the particle competition algorithm. Table 9.1 supplies details about the algorithms chosen for comparison matters. The genetic algorithm implemented in the Global Optimization Toolbox of MATLAB is used to optimize the parameters of the particle competition algorithm. Specifically, the λ parameter is optimized over the range $0.2 \leq \lambda \leq 0.8$ and its optimal values for the USPS, MNIST, and Letter Recognition data sets are 0.58, 0.60, 0.60, respectively. The number of particles to be inserted is determined according to the previous analysis, i.e., we choose the number of particles that maximizes the quantity $\langle R(t) \rangle$ measure, which are 10, 10, and 26 for the USPS, MNIST, and Letter Recognition data sets, respectively.

Table 9.2 presents the data clustering rate reached by the particle competition method and the aforementioned competing algorithms. Some of these results are readily extracted from [33] and [23]. Within this table, we have provided the Average Rank of each algorithm, which is calculated as follows: (1) for each data set we rank the algorithms according to their average performance (average data clustering accuracy), i.e., the best algorithm is ranked as 1, the second best one is ranked as 2, and so on; (2) for each algorithm, the Average Rank is given by the average value of its rank achieved in all the data sets. As we can verify by looking at the Average Rank column, the stochastic competition algorithm has reached one of the best positions, showing the effectiveness of the particle competition scheme.

In order to further verify the robustness of the particle competition method, we inspect the samples that compose a same cluster. Specifically, Figs. 9.16 and 9.17 show some samples of the clusters representing the pattern "2" and "5", respectively, of the MNIST data set. These samples are captured using the following strategy: we

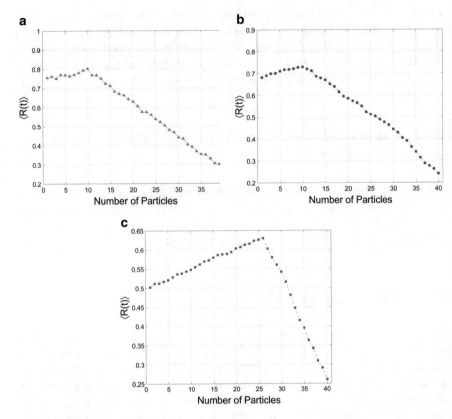

Fig. 9.15 Determination of the optimal number of particles K (the optimal number of clusters) in real-world data sets. The number of classes that each data set originally possesses are: (**a**) the USPS data set has 10 clusters (each cluster corresponding to a number from "0" to "9"); (**b**) the MNIST data set has 10 clusters (each cluster corresponding to a number from "0" to "9"); and (**c**) the Letter Recognition data set has 26 clusters (each cluster corresponding to a letter from the English alphabet ("A" to "Z")). 20 independent runs are performed and the average value is reported. Reproduced from [37] with permission from Springer

Table 9.1 Description of the competing state-of-the-art data clustering techniques

Technique	Reference
Gaussian mixture Model (GMM)	[5]
K-Means	[27]
Locally consistent Gaussian mixture Model (LCGMM)	[23]
Spectral clustering algorithm with normalized cut (Ncut)	[34]
Ncut embedding all (NcutEmb^{All})	[33]
Ncut embedding maximum (NcutEmb^{Max})	[33]

Reproduced from [37] with permission from Springer

Table 9.2 Data clustering accuracy reached by the particle competition technique and the competing methods listed in Table 9.1

	USPS	MNIST	Letter recognition	Avg. rank
LCGMM	73.83	73.60	93.03	2.33
GMM	67.30	66.60	91.24	5.33
K-Means	69.80	53.10	87.94	6.33
NCut	69.34	68.80	88.72	5.67
NCutEmb^{All}	72.72	75.10	90.07	3.67
NCutEmb^{Max}	72.97	75.63	90.59	2.67
Particle competition technique	80.46	74.53	91.37	2.00

For the stochastic methods, such as the particle competition method, thirty independent runs were performed and the corresponding mean is provided. Reproduced from [37] with permission from Springer

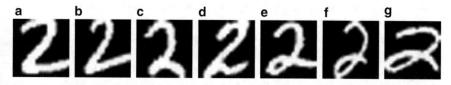

Fig. 9.16 A broad set of samples that were classified as being member of the cluster representing the pattern "2". Note that samples that are adjacent are similar with regard to the weighted eigenvalue dissimilarity function. The transitions from the sample (**a**)–(**g**) were captured from the maximum geodesic distance between two vertices in the cluster representing pattern 2. In this case, the diameter of such cluster is 17. We have only provided 7 representative samples above. Reproduced from [37] with permission from Springer

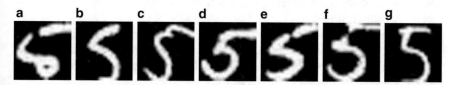

Fig. 9.17 A broad set of samples that were classified as being member of the cluster representing the pattern "5". Likewise the previous figure, adjacent samples are more similar to each other than distance samples. Reproduced from [37] with permission from Springer

compute the vertices that compose the maximum geodesic distance of the cluster representing each pattern (cluster diameter). Now, we select a representative subset of vertices composing the cluster diameter trajectory for illustrative purposes. In these figures, samples that are adjacent are more similar than those distant from one to another. On the basis of this analysis, we conclude that the graph representation has successfully captured several variations of these number patterns each of which in a single representative cluster, showing the robustness of the model.

9.6   Chapter Remarks

In this chapter, a rigorous definition of a competitive learning scheme in complex networks has been studied, whose foundations are biologically inspired by the competition process taking place in many nature and social systems. In this model, several particles navigate in the network to explore their territory and, at the same time, attempt to defend their territory from rival particles. If a particle frequently visits a specific vertex, it occurs that the domination level of the visiting particle on that vertex is strengthened. Concurrently to that, the domination levels of all of the other particles on the same vertex are weakened. In the long-run dynamic, each particle is expected to be confined within a community of the network.

The particle competition model is nonlinear and stochastic. Owing to the mathematical formality that the model is built upon, theoretical and empirical analyses have been conducted to better understand the underlying properties of the competitive model. A convergence analysis has shown that the dynamical system presents structural stability rather than asymptotic stability. This is a welcomed characteristic, since it better describes the uncertainty that revolves around real-world problems, which have noise and uncontrolled variables. In addition, due to this analysis, we have found that the model is a generalization of the process of single independent random walkers in a network. Specifically, we have shown that the model's behavior acts as multiple interacting walkers in a network. The interaction is molded in a competitive way, by using a probabilistic convex combination of random and preferential walks. Such generalization is realized by calibrating the values of the parameter λ and Δ of the system. If $\lambda = 0$, the model reduces to multiple non-interacting random walks; but, when $\lambda > 0$, the interaction among particles is turned on.

Furthermore, measures for detecting overlapping structures and for estimating the number of actual clusters or communities in a network have been discussed, whose calculations are embedded into the model's own algorithm. This permits their calculation to be performed in an efficient way.

Simulations have been carried out with the purpose of quantifying the robustness of the particle competition scheme on artificial and real-world data sets for the tasks of data clustering and community detection. Computer simulations have revealed that the model works well for community detection and for data clustering tasks. Finally, an application on handwritten digits and letters recognition has been provided and high clustering accuracies have been obtained. Moreover, we have analyzed the composition of the clusters formed in the MNIST data set and have verified that, within a specific cluster, several variations of the same pattern can be encountered, confirming the robustness of the model.

References

1. Allinson, N., Yin, H., Allinson, L., Slack, J.: Advances in Self-Organising Maps. Springer, New York (2001)
2. Amorim, D.G., Delgado, M.F., Ameneiro, S.B.: Polytope ARTMAP: pattern classification without vigilance based on general geometry categories. IEEE Trans. Neural Netw. **18**(5), 1306–1325 (2007)
3. Athinarayanan, R., Sayeh, M.R., Wood, D.A.: Adaptive competitive self-organizing associative memory. IEEE Trans. Syst. Man Cybern. Part A **32**(4), 461–471 (2002)
4. Bacciu, D., Starita, A.: Competitive repetition suppression (CoRe) clustering: a biologically inspired learning model with application to robust clustering. IEEE Trans. Neural Netw. **19**(11), 1922–1940 (2008)
5. Bishop, C.M.: Pattern Recognition and Machine Learning (Information Science and Statistics). Springer, New York (2007)
6. Carpenter, G.A., Grossberg, S.: Self-organization of stable category recognition codes for analog input patterns. Appl. Opt. **26**(23), 4919–4930 (1987)
7. Chen, M., Ghorbani, A.A., Bhavsar, V.C.: Incremental communication for adaptive resonance theory networks. IEEE Trans. Neural Netw. **16**(1), 132–144 (2005)
8. Çinlar, E.: Introduction to Stochastic Processes. Prentice-Hall, Englewood Cliffs (1975)
9. Deboeck, G.J., Kohonen, T.K.: Visual Explorations in Finance: With Self-Organizing Maps. Springer, New York (2010)
10. do Rêgo, R.L.M.E., Araújo, A.F.R., Neto, F.B.L.: Growing self-reconstruction maps. IEEE Trans. Neural Netw. **21**(2), 211–223 (2010)
11. Fortunato, S.: Community detection in graphs. Phys. Rep. **486**, 75–174 (2010)
12. Fu, X., Wang, L.: Data dimensionality reduction with application to simplifying rbf network structure and improving classification performance. IEEE Trans. Syst. Man Cybern., Part B: Cybern. **33**(3), 399–409 (2003)
13. Girvan, M., Newman, M.E.J.: Community structure in social and biological networks. Proc. Natl. Acad. Sci. USA **99**(12), 7821–7826 (2002)
14. Grossberg, S.: Competitive learning: from interactive activation to adaptive resonance. Cogn. Sci. **11**, 23–63 (1987)
15. Hull, J.J.: A database for handwritten text recognition research. IEEE Trans. Pattern Anal. Mach. Intell. **16**, 550–554 (1994)
16. Jain, L.C., Lazzerini, B., Ugur, H.: Innovations in ART Neural Networks (Studies in Fuzziness and Soft Computing). Physica, Heidelberg (2010)
17. Kaylani, A., Georgiopoulos, M., Mollaghasemi, M., Anagnostopoulos, G.C., Sentelle, C., Zhong, M.: An adaptive multiobjective approach to evolving ART architectures. IEEE Trans. Neural Netw. **21**(4), 529–550 (2010)
18. Knuth, D.E.: The Stanford GraphBase: A Platform for Combinatorial Computing. ACM, New York (1993)
19. Kohonen, T.: The self-organizing map. Proc. IEEE **78**(9), 1464–1480 (1990)
20. Kosko, B.: Stochastic competitive learning. IEEE Trans. Neural Netw. **2**(5), 522–529 (1991)
21. Lancichinetti, A., Fortunato, S., Radicchi, F.: Benchmark graphs for testing community detection algorithms. Phys. Rev. E **78**(4), 046,110(1–5) (2008)
22. Liu, D., Pang, Z., Lloyd, S.R.: A neural network method for detection of obstructive sleep apnea and narcolepsy based on pupil size and EEG. IEEE Trans. Neural Netw. **19**(2), 308–318 (2008)
23. Liu, J., Cai, D., He, X.: Gaussian mixture model with local consistency. In: AAAI'10, vol. 1, pp. 512–517 (2010)
24. López-Rubio, E., de Lazcano-Lobato, J.M.O., López-Rodríguez, D.: Probabilistic PCA self-organizing maps. IEEE Trans. Neural Netw. **20**(9), 1474–1489 (2009)
25. Lu, Z., Ip, H.H.S.: Generalized competitive learning of gaussian mixture models. IEEE Trans. Syst. Man Cybern., Part B: Cybern. **39**(4), 901–909 (2009)

26. Lusseau, D.: The emergent properties of a dolphin social network. Proc. R. Soc. B Biol. Sci. **270**(Suppl 2), S186–S188 (2003)
27. MacQueen, J.B.: Some methods for classification and analysis of multivariate observations. In: Proceedings of the fifth Berkeley Symposium on Mathematical Statistics and Probability, vol. 1, pp. 281–297. University of California Press, Berkeley (1967)
28. Meyer-Bäse, A., Thümmler, V.: Local and global stability analysis of an unsupervised competitive neural network. IEEE Trans. Neural Netw. **19**(2), 346–351 (2008)
29. Newman, M.E.J.: Fast algorithm for detecting community structure in networks. Phys. Rev. E **69**(6), 066,133 (2004)
30. Newman, M.E.J.: Modularity and community structure in networks. Proc. Natl. Acad. Sci. **103**(23), 8577–8582 (2006)
31. Príncipe, J.C., Miikkulainen, R.: Advances in Self-Organizing Maps - 7th International Workshop, WSOM 2009. Lecture Notes in Computer Science, vol. 5629. Springer, New York (2009)
32. Quiles, M.G., Zhao, L., Alonso, R.L., Romero, R.A.F.: Particle competition for complex network community detection. Chaos **18**(3), 033,107 (2008)
33. Ratle, F., Weston, J., Miller, M.L.: Large-scale clustering through functional embedding. In: Proceedings of the European conference on Machine Learning and Knowledge Discovery in Databases - Part II, European Conference on Machine Learning and Principles and Practice of Knowledge Discovery in Databases (ECML PKDD), pp. 266–281. Springer, New York (2008)
34. Shi, J., Malik, J.: Normalized Cut and Image Segmentation. Tech. rep., University of California at Berkeley, Berkeley (1997)
35. Silva, T.C., Zhao, L.: Stochastic competitive learning in complex networks. IEEE Trans. Neural Netw. Learn. Syst. **23**(3), 385–398 (2012)
36. Silva, T.C., Zhao, L.: Uncovering overlapping cluster structures via stochastic competitive learning. Inf. Sci. **247**, 40–61 (2013)
37. Silva, T.C., Zhao, L., Cupertino, T.H.: Handwritten data clustering using agents competition in networks. J. Math. Imaging Vision **45**(3), 264–276 (2013)
38. Sugar, C.A., James, G.M.: Finding the number of clusters in a data set: an information theoretic approach. J. Am. Stat. Assoc. **98**, 750–763 (2003)
39. Tan, A.H., Lu, N., Xiao, D.: Integrating temporal difference methods and self-organizing neural networks for reinforcement learning with delayed evaluative feedback. IEEE Trans. Neural Netw. **19**(2), 230–244 (2008)
40. Wang, Y., Li, C., Zuo, Y.: A selection model for optimal fuzzy clustering algorithm and number of clusters based on competitive comprehensive fuzzy evaluation. IEEE Trans. Fuzzy Syst. **17**(3), 568–577 (2009)
41. Xu, R., II, D.W.: Survey of clustering algorithms. IEEE Trans. Neural Netw. **16**(3), 645–678 (2005)
42. Zachary, W.W.: An information flow model for conflict and fission in small groups. J. Anthropol. Res. **33**, 452–473 (1977)

Chapter 10
Case Study of Network-Based Semi-Supervised Learning: Stochastic Competitive-Cooperative Learning in Networks

Abstract Information reaches us at a remarkable speed and the amount of data it brings is unprecedented. In many situations, only a small subset of data items can be effectively labeled. This is because the labeling process is often expensive, time consuming, and requires intensive human involvement. As a result, partially labeled data sets are more frequently encountered. In order to get a better characterization of partially labeled data sets, semi-supervised classifiers are designed to learn from both labeled and unlabeled data. It has turned out to be a new topic of machine learning research that has received increasing attention in the past years. In this chapter, the semi-supervised classification with focus on methods based on complex networks is explored. In special, the particle competition model that we have introduced in the previous chapter is adapted to this new learning paradigm. Specifically, this enhancement is achieved by introducing the idea of cooperation among the particles and by changing the inner mechanisms of the original algorithm so as to fit it into a semi-supervised environment. In contrast to the unsupervised learning model, where the particles are randomly spawned in the network because no prior analysis of the groups is available, the semi-supervised learning version does have some external knowledge by definition. This knowledge is represented by the labeled data items, usually offered as a small fraction of the entire data set. In this scenario, the objective is to propagate the labels from the labeled set to the unlabeled set. Likewise the previous chapter, a mathematical formalization of the model, as well as a theoretical analysis, is also provided. A great portion of this analysis is based on the model that we have studied in the last chapter. A validation is also presented linking the numerical and theoretical results. An application in imperfect data learning is also presented, where the particle competition model is employed to detect and prevent error propagation in the learning process due to noisy or wrongly labeled data.

10.1 A Quick Overview of the Chapter

As we have seen in Chap. 3, the semi-supervised learning differs from the unsupervised learning by the fact that the former has some external knowledge incorporated into the learning process, by means of pre-labeled data items. Moreover, semi-

© Springer International Publishing Switzerland 2016   291
T.C. Silva, L. Zhao, *Machine Learning in Complex Networks*,
DOI 10.1007/978-3-319-17290-3_10

supervised learning also differs from supervised learning because it uses both labeled and unlabeled data in the learning process.

In the initial part of this chapter, the semi-supervised particle competition model is presented [3, 11]. Once the basic concepts are properly introduced, we deal with the interesting problem of learning with imperfect data. In this case, the labeled data set is not totally reliable. Situations in which the training data is not perfectly reliable may arise when noises are embedded in the labeling source procedure or in the acquisition of the data items, or even when an external professor incorrectly labels data items (human error). Though being prone to all of these kinds of error sources, most semi-supervised learning algorithms assume a perfectly reliable training data. As we will see, their performances are largely impacted when that assumption is violated.

Due to the practicability in the real-world, the semi-supervised model of multiple particles is further enhanced to deal with this uncertainty in the training data. For that, the model is equipped with mechanisms to detect and to prevent wrongly labeled training data. These mechanisms only use information that the dynamical system of multiple particles generates. Therefore, the procedure of detection and prevention of imperfect data is embedded within the model. We show that the modified particle competition model can really provide good results even in environments where the training data reliability is low. We analyze how the model's accuracy rate behaves as we increase the percentage of noise or wrongly labeled items in the training data set. In this analysis, we find critical points that are characterized by border regions in which small increases in the noise in the labeled set produces large downfalls in the model's accuracy. We compare the critical points of the semi-supervised model of multiple particles with other competing techniques and show that the former can withstand much more noisy environments. We also show that, provided that the fraction of correctly labeled vertices is the majority, the model can in general correctly re-label the wrongly labeled vertices. This constraint is intuitive and confirms that the competitive model uses some kind of "natural selection" in the learning process that forces the majority to overwhelm the minority using the network topology.

10.2   Description of the Stochastic Competitive-Cooperative Model

In this section, we describe the semi-supervised version of the particle competition model in detail [11].

10.2.1   Differences of the Semi-Supervised and the Unsupervised Versions

The main difference in the semi-supervised version of the model of multiple interacting particles is that now each particle represents a labeled data item. The goal of each particle is to spread the associated label of its labeled vertex to other unlabeled vertices by visiting and dominating them in a competitive way. The labeled vertex that each particle represents is termed as the *home vertex* of that particle. Particles always start the dynamical process at their home vertices. In the reanimation procedure of particles, once exhausted, they always regress to their corresponding home vertices and not to random dominated vertices. Moreover, particles are guaranteed to always dominate their home vertices in such a way that the domination levels imposed on home vertices are not sensitive to visits of rival particles.

Figure 10.1 illustrates the initial condition of the semi-supervised version of the particle competition model. Note that the initial location of particles is deterministically established in accordance with their home vertices. Moreover, the number of particles equals the number of labeled data items.

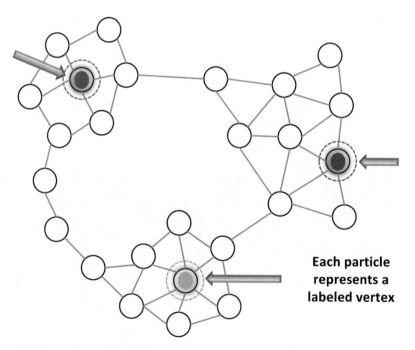

Each particle represents a labeled vertex

Fig. 10.1 Illustration of the initial conditions and the reanimation procedure of the semi-supervised model with multiple particles

This modification forces the model to work in a local label-spreading behavior, since each particle is now probabilistically bounded within a small region of the network. As a consequence, due to the competitive mechanism, each particle only visits a portion of vertices potentially having the same label as the particle's home labeled vertex. This concept can be roughly conceived as a "divide-and-conquer" effect embedded into the competitive scheme.

Recall that particles only have exploratory behavior due to the random movement rule when $\lambda = 0$. Conversely, they only present defensive characteristics due to the preferential movement rule when $\lambda = 1$. In addition, we have seen that a mixture of these two walking policies promotes better results for the model. Due to the deterministic behavior of the reanimation procedure, particles get probabilistically confined in regions potentially centered at their corresponding home vertices. The width of these regions is determined by the counterweighting factor λ. Figure 10.2 provides an intuitive schematic of this concept. Note that, as λ decreases, the larger are the regions that particles can potentially visit in that we are giving more importance to the exploratory in detriment to the defensive behavior. In this way, particles' dominated territories are expected to collide more often as λ decreases.

One interesting feature of the semi-supervised version is of the emergence of potential cooperation among particles. Frequently, we may have more than one labeled vertex from the same class. As a consequence, more than one particle may represent the same class or label of their respective home vertices. In this case, these

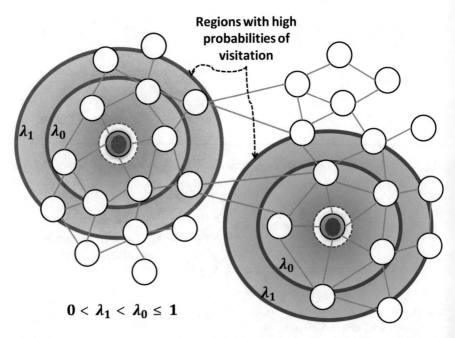

Fig. 10.2 Influence of the counterweighting factor λ in shaping regions with high probabilities of visitation

representative particles act together as a team because they propagate the same label to unlabeled vertices. In this way, several teams can potentially compete with each other to establish their class borders, while cooperating with their teammates.

10.2.2 Familiarizing with the Semi-Supervised Environment

In the semi-supervised learning scheme, denote $\mathscr{Y}$ as the set of possible discrete classes that can be predicted by the semi-supervised classifier. A set of data items $\mathscr{X} = \{x_1, \ldots, x_L, x_{L+1}, \ldots, x_{L+U}\}$ is supplied, in which each entry is a P-dimensional attribute vector of the form $x_i = (x_{i1}, \ldots, x_{iP})$. The first L data items are initially labeled and compose the labeled set $\mathscr{L}$. The remainder U data items are the unlabeled instances and comprise the unlabeled set $\mathscr{U}$. Note that $\mathscr{X} = \mathscr{L} \cup \mathscr{U}$. For each $x_i \in \mathscr{L}$, a label $y_i \in \mathscr{Y}$ is given. In contrast, no labels are supplied for data items in $\mathscr{U}$. The objective is to propagate the labels from $\mathscr{L}$ to $\mathscr{U}$, while preserving the data distributions. In practice, the proportion of unlabeled data items far surpasses the proportion of labeled data items, such that $U \gg L$ is observed in several practical occasions.

Since the particle competitive-cooperative model is a network-based technique, a network formation technique is employed to transform the vector-based data into a network. For this end, the data are mapped into a graph $\mathscr{G}$ using a network formation technique $g : \mathscr{X} \mapsto \mathscr{G} = \langle \mathscr{V}, \mathscr{E} \rangle$, in which $\mathscr{V} = \mathscr{L} \cup \mathscr{U}$ is the set of vertices and $\mathscr{E}$ is the set of edges. There are $V = |\mathscr{V}|$ vertices in the graph. Each vertex $v \in \mathscr{V}$ in the network corresponds to a data item $x \in \mathscr{X}$, so that $V = L + U$. Essentially, each vertex in $\mathscr{V}$ represents a data item in $\mathscr{X}$. The edges in $\mathscr{E}$ are created using a suitable network formation process, such as those explored in Chap. 4.

10.2.3 Deriving the Modified Competitive Transition Matrix

In this section, we focus on the technical differences of the unsupervised and semi-supervised learning transition matrices. If any part of the method has not been expressly indicated here, then it means that it is identical to the unsupervised transition matrix derived in Sect. 9.2.2.

The transition matrix assumes the same functional form as that in the unsupervised version, which, for convenience, we remember as follows:

$$\mathbf{P}^{(k)}_{\text{transition}}(t) \triangleq (1 - S^{(k)}(t)) \left[\lambda \mathbf{P}^{(k)}_{\text{pref}}(t) + (1 - \lambda) \mathbf{P}^{(k)}_{\text{rand}} \right] + S^{(k)}(t) \mathbf{P}^{(k)}_{\text{rean}}(t). \tag{10.1}$$

Basically, the technical differences are reflected on how each of these matrices comprising the transition matrix are defined. Specifically, the random and preferential terms do not suffer any modifications. The reanimation matrix, however, is adapted on account of two reasons:

- We must model the label diffusion process from labeled to unlabeled vertices in a local label-spreading behavior. For that, we cannot randomly transport particles from one place to another in the network as we would be creating labeling decisions that are non-smooth.
- We must now comport the idea of the existence of external information in the form of labels, which in turn are represented by labeled vertices. The idea is to use these labeled vertices, here termed as home vertices, as the proper destination for exhausted particles. As these labeled vertices are static in the network, we effectively force labeling decisions that are smooth.

Recall that each entry of $\mathbf{P}_{\text{rean}}^{(k)}(t)$ indicates the probability of bringing an exhausted particle $k \in \mathscr{K}$ back to its dominated territory. Here, we always transport the particle back to its home vertex, which is the vertex that particle k represents. Suppose that particle k is visiting vertex i when its energy becomes completely depleted. In this particular occasion, we transport the particle back in accordance with the following distribution:

$$\mathbf{P}_{\text{rean}}^{(k)}(i,j,t) \triangleq \begin{cases} 1, & \text{if } j = v_k \\ 0, & \text{otherwise} \end{cases}, \tag{10.2}$$

in which v_k indicates the home vertex of particle k, Therefore, matrix $\mathbf{P}_{\text{rean}}(t)$ only has non-zero entries for reallocations of particles to their respective home vertices. In computational terms, this can greatly enhance the process of deciding the next vertex that particle k will visit. For didactic purposes, Fig. 10.3 portrays a simple scenario of a reanimation taking place. In this case, the red or dark gray particle has its energy penalized, since it is visiting a vertex dominated by a rival particle. Supposing that its energy has been completely depleted, that particle becomes exhausted. Under these circumstances, the reanimation procedure of that particle is enabled, which will compel it to travel back to its home vertex so as to be properly recharged. Even though this is a simple mechanism, it can greatly increase the performance of the particle competitive-cooperative model, because it does not let particles go wander very far from their origions. Thus, the algorithm forces smoothness in the label diffusion process.

10.2.4   Modified Initial Conditions of the System

Recall that the internal dynamical state of system ϕ is composed of four terms: $p(t)$ is a random vector denoting the particles' locations at time t; $N(t)$ represents the number of visits that each of the vertices received up to time t; $E(t)$ indicates the

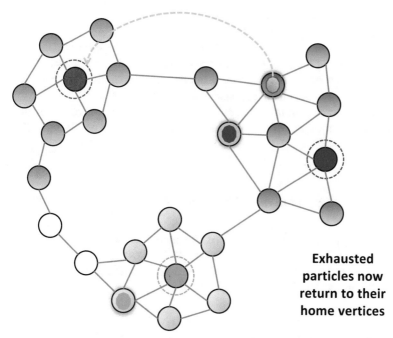

Exhausted particles now return to their home vertices

Fig. 10.3 The modified reanimation procedure performed by an exhausted particle. The surrounded vertices denote labeled instances. The color intensity in each vertex denotes the color of the particle imposing the highest domination level

particles' energy levels at time t; and $S(t)$ displays the particles' states at time t. In order to run the dynamical system ϕ, we need a set of initial conditions. In this way, we discuss how to fix the initial conditions for these four dynamical variables at $t = 0$.

For the initial particles' location $p(0)$, each particle is put at its corresponding home vertex.

We now discuss how to initialize matrix $\mathbf{N}(0)$. For those initially labeled vertices, we fix a permanent ownership to their representative particles as follows. Since the ownership is represented by the maximum actual domination level imposed on that vertex, we simply force the number of visits of the representative particle to its home vertex to be infinity at the beginning of the learning process. Thus, changes in the ownership of labeled vertices become impossible. In view of this scheme, we have that each entry of $\mathbf{N}(0)$ is now given by:

$$\mathbf{N}_i^{(k)}(0) = \begin{cases} \infty, & \text{if particle } k \text{ represents vertex } i \\ 1 + \mathbb{1}_{[p^{(k)}(0)=i]}, & \text{otherwise} \end{cases}, \qquad (10.3)$$

in which we apply (10.3) to every $(i, k) \in \mathcal{V} \times \mathcal{K}$. Note that the scalar 1 is used in the second expression of (10.3) so that unlabeled and unvisited vertices at time

$t = 0$ have their calculation well-defined, according to (9.7). Usually, more than one particle (a team) is generated to represent a set of pre-labeled examples of the same class. Each of them tries to dominate vertices independently. The cooperation among the particles of the same team happens only at the end of the process. In order to do so, for each vertex, we sum up the domination levels of all of the particles of the same team on it to obtain the aggregated domination level.

With respect to the initial particles' energy levels $E(0)$ and states $S(0)$, we maintain the configurations of the unsupervised learning model according to (9.25) and (9.26).

10.3   Theoretical Analysis of the Model

In this section, a theoretical analysis of the model is discussed. A numerical validation of the theoretical results is supplied. It is worth noting that only the main analytical differences in relation to theoretical analysis previously conducted on Sect. 9.3 (unsupervised version) are provided.

10.3.1   Mathematical Analysis

Since the dynamical system of the unsupervised and semi-supervised version are virtually the same, differing only on the initial distributions of the particle locations, the transition probability function is the same. In view of this, we rewrite it for convenience matters as follows:

$$
\begin{aligned}
P(\mathbf{X}(t+1) \mid \mathbf{X}(t)) = {}& \mathbb{1}_{[\mathbf{N}(t+1)=\mathbf{N}(t)+\mathbf{Q}_N(p(t+1))]} \\
& \times \mathbb{1}_{[S(t+1)=Q_S(E(t+1))]} \\
& \times \mathbb{1}_{[E(t+1)=E(t)+\Delta Q_E(p(t+1),\mathbf{N}(t+1))]} \qquad (10.4) \\
& \times \mathbf{P}_{\text{transition}}(\mathbf{N}(t), p(t)) \\
= {}& \mathbb{1}_{[\text{Compliance}(t)]}\mathbf{P}_{\text{transition}}(\mathbf{N}(t), p(t)).
\end{aligned}
$$

Following the reasoning applied in the analytical analysis of the unsupervised dynamical system, we are required to set feasible upper and lower limits for each of the random variables. The limits for $p(t), E(t), S(t)$ derived in the unsupervised version are integrally valid for the semi-supervised learning version.

Having in mind these considerations, we now derive these limits for the random variable $\mathbf{N}(t)$. The new initialization step indicated in (10.3) takes into account both labeled and unlabeled vertices, which invalidates Lemma 9.1 that only previews the existence of unlabeled vertices. Given this fact, we reformulate the aforementioned Lemma in the following.

Lemma 10.1. *The maximum reachable value of* $\mathbf{N}_i^{(k)}(t)$, $\forall (i,k) \in \mathcal{V} \times \mathcal{K}$, $t \in \mathbb{N}$, *is:*

- *If* $i \in \mathcal{U}$:

$$\mathbf{N}_{i_{\max}}^{(k)}(t) = \begin{cases} \left\lceil \frac{t+1}{2} \right\rceil + 1, & \text{if } t > 0 \text{ and } a_{ii} = 0 \\ t + 2, & \text{if } t > 0 \text{ and } a_{ii} > 0 \end{cases}. \tag{10.5}$$

- *If* $i \in \mathcal{L}$:

$$\bar{\mathbf{N}}_{i_{\max}}^{(k)}(t) = \infty. \tag{10.6}$$

in which $a_{ii} = 0$ *if there are no self-loops starting at vertex* i, *and* $a_{ii} > 0$ *otherwise.*

Proof. With regard to unlabeled vertices, i.e., which belong to $\mathcal{U}$, the proof supplied in Lemma 9.1 can be invoked *ipsis litteris*. This is valid because the movement policy of each particle remains the same in relation to the original unsupervised learning version.

With regard to labeled vertices, i.e., which belong to $\mathcal{L}$, this quantity can be inferred in a straightforward manner through the initial conditions of the system. According to (10.3), if i is a pre-labeled vertex and k is its representative particle, then $\mathbf{N}_i^{(k)}(t) = \infty, \forall t \geq 0$.   ∎

Since the upper and lower limits of the random variable $\mathbf{N}(t)$ have changed, the analysis of $\bar{\mathbf{N}}(t)$ needs some adjustments. Similarly to the previous case, Lemma 9.3 presented in the original version of the algorithm may only be applied to unlabeled vertices. In view of this, we reformulate this Lemma as follows:

Lemma 10.2. *The following assertions hold* $\forall (i,k) \in \mathcal{V} \times \mathcal{K}$:

- *If* $i \in \mathcal{U}$:

  (a) The minimum value of $\bar{\mathbf{N}}_i^{(k)}(t)$ *is:*

$$\bar{\mathbf{N}}_{i_{\min}}^{(k)}(t) = \frac{1}{1 + \sum_{u \in \mathcal{K} \setminus \{k\}} \mathbf{N}_{i_{\max}}^{(u)}(t)}. \tag{10.7}$$

  (b) The maximum value of $\bar{\mathbf{N}}_i^{(k)}(t)$ *is:*

$$\bar{\mathbf{N}}_{i_{\max}}^{(k)}(t) = \frac{\mathbf{N}_{i_{\max}}^{(k)}(t)}{\mathbf{N}_{i_{\max}}^{(k)}(t) + (K - 1)}. \tag{10.8}$$

- *If* $i \in \mathcal{L}$:

  (a) The minimum value of $\bar{\mathbf{N}}_i^{(k)}(t)$ *is:*

$$\bar{\mathbf{N}}_{i_{\min}}^{(k)}(t) = 0. \tag{10.9}$$

(b) The maximum value of $\bar{\mathbf{N}}_i^{(k)}(t)$ is:

$$\bar{\mathbf{N}}_{i_{\max}}^{(k)}(t) = 1. \tag{10.10}$$

Proof. With regard to unlabeled vertices, the proof supplied in Lemma 9.3 can be invoked *ipsis litteris.*

With regard to labeled vertices, Eq. (10.9) can be reached as follows: consider that particle k is not a representative from the labeled vertex i. However, by the initial conditions shown in (10.3), since vertex i is labeled by hypothesis, $\exists k' \in \mathcal{K}$: $\mathbf{N}_i^{(k')}(t) = \infty, \forall t \geq 0$. Now, as k does not represent i, via (10.3) again, we know that $\mathbf{N}_i^{(k)}(t)$ may only take on finite values $\forall t \geq 0$. Finally, applying (9.7) using this setup yields (10.9). Equation (10.10) can be achieved as follows: consider that particle k now represents the labeled vertex i. In this case, $\mathbf{N}_i^{(k)}(t) = \infty$. Considering that a labeled vertex may only be represented by one kind of particle, then all the remaining entries of $\mathbf{N}_i^{(k')}(t), k' \in \mathcal{K}, k' \neq k$ are finite. Using (9.7) under these circumstances, we arrive at (10.10). ∎

The final step before calculating the marginal distribution of the vertices' domination levels, i.e., $P(\bar{\mathbf{N}}(t))$, is to find all the possible irreducible fractions that an arbitrary entry of $\bar{\mathbf{N}}(t)$ can assume. Lemma 9.4 fails to provide us with enough information about the labeled vertices, since it only delimits the irreducible fractions for unlabeled vertices. Next, a reformulation of such Lemma is provided.

Lemma 10.3. *Denote* num/den *as an arbitrary irreducible fraction. Consider that the set $\mathcal{I}_t$ retains all the reachable values of $\bar{\mathbf{N}}_i^{(k)}(t), \forall (i,k) \in \mathcal{V} \times \mathcal{K}$, for a fixed t. Then, the elements of $\mathcal{I}_t$ are composed of all elements satisfying the following constraints:*

(i) With regard to unlabeled vertices:

   (a) The minimum element is given by the expression in (9.60).
   (b) The maximum element is given by the expression in (9.61).
   (c) All the irreducible fractions within the interval delimited by (a) and (b) such that:

       I. num, den $\in \mathbb{N}^*$.
       II. num $\leq \mathbf{N}_{i_{\max}}^{(k)}(t)$.
       III. den $\leq \sum_{u \in \mathcal{K}} \mathbf{N}_{i_{\max}}^{(u)}(t)$.

(ii) With regard to labeled vertices:

   (a) 0, if particle k does not represent vertex i.
   (b) 1, if particle k represents vertex i.

Proof. Regarding item (i): Straightforward from Lemma 9.4.

Regarding item (ii): (a) As vertex i is labeled, $\exists u \in \mathscr{K} : \mathbf{N}_i^{(u)}(t) = \infty$. In view of (9.7) and (10.3), we obtain $\bar{\mathbf{N}}_i^{(k)}(t) = 0$; (b) Similarly, using (9.7) and (10.3), we get $\bar{\mathbf{N}}_i^{(k)}(t) = 1$. ∎

Finally, the expression for calculating the domination matrix distribution remains the same as the one derived in the unsupervised version of the algorithm. In the following, we reiterate it for convenience:

$$P\left(\bar{\mathbf{N}}(t) = \mathbf{U} : \mathbf{U} \in \mathcal{M}_t\right) = \sum_{u=1}^{t} P\left(f(u\mathbf{N}(t)) = \mathbf{U}\right). \tag{10.11}$$

As $t \to \infty$, $P(\bar{\mathbf{N}}(t))$ provides enough information for classifying the unlabeled vertices. In this case, they are labeled according to the team of particles that is imposing the highest domination level. Since the domination level is a stochastic variable, the output of this model is fuzzy.

10.3.2 A Numerical Example

In this section, we show how the theoretical results derived in the previous section can be employed in a simple example. We limit the demonstration for a single iteration, namely for transition from $t = 0$ to $t = 1$. Consider the exemplificative network as a trivial 3-vertex regular network, identical to that in Fig. 9.10a. Consider that vertex 1 has been labeled as pertaining to class 1 and vertex 2, to class 2, i.e., $\mathscr{V} = \{1, 2, 3\}$, $\mathscr{L} = \{1, 2\}$, and $\mathscr{U} = \{3\}$. Clearly, we can see that the unlabeled vertex 3 possesses overlapping characteristics in relation to classes 1 and 2. We theoretically show this behavior through this illustrative example. Consider the arbitrary initial settings: we insert $K = 2$ particles into the network, i.e., $\mathscr{K} = \{1, 2\}$. Let the particle 1 represent vertex 1 and particle 2, vertex 2. In this setup, particle 1 propagates the label of vertex 1 and particle 2 diffuses labels of vertex 2. Suppose also that we have a certainty about the locations of the particles at $t = 0$, which satisfy the following distribution:

$$P\left(\mathbf{N}(0) = \begin{bmatrix} \infty & 1 \\ 1 & \infty \\ 1 & 1 \end{bmatrix}, p(0) = [1 \ 2], E(0), S(0)\right) = 1, \tag{10.12}$$

i.e., there is 100 % (certainty) that the particles 1 and 2 are generated at vertices 1 and 2, respectively. Observe that $\mathbf{N}(0)$, $E(0)$, and $S(0)$ are chosen such as to satisfy (10.3), (9.25), and (9.26), respectively; otherwise the probability should be 0, in view of (10.4).

From Fig. 9.10a, we can deduce the adjacency matrix $\mathbf{A}$ of the graph and, therefore, determine the transition matrix associated to the random movement term for a single particle. Applying (9.2), we get:

$$\mathbf{P}_{\text{rand}} = \begin{bmatrix} 0 & 0.50 & 0.50 \\ 0.50 & 0 & 0.50 \\ 0.50 & 0.50 & 0 \end{bmatrix}. \tag{10.13}$$

Given $\mathbf{N}(0)$, we can readily establish $\bar{\mathbf{N}}(0)$ with the aid of (9.7):

$$\bar{\mathbf{N}}(0) = \begin{bmatrix} 1 & 0 \\ 0 & 1 \\ 0.50 & 0.50 \end{bmatrix}. \tag{10.14}$$

Using (9.8) we are able to calculate the matrices associated to the preferential movement policy for each particle in the network:

$$\mathbf{P}_{\text{pref}}^{(1)}(0) = \begin{bmatrix} 0 & 0 & 1 \\ 0.67 & 0 & 0.33 \\ 1 & 0 & 0 \end{bmatrix}, \tag{10.15}$$

$$\mathbf{P}_{\text{pref}}^{(2)}(0) = \begin{bmatrix} 0 & 0.67 & 0.33 \\ 0 & 0 & 1 \\ 0 & 1 & 0 \end{bmatrix}. \tag{10.16}$$

To simplify calculations, let us assume $\lambda = 1$, so that (10.1) reduces to $\mathbf{P}_{\text{transition}}(0) = \mathbf{P}_{\text{pref}}^{(1)}(0) \otimes \mathbf{P}_{\text{pref}}^{(2)}(0)$ at time 0, which is a matrix with dimensions 9×9. Instead of building this matrix, we make use of Remark 9.1 to build the next particles localization vector $p(1)$ with the collection of two matrices 3×3, as given in (10.15) and (10.16). Note that, in the special case when $\lambda = 1$, the preferential movement matrix is the transition matrix itself, provided that all the particles are active, which indeed are at time 0, according to (9.26). For the first particle, one can see from (10.15) that, starting from vertex 1 (row 1), there can only be one next possible localization for particle 1, namely vertex 3. For the second particle, starting from the vertex 2 (row 2), one can state that the next localization of particle 2 can only be vertex 3, too. With that in mind, we have that:

$$P\left(\mathbf{N}(1) = \begin{bmatrix} \infty & 1 \\ 1 & \infty \\ 2 & 2 \end{bmatrix}, p(1) = [3\ 3], E(1), S(1) \mid \mathbf{X}(0) \right) = 1, \tag{10.17}$$

in which $\mathbf{X}(0)$ is given by (10.12). Furthermore, as we have fixed $\lambda = 1$, it is expected that the transition will be heavily dependent on the domination levels of the neighborhood vertices. Therefore, given that the labeled vertices constitute strong repulsive forces that act against rival particles, the preferential or defensive behavior of these particles will never adventure in these type of vertices. This provides a natural explanation for the reason that the state $p(1) = [3\ 3]$ is the only possible next particles localization vector.

Before doing the calculation of the marginal distribution $P(\mathbf{N}(1))$, we are required to find an upper limit for an arbitrary entry of a specific unlabeled vertex of the matrix $\mathbf{N}(1)$. This is readily evaluated from (10.5), which results in $\mathbf{N}_{i_{max}}^{(j)}(1) = 2$, $\forall i \in \mathscr{V}$, implying that we are only needed to take all numerical combinations of the matrix $\mathbf{N}(0)$ such that each entry may only take the values $\{1, 2\}$, since larger values would yield probability 0 according to Lemma 10.1. Moreover, we need to iterate through every feasible value of every entry of $E(0)$ and $E(1)$. In order to do so, we fix $\Delta = 0.25$, $\omega_{min} = 0$, and $\omega_{max} = 1$. With that, we are able to make use of Lemma 9.2, which yields $E(t) \in \{0, 0.25, 0.5, 0.75, 1\}$. The limits of the remaining system variables $S(0)$ and $S(1)$ are straightforward. In the present conditions, we have enough information to calculate the marginal distribution $P(\mathbf{N}(1))$, according to (9.57):

$$P\left(\mathbf{N}(1) = \begin{bmatrix} \infty & 1 \\ 1 & \infty \\ 2 & 2 \end{bmatrix}\right) = 1 \times 1 = 1. \tag{10.18}$$

As the last goal, the task is to determine the distribution $P(\bar{\mathbf{N}}(1))$. According to the specified steps in the previous section, we need to find all irreducible fractions that lie within the interval $[0, 1]$ with the constraints derived in the previous section. This means that we only have to consider entries of matrix $\bar{\mathbf{N}}(t)$ that contain elements of $\mathscr{I}_t$; the remainder $\bar{\mathbf{N}}(t)$ are infeasible and, thus, occur with probability 0. In view of the constraints previously enumerated, $\mathscr{I}_t = \{0, 1/4, 1/3, 1/2, 2/3, 3/4, 1\}$. It is worth commenting that the labeled vertices (vertices 1 and 2) can only assume the values $\{0, 1\} \subset \mathscr{I}_t$, as we have previously stated. Observing that we have the complete distribution of $\mathbf{N}(1)$, it is an easy task to apply (9.64), as follows:

$$P\left(\bar{\mathbf{N}}(1) = \begin{bmatrix} 1 & 0 \\ 0 & 1 \\ 0.5 & 0.5 \end{bmatrix}\right) = 1. \tag{10.19}$$

It is noteworthy to reinforce that the mapping between the probabilities of $\mathbf{N}(t)$ and $\bar{\mathbf{N}}(t)$ is not bijective: in this special simple case that we are studying, we did not have distinct $\mathbf{N}(t)$ that could generate $\bar{\mathbf{N}}(t)$, but as t increases, this is likely to happen quite frequently. This process is repeated for a sufficiently large t or until the system converges to a quasi-stationary state $\bar{\mathbf{N}}(t)$. A detailed look at the system's behavior that we have derived suggests that (10.19) holds for every $t \geq 1$, and particles 1 and 2 will visit vertex 3 with period 2. Hence, this shows the overlapping nature of vertex 3, as it can be naturally stated only by the topological structure of the graph. Ideally, for networks with presence of distinct classes, $\bar{\mathbf{N}}(t)$ varies as we iterate the dynamical system. We can then check the most probable classification of each of the vertices by looking at the domination levels of those particles with highest probability $P(\bar{\mathbf{N}})$ in the corresponding rows of each vertex.

For example, in Ref. [11], a simulation is performed to illustrate that the theoretical results really approximate the empirical behavior of the stochastic competitive model for a large number of independent runs of the algorithm. In addition, extensive numerical analyses are conducted to show the good performance of the particle competitive-cooperative model.

10.4 Numerical Analysis of the Model

In this section, we present simulation results in order to show the effectiveness of the semi-supervised particle competitive-cooperative model.

10.4.1 Simulation on a Synthetic Data Set

Here, we investigate the performance of the algorithm when applied to a network consisted of $V = 15$ vertices split into 3 unbalanced communities, as depicted in Fig. 10.4. $K = 3$ particles are inserted into the network at the initial positions $p(0) = [2\ 8\ 15]$, meaning the first particle (representing the red or "circle" class) starts at vertex 2, the second particle (representing the blue or "square" class) starts at vertex 8, and the third particle (representing the green or "triangle" class) starts at vertex 15. All the remaining vertices in the network are initially unlabeled (in the

Fig. 10.4 A simple networked data set. The *red* or *"circle"* class is composed by vertices 1–4, the *blue* or *"square"* class comprises the vertices 5–10, and the *green* or *"triangle"* class encompasses the vertices 11–15. Initially, only the vertices 2 (*red* or *"circle"* particle), 8 (*blue* or *"square"* particle), and 15 (*green* or *"triangle"* particle) are labeled

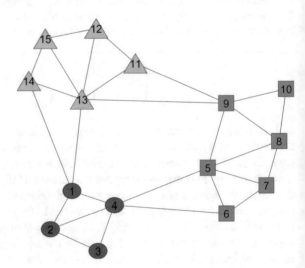

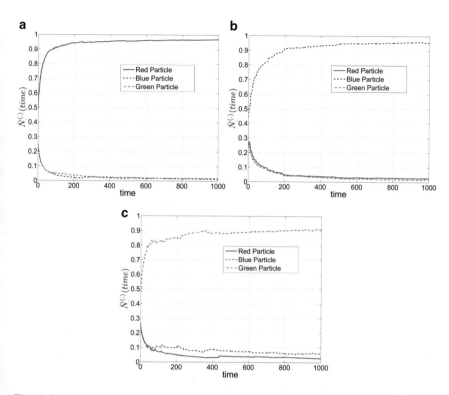

Fig. 10.5 Evolutional behavior of the average domination level imposed by the 3 particles on the existing classes in the network. (**a**) Red or "circle" class (vertices 1–4). (**b**) Blue or "square" class (vertices 5–10). (**c**) Green or "triangle" class (vertices 11–15)

figure, they are colored for the sake of easily identifying the classes). The competitive system is iterated until $t = 1{,}000$ and the predicted label for each of the unlabeled vertices is given by the particle's label that is imposing the highest domination level. Figure 10.5a–c show the evolutional behavior of the domination levels imposed by the three particles on the red or "circle" class, the blue or "square" class, and the green or "triangle" class, respectively. Specifically, from Fig. 10.5a, we can verify that red or "circle" particle dominates vertices 1–4 (red or "circle" class), due to the fact that the average domination level on these vertices approaches 1, whereas the average domination levels of the other two rival particles decay to 0. Considering Fig. 10.5b, c, we can use the same logic to confirm that the blue or "square" particle completely dominates the vertices 5–10 (blue or "square" class) and the green or "triangle" particle dominates vertices 11–15 (green or "triangle" class).

Table 10.1 Brief
meta-information of the UCI
data sets

Data set	# Instances	# Dimensions	# Classes
Heart	303	75	2
Heart-statlog	270	13	2
Ionosphere	351	34	2
Vehicle	946	18	4
House-votes	435	16	2
Wdbc	569	32	2
Clean1	476	168	2
Isolet	7797	617	26
Breastw	569	32	2
Australian	690	14	2
Diabetes	768	8	2
German	1000	20	2
Optdigits	5620	64	10
Sat	6435	36	7

10.4.2   Simulations on Real-World Data Sets

In this section, computer simulations on real-world data sets gathered from the UCI
Machine Learning Repository are presented. A brief meta-information of the 14
real-world data sets that are going to be studied here is supplied in Table 10.1. For a
detailed description, refer to [7].

With respect to the experimental setup, for each data set, 10 examples are
randomly selected to compose the labeled set and the labels of the remainder of
the examples are dropped. The labeled set is formed in such a way that each class
has at least a labeled example. It is worth observing that in a semi-supervised
learning task, the quantity of labeled examples is often too few to afford a valid cross
validation, and therefore hold-out tests are usually used for the evaluation process.
Also, for the purpose of comparing the performance of different algorithms, we
use two competitive semi-supervised data classification techniques: Transductive
SVM (TSVM) and low density separation (LDS). For a description and the default
parameters used, one can refer to [5].

Since the particle competitive-cooperative model relies on a networked environment, we need to construct a graph from the vector-based data sets shown in
Table 10.1. For this end, we use the k-nearest neighbor technique with $k = 3$. We
fix $\lambda = 0.6$ and $\Delta = 0.07$, which respects the parameter sensitivity analysis in
Sect. 9.2.6.

The simulation results are reported in Table 10.2. In this table, the average test
error and the corresponding standard deviation achieved by each algorithm are
supplied. We can note that the semi-supervised algorithm based on multiple particles
achieves better results most of the time.

The average rank attained by each of the algorithms is also provided. We estimate
the average rank as follows: (i) for each data set, the algorithms are ranked according

Table 10.2 Test errors (%) with 10 labeled training points and the corresponding average rank and standard deviation of each technique

Data set	TSVM	LDS	Proposed technique
Heart	27.4 ± 10.4	22.9 ± 9.6	**21.3 ± 9.9**
Heart-statlog	26.1 ± 5.9	21.7 ± 6.1	**20.5 ± 5.4**
Ionosphere	**23.9 ± 8.2**	24.1 ± 10.9	24.7 ± 9.0
Vehicle	36.8 ± 7.8	33.7 ± 8.5	**31.8 ± 8.8**
House-votes	16.0 ± 5.3	11.6 ± 4.0	**11.4 ± 3.7**
Wdbc	**11.1 ± 3.7**	15.0 ± 8.7	11.9 ± 5.1
Clean1	46.7 ± 4.8	43.2 ± 3.7	**40.2 ± 2.9**
Isolet	13.3 ± 9.5	**8.0 ± 11.4**	12.7 ± 8.8
Breastw	11.1 ± 8.8	**9.6 ± 7.6**	10.5 ± 9.4
Australian	**31.4 ± 11.4**	34.0 ± 14.5	31.6 ± 12.2
Diabetes	34.2 ± 4.6	33.8 ± 4.8	**32.1 ± 4.6**
German	36.5 ± 5.1	**35.3 ± 4.2**	35.9 ± 4.3
Optdigits	8.6 ± 7.6	**3.6 ± 11.1**	5.4 ± 8.9
Sat	13.5 ± 10.8	**5.8 ± 14.2**	9.3 ± 10.1
Average rank	2.6	1.9	1.6

The experiments are repeated 30 times for each labeled set and the average test error and standard deviations are recorded. The smallest test errors for each data set are in bold

to their average performance, i.e., the best algorithm (the smallest test error) is ranked as 1st, the second best one is ranked as 2nd, and so on; and (ii) for each algorithm, the average rank is given by the average value of its ranks scored on all the data sets.

With the purpose of examining these simulation results in a statistical manner, the procedure outlined in [6] is adopted. The methodology described therein uses the calculated average rank of each algorithm for the statistical inference. Specifically, the Friedman Test is used to check whether the measured ranks are significantly distinct from the mean value of the ranks. In this case, the mean value of the ranks is 2, since there are 3 algorithms. The null-hypothesis considered here is that all the algorithms are equivalent, so their ranks should be the same. Here, we fix a significance level of 0.05. For our experiments, according to [6], we have that $N = 14$ and $k = 3$, resulting in a critical value given by $F(2, 26) \approx 3.37$, in which the two arguments are derived from the degrees of freedom defined as $k - 1$ and $(N - 1)(k - 1)$, respectively. In our case, we get a value $F_F \approx 5.32$ that is higher than the critical value, so the null-hypothesis is rejected at a 5 % significance level.

As the null hypothesis is rejected, one can advance to post-hoc tests which aim at verifying the performance of the proposed algorithm in relation to others. For this task, we opt to use the Bonferroni-Dunn Test, for which the control algorithm is fixed as the proposed technique. According to [6], one should not make pairwise comparisons when we test whether a specific method is better than others. Basically, the Bonferroni-Dunn Test quantifies whether the performance between an arbitrary algorithm and the reference is significantly different. This is done by verifying whether the corresponding average ranks of these algorithms differ by at least a critical difference (CD). If they do differ that much, then it is said that the better ranked algorithm is statistically superior to the worse ranked one. Otherwise, they do not present a significant difference for the problem at hands. Thus, if we perform the evaluation of the CD for our problem, we encounter CD $\approx$ 0.8. The average rank of the proposed method is 1.6. By virtue of that, if any rank does lie in the interval 1.6±0.8, the control algorithm and the compared algorithms are statistically equivalent. We conclude that our algorithm is superior to Transductive SVM for the simulations performed on these data sets. However, the comparison of the LDS to the control algorithm does not surpass the CD, meaning that the differences among them are statistically insignificant.

10.5   Application: Detection and Prevention of Error Propagation in Imperfect Learning

In this section, we tackle the problem of learning with imperfect data, i.e., some of the labeled samples are incorrectly labeled, via the competitive model described in the previous section.

Section 10.5.1 motivates the importance and real practicability of detecting and preventing error propagation in semi-supervised tasks in imperfect data training data. Section 10.5.2 enhances the particle competitive-cooperative model to deal with imperfect learning by providing a detection mechanism of possible wrong labeled instances. Section 10.5.3 shows a mechanism to prevent error propagation by flipping labels from those vertices detected as possibly wrong labeled by the detection module. Section 10.5.4 formally presents the modified particle competitive-cooperative model to withstand imperfect training data. Section 10.5.5 investigates the sensitivity of the model's parameters related to detecting and preventing error propagation. Finally, Sect. 10.5.6 tests the error detection and prevention mechanisms on synthetic and real-world data.

10.5.1   Motivation

The quality of the training data is a fundamental issue in machine learning. It becomes more critical in semi-supervised learning, because fewer labeled data are available and errors (wrong labels) may easily propagate to a portion of or to the entire data set. Up to now, there are still few works devoted to studying semi-supervised learning from imperfect data [1, 2, 8]. Usually, in machine learning, the input label information of the training data set is supposed to be completely reliable. Intuitively, this is not always true and mislabeled samples are commonly found in the data sets due to instrumental errors, corruption from noise, or even human mistakes in the labeling process. If these kinds of wrong labels are used to further classify new data (in the supervised learning case) or are propagated to the unlabeled data (in the semi-supervised learning), severe consequences may occur. Therefore, designing mechanisms to prevent error propagation is important in the machine learning area. Specifically, the prevention of error propagation can benefit the learning systems from two complementary aspects:

i. Improvement of the performance of the learning system, permitting the system to learn from errors;
ii. Avoidance of a system's catastrophe by limiting the spread of wrong labels from imperfect training data.

In the next section, a mechanism for preventing error propagation embedded in the particle competition-cooperation model is presented [4, 10].

10.5.2 Detecting Imperfect Training Data

The idea of mislabeled vertex identification is described in the following. In the competition-cooperation model, for each labeled vertex, a representative particle is generated. For simplicity, we here use the term *correctly labeled particle* to denote the representative particle of a correctly labeled vertex and the *mislabeled particle* to represent the representative particle of a mislabeled vertex. In this way, the vertices in the vicinity of mislabeled vertices are expected to be in constant competition among the correctly labeled and the mislabeled particles. Therefore, the mislabeled particles will be stranded in the small region centered at the mislabeled vertex. Since the number of mislabeled particles in each region is generally much smaller than the number of correctly labeled ones, the surrounding region of the mislabeled vertex tends to be heavily dominated by the correctly labeled particle team. By virtue of the combination of random and preferential walking rules, particles will eventually try to venture far away from their home vertices. Once a mislabeled particle goes far away from its home vertex, it has a high probability to get exhausted. Hence, the number of times that a particle becomes exhausted is a good indicator of whether or not the home vertex that it represents is mislabeled. If the associated particle is constantly getting exhausted, it is possibly representing an imperfect labeled vertex. Otherwise, it is probably representing a correctly labeled vertex.

In order to detect possible mislabeled vertices, consider the random vector $D(t) = [D^{(1)}(t), \ldots, D^{(K)}(t)]$, in which the k-th entry, $D^{(k)}(t)$, stores the number of times that particle k has become exhausted up to time t. In view of this, the update rule of each entry of $D(t)$ is expressed by:

$$D^{(k)}(t) = D^{(k)}(t-1) + S^{(k)}(t), \qquad (10.20)$$

in which $S^{(k)}(t)$ is the boolean-valued variable that indicates whether particle k is active or exhausted at time t. In brief, it yields 1 if particle k is exhausted at time t

and 0, otherwise. On account of that, Eq. (10.20) simply adds 1 or remains with the same summation, depending on the state of particle k at the current time.

In order to check whether or not a particle is getting exhausted more times than the others, we use the average number of times that particles get exhausted as a statistical descriptor/threshold. The average number of times that particles get exhausted in the network is expressed by:

$$\langle D(t) \rangle = \frac{1}{K} \sum_{u=1}^{K} D^{(u)}(t). \tag{10.21}$$

In view of the random variable introduced in (10.21), any particle $k \in \mathcal{K}$ such that:

$$D^{(k)}(t) \geq (1 + \alpha) \langle D(t) \rangle \tag{10.22}$$

holds is considered to be getting exhausted more times than the other particles. Therefore, such particle is a great candidate of representing an incorrectly labeled or imperfect home vertex. The parameter $\alpha \in [-1, \infty)$ is a confidence value that indicates the percentage above the average value $\langle D(t) \rangle$ that must occur in order to a particle to be conceived as a representative of an incorrectly labeled vertex. A small α tends to classify more vertices as incorrectly labeled ones than a large value. In the extreme case, when $\alpha \to \infty$, then the model reduces to its original form, i.e., it does not detect nor prevent incorrectly labeled vertices from propagating incorrect labels.

At the beginning of the competitive process, a very small portion of the unlabeled data is expected to be dominated by the particles. In this way, a correctly labeled particle may accidentally get exhausted and (10.22) turns out to be true. In an attempt to prevent this false positive, a weighted function in (10.22) is introduced, which penalizes (10.22) at the beginning of the competition process and eliminates the effect of that penalty when t is large. In this way, the weighted version of (10.22) becomes:

$$(1 - e^{-\frac{t}{\tau}}) D^{(k)}(t) \geq (1 + \alpha) \langle D(t) \rangle, \tag{10.23}$$

in which $\tau \in (0, \infty)$ is the time constant of the exponential decaying function.

Next, we analyze the minimum number of times that a particle must get exhausted in order to (10.23) to hold. For this end, we plot the minimum $D^{(k)}(t)$, denoted as $D_{\min}^{(k)}(t)$, in a such a way that (10.23) holds. Mathematically, it satisfies the following expression:

$$D_{\min}^{(k)}(t) = \frac{(1 + \alpha) \langle D(t) \rangle}{(1 - e^{-\frac{t}{\tau}})}. \tag{10.24}$$

For the sake of clarity, let us suppose that the dynamical process produces $\langle D(t) \rangle = 1, \forall t \geq 0$. Fix $\alpha = 0$ for simplicity. With respect to the non-weighted version, even when t is very small, a $D^{(k)}(t) = 1$ is sufficient to (10.22) to be satisfied. Therefore, if a correctly labeled vertex happens to reach the "exhausted" state at the beginning of the competitive process, it is immediately marked as a possible wrongly labeled vertex. On the other hand, the weighted version penalizes $D^{(k)}(t)$ when t is small, as (10.23) reveals. Therefore, it is unlikely that labeled vertices are classified as wrongly labeled for small values of t. However, for a sufficient large t, this penalization ceases and (10.23) asymptotically approximates (10.22). In particular, when $t \to \infty$, one has:

$$\lim_{t \to \infty} (1 - e^{-\frac{t}{\tau}}) D^{(k)}(t) \geq \lim_{t \to \infty} (1 + \alpha) \langle D(t) \rangle \Rightarrow$$

$$D^{(k)}(\infty) \lim_{t \to \infty} (1 - e^{-\frac{t}{\tau}}) \geq (1 + \alpha) \langle D(\infty) \rangle \Rightarrow$$

$$D^{(k)}(\infty) \geq (1 + \alpha) \langle D(\infty) \rangle, \tag{10.25}$$

i.e., (10.23) reduces to (10.22). Finally, parameter τ is used to control the decaying speed of the exponential function. Looking in isolation to this function, a small τ yields a large negative derivative for this function and a large τ produces a small negative derivative for this function. In other words, the speed of decaying increases as τ decreases.

10.5.3 Preventing Label Propagation from Imperfect Training Data

In the previous section, we have presented a method for detecting possible wrongly labeled vertices by means of using the information generated by the competitive process itself. Now, whenever we detect a possible imperfect labeled vertex, we need to take actions in order to prevent it from propagating wrong labels throughout its neighborhood.

In an imperfect labeled vertex, the associated particle is expected to constantly getting exhausted, as it is probably in a region with several other labeled data items from rival teams of particles. As such, the neighborhood of the imperfect labeled vertex receives a large quantity of visits by other particles, in such a way that competition is always taking place. In order to correct for the imperfectness of the training labeled data, a natural approach therefore is to drop the label from that home vertex and reset it accordingly to the most dominant class in the neighborhood. By using a local approach, we maintain the smoothness assumption of the model. Figure 10.6 portrays a schematic of this relabeling process that prevents error propagation in the model. Note that the neighborhood is mostly dominated by the red class in such a way that the home vertex has its label flipped to the red

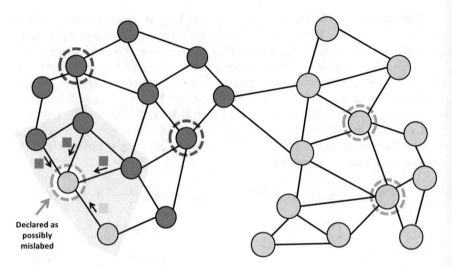

Fig. 10.6 Schematic of the mechanism of error propagation prevention. Surrounded vertices denote home vertices, which correspond to the labeled data. The *left*-most *blue* labeled item is declared as possibly mislabeled. In this case, the label is reset accordingly to the most dominant class in the neighborhood

instead of the original blue class. In this way, the training labeled data gets reshaped such as to have smoother properties. Note that we never use the information of the possibly imperfect labeled vertex, as the detection module warns that its label may be incorrect.

We now formalize that idea. Suppose vertex i has been considered as a possible imperfect labeled vertex at time t, meaning that (10.23) holds. In view of this, vertex i is going to have its random vector $\mathbf{N}_i(t)$ altered, in such a way to reflect how the neighborhood is being dominated at time t. We simply restart $\mathbf{N}_i(t)$ as the average number of visits received by its neighbors. Mathematically, for all $k \in \mathscr{K}$, we have:

$$\mathbf{N}_i^{(k)}(t) = \frac{1}{|\mathscr{N}(i)|} \sum_{j \in \mathscr{V}} \mathbf{A}_{ij} \mathbf{N}_j^{(k)}(t)$$

$$= \frac{1}{|\mathscr{N}(i)|} \sum_{j \in \mathscr{N}(i)} \mathbf{N}_j^{(k)}(t), \tag{10.26}$$

in which $\mathscr{N}(i)$ is the set of neighbors of the imperfect labeled vertex i and $|\mathscr{N}(i)|$ is the corresponding number of neighbors. This idea can be further extended to encompass not only the direct neighborhood but also indirect levels of neighborhoods of the home vertex. In this chapter, we use the simple approach of relabeling according to the immediate neighborhood, which is the most conservative option in terms of smoothness in the labeling decision.

One can observe that the difference between the modified model and the original model is that the former is capable of relabeling labeled vertices, while the latter is not. This new feature is processed when (10.26) is applied.

10.5.4   Definition of the Modified Learning System to Withstand Imperfect Data

With the mechanism introduced before, the dynamical system of the original particle competitive-cooperative model presented in Sect. 10.2 is modified as follows:

$$
\mathbf{X}(t) = \begin{bmatrix} p(t) \\ \mathbf{N}(t) \\ E(t) \\ S(t) \\ D(t) \end{bmatrix}. \tag{10.27}
$$

If $\mathrm{wrong}(k, t) = (1 - e^{-\frac{t}{\tau}})D^{(k)}(t) \geq (1 + \alpha)\langle D(t) \rangle$, then the new competition-cooperation system that supports detection and prevention of incorrectly labeled vertices is given by:

$$
\phi : \begin{cases} p^{(k)}(t+1) = j, \quad j \sim \mathbf{P}^{(k)}_{\mathrm{transition}}(t) \\[4pt] \mathbf{N}^{(k)}_i(t+1) = \mathbb{1}_{[\mathrm{wrong}(k,t)]} \left[\frac{1}{|\mathcal{N}(i)|} \sum_{j \in \mathcal{N}(i)} \mathbf{N}^{(k)}_j(t) \right] \\[6pt] \qquad\qquad\quad + \mathbb{1}_{[\neg\mathrm{wrong}(k,t)]} \left[\mathbf{N}^{(k)}_i(t) + \mathbb{1}_{[p^{(k)}(t+1)=i]} \right] \\[6pt] E^{(k)}(t+1) = \begin{cases} \min(\omega_{\max}, E^{(k)}(t) + \Delta), \text{if } \mathrm{owner}(k, t) \\ \max(\omega_{\min}, E^{(k)}(t) - \Delta), \text{if } \neg\mathrm{owner}(k, t) \end{cases} \\[10pt] S^{(k)}(t+1) = \mathbb{1}_{[E^{(k)}(t+1)=\omega_{\min}]} \\[4pt] D^{(k)}(t+1) = D^{(k)}(t) + S^{(k)}(t+1) \end{cases} \tag{10.28}
$$

We see that the update rule related to the number of visits (2nd expression) now consists of two terms: the term which is employed for vertices that have been detected to be wrongly labeled (first term) and the term for vertices that are not considered wrongly labeled (second term).

10.5.5   Parameter Sensitivity Analysis

Likewise we have performed for the original particle competitive-cooperative model in Sect. 9.2.6, here we focus on understanding the roles of parameters α and τ that deal with the error detection and prevention mechanisms in imperfect data environments. Following the same setup in Sect. 9.2.6, we also use the benchmark of Lancichinatti et al. [9] with $V = 5,000$ vertices, a network connectivity of $\bar{k} = 8$, and intercommunity mixture of $\mu = 0.3$. Essentially, the benchmark process consists in varying the mixing parameter μ and evaluating the resulting model's accuracy rate.

To test for robustness against imperfect data, the benchmark of Lancichinatti et al. is altered [12]. Once the networks are generated, we label a fraction of the vertices using a stratified uniform distribution to compose the labeled training data. Now, we purposely flip some labels of the labeled data so as to introduce noise or imperfectness. In the simulations, the labeled set is fixed as 10% of the size of the data set. Finally, we deliberately flip 30% of the correct labels to incorrect labels $q = 0.3$ in a stratified manner, so as to maintain the proportion of labeled samples at each class.

10.5.5.1   Impact of α

Parameter α is employed in the module of detecting imperfect labeled data. In essence, it determines how much $D^{(k)}(t), k \in \mathcal{K}$, must deviate from $\langle D(t) \rangle$ in order to the corresponding labeled vertex to be declared as mislabeled. Observe that, when $\alpha = -1$, the detection process always accuses labeled vertices as mislabeled, because, in accordance with (10.24), one has:

$$D^{(k)}(t) \geq \lim_{\alpha \to -1} \frac{(1+\alpha)\langle D(t) \rangle}{(1 - e^{-\frac{t}{\tau}})}$$
$$\implies D^{(k)}(t) \geq 0. \tag{10.29}$$

Taking into account that the domain of $D^{(k)}(t)$ is the interval $[0, \infty)$, then (10.29) is always satisfied. Therefore, (10.26) is applied to every vertex in the network for any $t \geq 0$, i.e., the detection procedure reduces to this simple strategy of locally verifying the label validity. Now, when α assumes larger values, the competition dynamics are incorporated to the detection scheme and nonlinear interactions are taken into account in the detection scheme. As α is increased, one can see that solution space of (10.24) becomes more distant from the origin, meaning that (10.24) is more difficult to be satisfied. On the extreme case, when $\alpha \to \infty$, one has that:

$$D^{(k)}(t) \geq \lim_{\alpha \to \infty} \frac{(1+\alpha)\langle D(t) \rangle}{(1 - e^{-\frac{t}{\tau}})} \implies \tag{10.30}$$

$$D^{(k)}(t) \geq \infty, \tag{10.31}$$

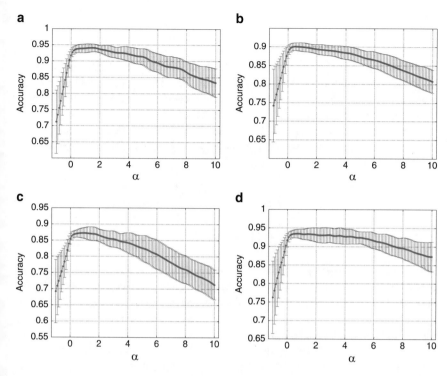

Fig. 10.7 Accuracy rate vs. α. We fix $\tau = 40$. Taking into account the steep peek that is verified and the large negative derivatives that surround it, one can see that the parameter α is sensible to the overall model's performance. Results are averaged over 30 simulations. Reproduced from [12] with permission from Elsevier. (**a**) $\gamma = 2$ and $\beta = 1$. (**b**) $\gamma = 2$ and $\beta = 2$. (**c**) $\gamma = 3$ and $\beta = 1$. (**d**) $\gamma = 3$ and $\beta = 2$

that is only satisfied when $t \to \infty$, which is empirically unattainable. Hence, (10.31) never holds for a finite t. As a result, the detection scheme is virtually turned off. In other terms, the particle competition algorithm reduces to its original form proposed in Sect. 10.2.

Taking into account that analysis, Fig. 10.7a, d display how the accuracy rate of the model behaves as we vary α from -1 to 10 in the networks constructed using the methodology described in [9] with different values of γ and β. As one can verify from the figures, this parameter is sensitive to the outcome of the technique. Oftentimes, the optimal accuracy rate is achieved when a mixture of random and preferential walks occurs. Using a conservative approach, for $0 \leq \alpha \leq 3$, the model gives decent accuracy results when applied to networks with communities.

10.5.5.2   Impact of τ

Parameter τ is employed in detection module of the error propagation mechanism. It assumes the range $\tau \in (0, \infty)$ and is responsible for adjusting the speed of the penalizing function so as to prevent vertices from being signalized as wrongly labeled ones at the beginning of the stochastic process. Next, we study the behavior of the algorithm for small and large τ in a theoretical and empirical manner.

When parameter τ assumes small values, the exponential decaying function has large-valued derivatives, meaning that its decaying speed is faster compared to larger values of τ. In the extreme case, i.e., $\tau \to 0$, one has:

$$D^{(k)}(t) \geq \lim_{\tau \to 0} \frac{(1 + \alpha)\langle D(t)\rangle}{(1 - e^{-\frac{t}{\tau}})} \implies \tag{10.32}$$

$$D^{(k)}(t) \geq (1 + \alpha)\langle D(t)\rangle, \tag{10.33}$$

in which we have used the fact that $\lim_{\tau \to 0}(1 - e^{-\frac{t}{\tau}}) = 1$, since $e^{-\frac{t}{\tau}} \to 0$ provided that $\tau \to 0$ and t is finite. This shows that the model's behavior is dictated by the value of α, which has been studied in the previous section. This means that, in this special case, the penalizing function ceases to exist, because it decays so fast to be considered relevant in the learning process.

When τ assumes larger values, then the decaying speed of the exponential function reduces accordingly. By virtue of that, we have that:

$$D^{(k)}(t) \geq \lim_{\tau \to \infty} \frac{(1 + \alpha)\langle D(t)\rangle}{(1 - e^{-\frac{t}{\tau}})} \implies \tag{10.34}$$

$$D^{(k)}(t) \geq \frac{(1 + \alpha)\langle D(t)\rangle}{(1 - \lim_{\tau \to \infty} e^{-\frac{t}{\tau}})} \implies \tag{10.35}$$

$$D^{(k)}(t) \geq \infty. \tag{10.36}$$

In this case, the denominator of (10.36) approaches 0 in a quick manner, since $e^{-\frac{t}{\tau}} \to 1$ provided that $\tau \to \infty$ and t is finite. This reveals that the detection scheme is turned off, since there is no reachable solution for (10.36) when t remains finite. Therefore, in this case, the model reduces to its original form.

Bearing in mind these considerations, Figs. 10.8a, d portray the accuracy rate attained by the algorithm for distinct values of τ. We can conclude that, for intermediate values of τ, namely $30 \leq \tau \leq 60$, the model is not sensitive to τ when applied to networks with communities.

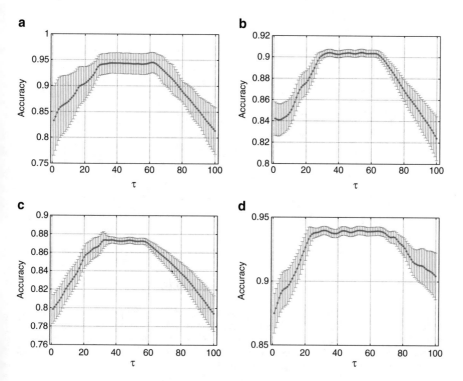

Fig. 10.8 Accuracy rate vs. τ. We fix $\alpha = 0$. Taking into account the large steady region that is verified, one can see that the parameter τ is not very sensible to the overall model's performance if one correctly uses it. Results are averaged over 30 simulations. Reproduced from [12] with permission from Elsevier. (**a**) $\gamma = 2$ and $\beta = 1$. (**b**) $\gamma = 2$ and $\beta = 2$. (**c**) $\gamma = 3$ and $\beta = 1$. (**d**) $\gamma = 3$ and $\beta = 2$

10.5.6 Computer Simulations

In this section, computer simulations are performed to show the robustness of the particle competition-cooperation model for semi-supervised learning in an error-prone environment. We use synthetic and real-world data sets to check the model's performance.

10.5.6.1 Synthetic Data Sets

We first show the behavior of the particle competition algorithm on artificial networks using the Girvan-Newman's benchmark, which we discussed in Sect. 6.2.4. Figure 10.9 depicts the model's accuracy rate as a function of the proportion of mislabeled instances or imperfect data q for three different sizes of the labeled set.

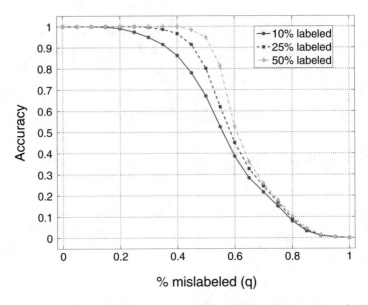

Fig. 10.9 Behavior of the model's accuracy rate of the model vs. the proportion of mislabeled instances or imperfect data q, when random clustered networks with constant mixture are used. The networks are generated using $V = 10,000$ vertices, $M = 16$ communities, and a intercommunity mixture of $z_{out}/\langle k \rangle = 0.3$. 100 independent runs are performed and the average value is reported. Reproduced from [12] with permission from Elsevier

From this figure, we can observe three different regions that are separated by two critical points of interest:

- When q is small, the effect of mislabeled vertices on the accuracy rate of the algorithm is minimal. Visually, this is translated by the plateau region in the plot depicted in Fig. 10.9 with basis near the 100 % accuracy rate. This behavior can be explained by the competition that happens in the dynamical system as it evolves in time. Since the majority of the labeled samples is correctly labeled, the propagation of the correctly labeled samples literally overwhelms that of mislabeled samples. As a consequence, the model's performance is slightly altered.
- When q's value is neither small nor large, the accuracy rate decreases. The first critical point indicates that the label propagation originated from imperfect labeled vertices start to overwhelm that of correctly labeled data. From the same figure, we see that such a phenomenon is also dependent on the size of the labeled set. As the labeled set gets bigger, the algorithm becomes more robust in error-prone environments.
- When q is large, the accuracy rate enters a new steady state with low accuracy rate values that start from the second critical point onwards. At this point, the propagation of wrongly labeled vertices by misrepresented particles completely overwhelms the particles that are spreading correct labels.

Table 10.3 Description of the semi-supervised data classification techniques used in the error propagation analysis

Abbreviation	Technique	Reference
LP	Linear propagation	Zhou et al. [14]
LNP	Linear neighborhood propagation	Wang and Zhang [13]
Original PCCM	Original particle competitive-cooperative method	Sect. 10.2
Modified PCCM	Modified particle competitive-cooperative method	Sect. 10.5

10.5.6.2 Real-World Data Sets

In this section, the algorithm is applied to real-world data sets in an imperfect labeled data environment. For comparison matters, a set of competing semi-supervised learning techniques is also employed, which is summarized in Table 10.3.

With respect to the model selection procedure, all of the parameters are tuned in accordance with the best accuracy rate reached by the algorithms. The model selection is conducted as follows:

- *LP*: σ is selected over the discretized interval $\sigma \in \{0, 1, \ldots, 100\}$ and α is fixed to $\alpha = 0.99$ (the same setup as [14]);
- *LNP*: k is evaluated over the discretized interval $k \in \{1, 2, \ldots, 100\}$, and σ, as well as α, are selected using the same process employed in the LP parameters selection;
- *Original and Modified PCCM*: the data are first transformed into a network representation using the k-nearest neighbor network formation technique. For this purpose, k is chosen over the discretized interval $k \in \{1, 2, \ldots, 10\}$. For the model's parameters, we test λ in the interval $\{0.20, 0.22, \ldots, 0.80\}$. We fix $\Delta = 0.1$. The selection of these parameters and the candidate range for λ are in accordance with the guidelines provided in the parameter sensitivity analysis supplied in Sect. 9.2.6.

Now we apply the LP, LNP, the original and modified PCCMs on two data sets from the UCI Machine Learning Repository [7]: Iris and Letter Recognition. The former is composed of three equal-sized classes, each of which comprising 50 samples, totalizing 150 samples. The latter is composed of 20,000 samples divided into 26 unbalanced classes, each representing a different letter of the English alphabet. Thus, the Letter Recognition data set can be considered as a large-scale data set.

Figure 10.10a, b show the behavior of the test error vs. the proportion of mislabeled samples q. One can verify that, as q grows, intuitively all of the algorithms' performances start to decline, producing larger test errors. However, the modified PCCM is able to outperform the compared algorithms, by virtue of the detection and prevention mechanisms embedded into the competitive model. In an error-prone environment, we can conceive the algorithm as having two types of competitions taking place simultaneously: competition of particles spreading

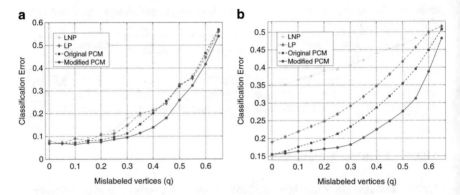

Fig. 10.10 Behavior of the test error as a function of the proportion of imperfect data on two real-world data sets. 100 independent runs are performed and the average value is reported. Reproduced from [12] with permission from Elsevier. (**a**) Iris data set. (**b**) Letter Recognition data set

correct and incorrect labels. Since the competition is always taking place indirectly,[1] then these two types of label diffusion processes are always in opposition. In practical situations, it is fair to assume that the number of correctly labeled samples is usually larger than that of incorrect labels, in such a way that the diffusion process represented by those particles that spread correct labels will eventually overwhelm or win the competition against the label propagation originated by the imperfect training data.

10.6   Chapter Remarks

This chapter presents a semi-supervised learning technique that uses mechanisms of competition and cooperation among particles, which runs in a networked environment. In this model, several particles, each of which representing a class, navigate in the network to explore their territory and, at the same time, attempt to defend their territory against rival particles. If several particles propagate the same class label, then a team is formed, and a cooperation process among these particles occurs.

Wrong label propagation is a fundamental question in machine learning because mislabeled samples are commonly found in the data sets due to several factors, such as instrumental errors, corruption from noise, or even human mistakes. In autonomous learning systems, errors are much easier to be propagated to the whole data set due to the absence or few external intervention, which makes the situation more critical.

---

[1] The indirect competition among particles occurs by the accumulated domination levels of each vertex and by the particles' movement policy.

To deal with that situation, in this chapter, a method is introduced for detecting and preventing error propagation embedded in the semi-supervised learning technique. The error detection mechanism is realized by weighting the total number of times a particle has become exhausted to a thresholded value, which is dependent and vary in time. When the dynamical competitive system begins, there is a penalizing factor which prevents the detection of false positives. This has been introduced in order to diminish the dependency of the error propagation model on the initial locations of the labeled samples (transient part of the dynamics). As the system evolves, this penalization ceases to exist and the plain domination level that each vertex has is used in the error propagation inference. Once a vertex is declared as mislabeled, the particle competition technique resets its domination levels as the average value of its neighborhood, so as to conform to the cluster and smoothness assumptions.

References

1. Amini, M.R., Gallinari, P.: Semi-supervised learning with explicit misclassification modeling. In: Proceedings of the 18th International Joint Conference on Artificial Intelligence (IJCAI), pp. 555–560. Morgan Kaufmann, San Francisco (2003)
2. Amini, M.R., Gallinari, P.: Semi-supervised learning with an imperfect supervisor. Knowl. Inf. Syst. 8(4), 385–413 (2005)
3. Breve, F., Zhao, L., Quiles, M.G., Pedrycz, W., Liu, J.: Particles competition and cooperation in networks for semi-supervised learning. IEEE Trans. Data Knowl. Eng. 24, 1686–1698 (2010)
4. Breve, F.A., Zhao, L., Quiles, M.G.: Particle competition and cooperation for semi-supervised learning with label noise. Neurocomputing 160, 63–72 (2015)
5. Chapelle, O., Schölkopf, B., Zien, A. (eds.): Semi-supervised learning. Adaptive Computation and Machine Learning. MIT, Cambridge (2006)
6. Demšar, J.: Statistical comparisons of classifiers over multiple data sets. J. Mach. Learn. Res. 7, 1–30 (2006)
7. Lichman, M.: UCI Machine Learning, Repository University of California, Irvine, School of Information and Computer Sciences, University of California, Irvine (2013)
8. Hartono, P., Hashimoto, S.: Learning from imperfect data. Appl. Soft Comput. 7(1), 353–363 (2007)
9. Lancichinetti, A., Fortunato, S., Radicchi, F.: Benchmark graphs for testing community detection algorithms. Phys. Rev. E 78(4), 046110(1–5) (2008)
10. Silva, T.C., Zhao, L.: Detecting error propagation via competitive learning. Procedia Comput. Sci. 13, 37–42 (2012)
11. Silva, T.C., Zhao, L.: Network-based stochastic semisupervised learning. IEEE Trans. Neural Netw. Learn. Syst. 23(3), 451–466 (2012)
12. Silva, T.C., Zhao, L.: Detecting and preventing error propagation via competitive learning. Neural Netw. 41, 70–84 (2013)
13. Wang, F., Zhang, C.: Label propagation through linear neighborhoods. IEEE Trans. Knowl. Data Eng. 20(1), 55–67 (2008)
14. Zhou, D., Bousquet, O., Lal, T.N., Weston, J., Schölkopf, B.: Learning with local and global consistency. In: Advances in Neural Information Processing Systems, vol. 16, pp. 321–328. MIT, Cambridge (2004)

Index

© Springer International Publishing Switzerland 2016
T.C. Silva, L. Zhao, *Machine Learning in Complex Networks*,
DOI 10.1007/978-3-319-17290-3

Printed in the United States
By Bookmasters